Author Cocitation Analysis: Quantitative Methods for Mapping the Intellectual Structure of an Academic Discipline

Sean Eom
Southeast Missouri State University, USA

Information Science REFERENCE
INFORMATION SCIENCE REFERENCE
Hershey · New York

Director of Editorial Content: Kristin Klinger
Director of Production: Jennifer Neidig
Managing Editor: Jamie Snavely
Assistant Managing Editor: Carole Coulson
Typesetter: Kim Barger
Cover Design: Lisa Tosheff
Printed at: Yurchak Printing Inc.

Published in the United States of America by
Information Science Reference (an imprint of IGI Global)
701 E. Chocolate Avenue, Suite 200
Hershey PA 17033
Tel: 717-533-8845
Fax: 717-533-8661
E-mail: cust@igi-global.com
Web site: http://www.igi-global.com/reference

and in the United Kingdom by
Information Science Reference (an imprint of IGI Global)
3 Henrietta Street
Covent Garden
London WC2E 8LU
Tel: 44 20 7240 0856
Fax: 44 20 7379 0609
Web site: http://www.eurospanbookstore.com

Library of Congress Cataloging-in-Publication Data

Eom, Sean B.

Author cocitation analysis : quantitative methods for mapping the intellectual structure of an academic discipline / by Sean Eom.

p. cm.

Includes bibliographical references and index.

Summary: "This book introduces an alternative approach to conducting author cocitation analysis (ACA) without relying on commercial citation databases such as index ISI citation index. It is based on a custom bibliographic database and cocitation matrix generation systems specifically developed to use the custom database"--Provided by publisher.

ISBN 978-1-59904-738-6 (hardcover) -- ISBN 978-1-59904-740-9 (ebook)

1. Bibliometrics. 2. Bibliography--Databases. 3. Multivariate analysis. 4. Research--Statistical methods--Data processing. I. Title. II. Title: Author co-citation analysis.

Z669.8.E55 2008

010.72--dc22

2008030770

British Cataloguing in Publication Data
A Cataloguing in Publication record for this book is available from the British Library.

Table of Contents

Section V:
Applications in the MIS Area

Preface

Author cocitation analysis is a subfield of informetrics. Informetrics is a broader term that encompasses electronic communication of media including the Internet and World Wide Web, books, and journals. Informetrics is defined as "the study of the quantitative aspects of information in any form, not just records or bibliographies, and in any social group, not just scientists" (Tague-Sutcliffe, 1992). The development of the Internet has expanded the scope of bibliometrics into electronic communication media. These new areas are often called Webometrics, cybermetrics, technometrics, or scientometrics.

The terms bibliometrics, librametry scientometrics, and informetrics are frequently used interchangeably. Even in the late 1980s, all three terms were not clearly distinguishable from one another. The chaotic state of terminologies existed until the late 1980s. Now, the library and information science area seems to have accepted "informetrics" (Björneborn & Ingwersen, 2004; Wormell, 1998) as the umbrella term enveloping all subfields to study all the quantitative aspects of various information resources including journals, books, and information resources on the Web and the Internet.

This book focuses on a small spot regarding the study of informetrics and author cocitation analysis. The huge body of knowledge that exists today is the result of a cumulative research tradition. Researchers build on each other and their own previous work. Definitions, topics, and concepts are shared and interesting lines of inquiry need to be continuously followed up. In this process of knowledge creation, it is necessary to identify, examine, and trace the intellectual linkage to each other in a given academic field as a basis of assessing the current state of its field to guide future development. These intellectual linkages can be systematically examined by means of counting and analyzing the various facets of intellectual activity outputs in the form of written communications.

Over the past 80 years, the way we count and analyze the citation frequency has dramatically changed from the early manual transcribing and statistical computation of citation data to computer-based citation data creation and its manipulation. The term statistical bibliography was coined by Hulme (1923) as a research tool for examining the intellectual development and structure of an academic discipline. Since then, we have seen continuous development in the field of bibliometrics. The principal method of bibliometrics is citation analysis through counting and analyzing the citation frequencies. The most important milestone in the development of citation analysis was established by Garfield. He presented an idea for the management of scientific information using a comprehensive citation index in 1955 and three years later founded the Institute for Scientific Information (ISI) (Garfield, 1955). For a detailed description of theory and application of citation indexing, see (Garfield, 1979). A citation index is a listing of all referenced or cited source items published in a given time span associated with the citing articles. The Web version of citation index appeared in 1997 is Web of Science®. The Web of Science provides access to multidisciplinary citation index information from approximately 8,700 high impact research journals in the world.

Due to the rich information resources available today such as Web of Science®, bibliometric analysis researchers can easily access rich bibliographic information using the World Wide Web. There are two important recent developments in author cocitation analysis: The use of Pearson correlations coefficients, r, as a similarity measure and several new developments in ACA visualization tools such as Pathfinder networks (White, 2003), AuthorLink (Lin, White, & Buzydlowski, 2003), and VxInsight (Boyack, Wylie, & Davidson, 2002). Although there are some developments in applying common bibliometric methods to Web co-link analysis (Zuccala, 2006), Chapter I briefly discusses only two streams of developments in the ACA area.

THE AUDIENCE OF THIS BOOK

This book is for graduate students and researchers in any academic discipline who want to learn the research techniques and tools to delineate the intellectual structure of various academic disciplines, compare cumulative research traditions, demonstrate theoretical differences between competing approaches, and to trace a paradigm shift in various academic disciplines over time. Author cocitation analysis (ACA) is one of research methodologies that transcends the individual field of inquiry. Despite its usefulness and capabilities that reveal a larger vista hidden in the bibliographic databases, ACA has not been a popular research tool in some academic disciplines including management information systems. For example, in the area of management information systems, there are a total of 2,744 individuals listed in the database of MIS faculty directory. This service was developed and is operated by the Information and Decision Sciences Department and the MIS Research Center of the Carlson School of Management at the University of Minnesota. Of these 2,744, less than 10 researchers have conducted and published ACA research over the past four decades.

This book aims to open the vast expanse of wasteland. Considering the limited exposure of this research methodology to our area, this book covers all essential ACA topics for graduate students and researchers who want to learn the basics of ACA as well as recent developments in ACA such as controversial debates on proximity measurer, diagonal values in cocitation frequency matrix, visualization tools and techniques. The basics of ACA include how to retrieve cocitation frequency counts from online commercial bibliographic databases and how to build custom databases using spreadsheet and database management systems. The basics also include the step-by-step procedures of ACA using the factor, cluster, and multi-dimensional scaling procedures.

THE OBJECTIVE AND CONTRIBUTIONS OF THIS BOOK

This book introduces an alternative approach to conducting author cocitation analysis (ACA) without relying on commercial citation databases such as index ISI citation index. It is based on a custom bibliographic database and cocitation matrix generation systems specifically developed to use the custom database. The alternative approach can be an effective research tool overcoming several weaknesses of the commercial online data-based ACA research. The custom data-based ACA is not a replacement for the commercial data-based ACA. These two approaches are complementary to each other. Our approach clearly has advantages but its critical drawback is the time and effort needed to build the database.

First, the approach we are introducing here has the capability to access the non-primary authors of cited references. The non-primary authors refer to all authors other than the first author. The inability to access non-primary authors is a critical shortcoming of ACA research utilizing the commercial databases. Theoretically, the contributions made by non-primary authors must be counted when examining the intellectual structure of an academic discipline.

Second, strict criteria can be applied to the selection of citing articles. A researcher does not always write articles in a specialized field throughout his/her lifetime. Research interests can shift from one subspecialty area to other areas within an academic discipline. Custom bibliographic databases can be built to include only writings in a specific field. Custom database requires hard labor and a time-consuming process from selecting citing articles, entering cited references from the citing articles, and to maintaining the databases.

Third, the alternative approach effectively identifies the intellectual structure of an academic field and its reference disciplines more accurately as well. All previous ACA studies, except the ones conducted by Eom and his colleagues (Eom, 1996, 1998a, 1998b, 2002; Eom & Farris, 1996; Eom, Lee, & Kim, 1993), failed to identify the reference disciplines of an academic field. The reason for the failure was the method used to select authors to use for ACA. The method starts with a predetermined list of authors selected by the subjective judgments of researchers. It is impractical for ACA researchers to include all authors in the reference disciplines of an academic field prior to conducting ACA analysis. If ACA researchers somehow managed to include authors in the reference disciplines of an academic field, ACA would produce empirical maps of prominent authors selected by the researchers. However, with the approach introduced in this book, ACA becomes an exploratory tool. It can dig up the roots (reference disciplines), locate the trunk (foundations of an academic discipline), and sift through branches (subspecialties) of a tree (an academic discipline). The critical element that makes ACA an exploratory tool is the custom bibliographic databases and the author selection method of screening entire databases to finalize the author set for ACA analysis. This can be called the bottom-up approach. The majority of , if not all, ACA studies using commercial databases are based on the top-down approach – selecting authors applying the subjective judgments prior to ACA analysis. The end result of the top-down approach is simply clustering the subjective author set into several subgroups. With this approach, ACA is inherently a limited tool for identifying the changing structure of an academic field and tracing emerging/fading scholars.

Fourth, the custom databases can be built to include only writings in a specific domain/subspecialty. For example, if anyone wants to study the intellectual structure and main themes and reference disciplines used by the researchers who attended only the International Conference on Information Systems (ICIS), the existing commercial database cannot be used. The only way is to build a custom database from the proceedings of the ICIS. Building custom databases requires hard labor, making it a time-consuming process. However, there are important advantages in using custom databases. The ISI social science citation index includes bibliographic information, author abstracts, and cited references found in more than 1,700 scholarly social science journals covering more than 50 disciplines. To identify the intellectual structure of the decision support systems area, social science citation index-based research could possibly reach inaccurate results due to the technical limitations of ISI citation index files. In this case, building custom databases could be an effective approach.

Fifth, this book describes step-by-step ACA procedures for novice SAS users as well as SPSS. The SAS® system is an integrated system of software that provides complete control over data access, management, analysis, and presentation. The SPSS® is a statistical and data management package for analysts and researchers. This book provides explicit instructions to build bibliographic databases, compile a cocitation matrix, prepare SAS input files, and interpret the results. This book provides the reader with a useful, instructional guideline to conduct ACA research regardless of the bibliographic databases used; in-house databases or commercial citation databases. With commercial citation databases, the cocitation matrix can be easily retrieved to create the bibliographic databases. After the retrieval of author cocitation counts, many steps and procedures must still be followed to accomplish the goals of ACA as shown in Figures 1 and 2 (Chapter VII). Each and every step is an unstructured process for those inexperienced researchers. This book is intended to help them conduct ACA research.

This book can also be useful for those who are not familiar with the three multivariate statistical techniques (factor analysis, cluster analysis, and multidimensional scaling). The book shows the entire procedure to prepare SAS data files, process them, and analyze the outputs. Some of the chore activities must be learned from trial and error, which is often time-consuming and frustrating. Even to those who are not ACA researchers, the book provides useful tips on each process of research using multivariate techniques. Although I have included the basics of SAS and SPSS programs for three multivariate statistical analysis techniques (factor analysis, cluster analysis, and multidimensional scaling), this introduction is not intended to give a comprehensive one-step guideline for ACA students. It is an introduction of multivariate statistical techniques using the SAS and SPSS systems to analyze cocited author counts. With this introduction, ACA students are in a better position to study SAS and SPSS language and procedures; SAS graph software, and SAS/STAT users' guide. The sample SAS and SPSS programs in the book are working programs that can be used with different data sets.

THE STRUCTURE OF THE BOOK

The book consists of five sections: Foundations, Fundamental Issues in ACA Online Data Retrieval, Alternative Approaches of Building Custom Databases, ACA Procedures, and ACA Applications.

Foundations

The first section, which includes one chapter, is concerned with the foundation of ACA. Chapter I provides readers with a big picture of bibliometrics and introduces ACA as a subfield of bibliometrics. Author cocitation analysis (ACA) is a branch of bibliometrics. Bibliometrics/informetrics is one of the older areas of information science research. This chapter briefly overviews the bibliometrics, including the basic concepts, scopes, and study area of bibliometrics. The area of study covers bibliometric distribution, citation, and cocitation analyses, and library use studies. The study of bibliometric distribution led to the invention of Lotka's law of scientific productivity, Bradford's law of core scatter in journals, and Zipf's law of word occurrence. The researchers in the citation and co-citation areas identify the pattern of how published documents are cited over time using many different approaches such as bibliometric coupling, document cocitation analysis, author cocitation analysis, and co-ward analysis. The last section briefly discusses the assumptions, purposes, benefits, limitations, and criticisms of ACA.

Fundamental Issues in ACA Online Data Retrieval

Section II consists of 3 chapters. Chapter II introduces the basics of the Institute for Scientific Information online data retrieval, using the Web of Science and Dialog Classic. The Web of Science provides access to multidisciplinary citation index information from approximately 8,700 high impact research journals in the world. Users can navigate to electronic full-text journal articles with complete bibliographic data, cited reference data, and direct links to the full text. A citation index, developed by ISI, is an alphabetical listing by author, of all the references found in footnotes and bibliographies of the journals covered in the index. This chapter overviews three search options: general search, cited reference search, and advanced search. The following section provides some useful information about the entire procedure to retrieve cocitation frequency counts using Dialog Classic and the free ONTAP® (ONline Training And Practice) site. This chapter points out several technical limitations of the ISI online citation index databases including multiple authorship: all citation index files permit retrieving records only by the last name and initials of the first author only. Another limitation is name-homographs: SSCI indexes only author's last name and initials. Consequently, citation records by an author of the same last name and initials may not be authored by the same author. Another limitation is synonyms: the same author's initials are recorded in many different ways. Some examples of synonym are Keen, P., or Keen, P. G. W., Lee, S. or Lee, S. M.

Chapter II introduces the first of the two fundamental and long standing issues in ACA using ISI online databases. The majority of ACA research has relied on the Institute for Scientific Information (ISI) citation databases. ISI convention allows only the retrieval of papers citing works of which the author is the first or sole author. Non-primary authors (authors whose name appear in second or a later position) will not be counted when assembling a cocitation frequency matrix. This chapter empirically examines the impact of the ISI convention on the results of ACA. Virtually all ACA studies use Thomson's ISI citation indexes that use the first author to retrieve the cocitation counts. Therefore, this has been a methodological issue in ACA study. First, literature survey is conducted to review what has been done to deal with this issue. Second, based on the survey of literature, we further argue that previous research has addressed and shed light on some parts of methodological issues. However, it had failed to address issues such as to what extent the use of a different approach has resulted in different outcomes in terms of an actual intellectual structure of a given academic discipline. Using our data and cocitation matrix generation systems, we compare the differences in the process and outcomes of using different cocitation matrices. Three conclusions can be reached based on our study. First, an all author-based ACA is better than first author-based ACA to capture all influential researchers in a field. Second, it identifies more subspecialties. Finally, an all author-based ACA and first author-based ACA produce little differences in stress values.

Chapter IV investigates the second of the two fundamental and long standing issues in ACA using ISI online databases. Diagonal values in the cocitation frequency counts matrix have been considered a fundamental issue in ACA study. Diagonal values are the cocitation frequency counts between the author and himself/herself. Finding the exact values of diagonal values in the co-citation matrix requires the manual procedure of examining the total number of contributions including journal articles, books, proceedings, and so forth. For that reason, ACA researchers suggested many different approaches to fill the diagonal cells in the cocitation matrix. They include the mean cocitation count, missing values, zeroes, highest off-diagonal counts, adjusted off-diagonal values, and the number of times cocited with himself/herself. The majority of ACA researchers prefer to use either the adjusted value approach by adding the three highest off-diagonal values and dividing by two or the missing value approach. This chapter empirically examines the impact of these different approaches on the ACA outcomes. Based on the results of this study, if the pure cocitation counts are not used, the next best alternatives are as follows. They are the missing value approach, mean cocitation value approach, and the highest off-diagonal value approach in the order of the highest total variance explained.

Alternative Approaches: Building Custom Databases

The third section of the book presents two other alternative approaches in Chapters V and VI that can be used to retrieve cocitation counts in lieu of using the ISI citation index files and Dialog Classic. Chapter V introduces the first, using a popular database management system, of the two alternative approaches to overcome the technical limitations associated with online cocitation counts retrieval using Dialog Classic and citation index files. Certainly Dialog Classic is an attractive alternative because the user is using the readily available bibliographic databases and retrieval software. The majority of ACA researchers have used ISI databases and Dialog Classic to retrieve cocitation counts. However, this approach has some technical limitations as discussed earlier. They include the issue of *Multiple Authorship, Name-Homographs, and Synonyms.* This chapter introduces an alternative approach to retrieving co-citation counts from the custom databases through the system we have designed and implemented. Custom database and retrieval systems need time and investment for development, but they can manage most of the technical limitations discussed. This chapter introduces the fox-base approach, the first of the two, in developing custom databases and the cocitation matrix generation system. The first part is concerned with the design of databases. The second part describes the cocitation retrieval system. We also discuss how our system can eliminate or minimize the technical limitations of the Thomson ISI database and Dialog Classic Software system.

Chapter VI introduces the second alternative approach using a spreadsheet program, Microsoft Excel. McIntire (2007) invented this approach as part of his Master's thesis at the University of Columbia. His thesis is based on the International Textile and Apparel Association (ITAA) publication database. The motive for the design of database and cocitation counts system was simply that the ISI citation index files do not include the specific journal in the textile and apparel area. The chapter shows the design of databases and retrieval of cocitation counts using the spreadsheet based cocitation counts generation system.

ACA Procedures

Section IV deals with the procedures of ACA analysis and consists of 5 chapters. Chapter VII overviews several important steps in author cocitation analysis. ACA consists of the six major steps beginning with the selection of author sets for further analysis, collection and statistical analysis of the cocitation frequency counts, and the validation and interpretation of statistical outputs.

The remaining 4 chapters (VIII through XI) focus on statistical procedures using the SAS and SPSS systems. Chapter VIII describes principal component analysis using the factor procedure of the SAS system. The first section of the chapter begins with the definition of factor analysis. It is the statistical techniques whose common objective is to represent a set of variables in terms of a smaller number of hypothetical variables (factor). We also present many different approaches of preparing datasets including importing from external sources, manual data inputs, and in-file statements. We discuss each of the key SAS statements including DATA, INPUT, CARDS, PROC, and RUN. In addition, we examine several option statements to specify the following: method for extracting factors, number of factors, rotation method, and displaying output options.

Chapter IX describes the distance and cluster procedures of the SAS system. Cluster analysis is a data reduction technique for grouping various entities (e.g. individuals, variables, objects) into clusters so that the entities in the same cluster have more similarities with each other with respect to some predetermined selection criteria. The first section of this chapter explains the creation of a distance matrix, which is the input to the cluster procedure. The second part of this chapter focuses on the PROC CLUSTER statement which sets out the CLUSTER procedure steps. This chapter includes the discussions of generations of a distance matrix, the PROC CLUSTER Statement, and interpreting results of cluster analysis.

Chapter X presents multidimensional scaling (MDS) procedures in the SAS system. MDS is a class of multivariate statistical techniques/procedures to produce two or three dimensional pictures of data (geometric configuration of points) using proximities among any kind of object as input. Three SAS procedures (MDS, PLOT, and G3D) are necessary to convert the author cocitation frequency matrix to two or three dimensional pictures of data. The distance matrix produced earlier by using xmacro.sas and distnew.sas programs in SAS version 8 or the DISTANCE procedure in version 9 is converted to a coordinate matrix, to produce two-dimensional plots and annotated three-dimensional scatter diagrams. This chapter also discusses how to label data points on a plot. The annotate facility in the SAS system produces figures with the name of the author on each data point. The PROC MDS procedure includes many of the features of the ALSCAL procedure.

Chapter XI briefly introduces the use of SPSS version 15.0 to conduct ACA analysis. The SPSS accepts data files in many different formats including spreadsheets, database files, tab-delimited, and other types of ASCII text files. Assuming that cocitation frequency counts are stored in a spreadsheet file in Excel, we demonstrate each step of ACA analysis to produce outputs using factor, cluster, and multi-dimensional scaling analyses.

ACA Applications in the MIS Area

Section V introduces an ACA study in the management information systems area to demonstrate some concepts that cannot be adequately explained with the smaller dataset used in prior chapters. Throughout this book, we use a small data set to demonstrate the step-by-step procedures of converting the dataset to the final ACA outputs. Advantages of using such a small number of variables include a clearer understanding of data preparation steps and an easier interpretation of outputs. On the other hand, a smaller data set may make it difficult to fully demonstrate the problems that can arise with a large number of variables such as scree plot, finding the optimal number of factors based on the factor interpretation, and so forth.

Section V has 2 chapters. Chapter XII infers the intellectual structure of the decision support systems (DSS) field by means of an empirical assessment of the DSS literature from 1969 to 1989. Three multivariate data analysis tools (e.g. factor analysis, multidimensional scaling, and cluster analysis) are applied to an author cocitation frequency matrix derived from a large database file of comprehensive DSS literature over the same period. Seven informal clusters of DSS research subspecialties and reference disciplines were uncovered. Four of them represent DSS research subspecialties—foundations, group DSS, model/data management, and individual differences. Three other conceptual groupings define the reference disciplines of DSS—organizational science, multiple criteria decision making, and artificial intelligence. DSS is a very young academic field that is still growing. DSS has entered the era of growth after 20 years of research. During the 1990s, DSS research was further grounded in a diverse set of reference disciplines. Furthermore, it is in the active process of solidifying its domain and demarcating its reference disciplines.

The last chapter of the book, Chapter XIII, extends an earlier benchmark study (Eom, 1995), which examined the intellectual structure, major themes, and reference disciplines of decision support systems (DSS) over the last two decades (1969-1990). Factor analysis of an author cocitation matrix over the period of 1990 through 1999 extracted 10 factors, representing 6 major areas of DSS research: group support systems, DSS design, model management, implementation, and multiple criteria decision support systems, and 5 contributing disciplines: cognitive science, computer supported cooperative work, multiple criteria decision making, organizational science, and social psychology. We have highlighted several notable trends and developments in the DSS research areas over the 1990s.

REFERENCES

Björneborn, L., & Ingwersen, P. (2004). Toward a Basic Framework for Webometrics. *Journal of the American Society for Information Science and Technology, 55*(14), 1216-1227.

Boyack, K. W., Wylie, B. N., & Davidson, G. S. (2002). Domain Visualization Using Vxinsight for Science and Technology Management. *Journal of the American Society for Information Science and Technology, 53*(9), 764-774.

Eom, S. B. (1995). Decision Support Systems Research: Reference Disciplines and a Cumulative Tradition. *Omega: The International Journal of Management Science, 23*(5), 511-523.

Eom, S. B. (1996). Mapping the Intellectual Structure of Research in Decision Support Systems through Author Cocitation Analysis (1971-1993). *Decision Support Systems, 16*(4), 315-338.

Eom, S. B. (1998a). The Intellectual Development and Structure of Decision Support Systems (1991-1995). *Omega, 26*(5), 639-658.

Eom, S. B. (1998b). Relationships between the Decision Support System Subspecialties and Reference Disciplines: An Empirical Investigation. *European Journal of Operational Research, 104*(1), 31-45.

Eom, S. B. (2002). *Decision Support Systems Research (1970-1999): A Cumulative Tradition and Reference Disciplines*. Lewiston, New York: Edwin Mellen Press.

Eom, S. B., & Farris, R. (1996). The Contributions of Organizational Science to the Development of Decision Support Systems Research Subspecialties. *Journal of the American Society for Information Science, 47*(12), 941-952.

Eom, S. B., Lee, S. M., & Kim, J. K. (1993). The Intellectual Structure of Decision Support Systems (1971-1989). *Decision Support Systems, 10*(1), 19-35.

Garfield, E. (1955). Citation Indexes for Science. *Science, 122*, 108-111.

Garfield, E. (1979). *Citation Indexing: Its Theory and Application in Science, Technology, and Humanities*. New York: Wiley.

Hulme, E. W. (1923). *Statistical Bibliography in Relation to the Growth of Modern Civilization*. London: Grafton.

Lin, X., White, H. D., & Buzydlowski, J. (2003). Real-Time Author Co-Citation Mapping for Online Searching. *Information Processing & Management, 39*(5), 689.

McIntire, J. S. (2007). *The Clothing and Textile Research Base: An Author Cocitation Study*. Unpublished Master's Thesis, University of Missouri, Columbia, Columbia, Missouri.

Tague-Sutcliffe, J. (1992). An Introduction to Informetrics. *Information Processing & Management, 28*(1), 1-3.

White, H. D. (2003). Pathfinder Networks and Author Cocitation Analysis: A Remapping of Paradigmatic Information Scientists. *Journal of the American Society for Information Science and Technology, 54*(5), 423-434.

Wormell, I. (1998). Informetrics: An Emerging Subdiscipline in Information Science. *Asian Libraries, 7*(10), 257-268.

Zuccala, A. (2006). Author Cocitation Analysis Is to Intellectual Structure as Web Colink Analysis Is To...? *Journal of the American Society for Information Science and Technology, 57*(11), 1487-1502.

Acknowledgment

I would like to acknowledge the help of three anonymous reviewers involved in the collation and review process of the book, without whose support the project could not have been satisfactorily completed.

Special thanks also go to the publishing team at IGI Global, whose contributions throughout the whole process from inception of the initial idea to final publication have been invaluable. In particular to Jan Travers (vice president of editorial) and Heather Probst (assistant executive editor) who have been assigned to work with me. They continuously provided feedbacks and encouragements via e-mail and telephone calls for keeping the project on schedule.

Special thanks to my daughter, Caroline Eom, who has done an excellent job in proofreading the entire manuscript of this book and several other previous books.

Sean Eom, PhD

Cape Girardeau, Missouri, USA

June 2008

Section I
Foundations

Chapter I
An Introduction to Bibliometrics and Informetrics

ABSTRACT

Author cocitation analysis (ACA) is a branch of bibliometrics. Bibliometrics/informetrics is one of the older areas of library and information science. The terms bibliometrics, scientometrics, and informetrics are often used synonymously. This chapter briefly overviews bibliometrics, including basic concepts, scopes, and study areas of bibliometrics. The areas of study cover bibliometric distribution, citation and cocitation analyses, and library use studies. The study of bibliometric distribution led to the invention of Lotka's law of scientific productivity, Bradford's law of core scatter in journals, and Zipf's law of word occurrence. The researchers in the citation and co-citation areas identify the pattern of how published documents are cited over time using many different approaches such as bibliometric coupling, document cocitation analysis, author cocitation analysis, and co-word analysis. This chapter also discusses assumptions, purposes, benefits, limitations, and criticism of ACA. The last section of this chapter includes discussions of several developments in informetrics and ACA. Since the late 1990s, a new subset of informetrics, webometrics/cybermetrics, has become part of the main stream library and information science research area. In ACA, there had been a series of intense debates on the use of Pearson correlations coefficients, r, as a similarity measure along with several new developments in ACA visualization tools such as Pathfinder networks (Howard D. White, 2003b), AuthorLink (Lin, White, & Buzydlowski, 2003), and VxInsight (Boyack, Wylie, & Davidson, 2002).

INTRODUCTION

The library and information science (LIS) field consists of informetrics, bibliometrics, scientometrics, cybermetrics, and webometrics. The terms bibliometrics, librametry scientometrics, and informetrics are frequently used interchangeably. Even in the late 1980's, all these terms were not clearly distinguishable each other. Wormell described the chaotic state of terminologies and the acceptance of the term "informetrics" this way (Wormell, 1998, p. 258):

The individual identities of the subfields "bibliometrics", "informetrics", "scientometrics" and "technometrics" are unfortunately not very clear, and there is chaos in the terminology. At the 1987 international conference some thoughts were given to changing the name of the discipline to "informetrics", and since the late 1980s there has been some support for use of this term. But alongside or parallel with this, both "bibliometrics" and "scientometrics" are frequently used terms. The field is becoming a scientific discipline including all the statistical and mathematical aspects connected with library, documentation and information problems with strong links to the theoretical aspects of information retrieval.

As Figure 1 shows, nowadays, informetrics is a broader term that encompasses electronic communication of media including the Internet and World Wide Web, books, and journals. Informetrics is defined (Tague-Sutcliffe, 1992) as "the study of the quantitative aspects of information in any form, not just records or bibliographies, and in any social group, not just scientists." The development of the Internet has expanded the scope of bibliometrics into electronic communication media.

Figure 1. Relationships among many subfields of library and information science fields

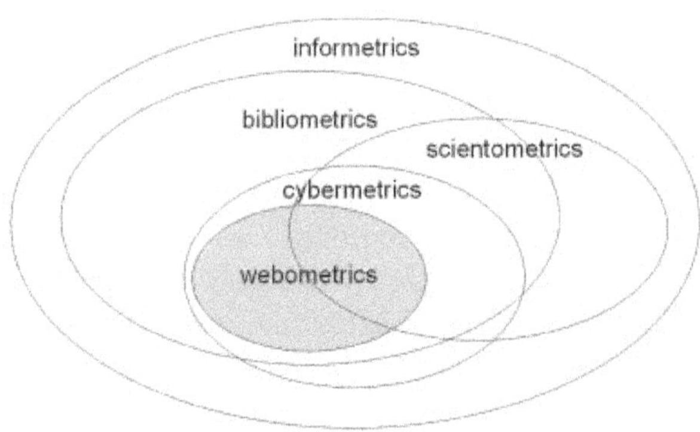

Source: Björneborn & Ingwersen (2004, p. 1217).

These new areas are often called webometrics and cybermetrics. Scientometrics is the application of quantitative tools to the study of scientific communications (Loet Leydesdorff, 2001).

This chapter aims to provide a bird's-eye view of bibliometrics/informetrics. This chapter provides basic concepts and the scope of bibliometrical studies. The last section discusses the purposes, assumptions, and limitations of ACA .

WHAT IS BIBLIOMETRICS (STATISTICAL BIBLIOGRAPHY)?

The term statistical bibliography was coined by E. Wyndham Hulme (1923). The purposes of statistical bibliography are:

1. to shed light on the processes of written communication and of the nature and course of development of a discipline (in so far as this is displayed through written communication), by means of counting and analyzing the various facets of written communications (Prichard, 1969).
2. the assembling and interpretation of statistics relating to books and periodicals ... to demonstrate historical movements, to determine the national or universal research use of books and journals, and to ascertain in many local situations the general use of books and journals (Raisig, 1962).

Pritchard (1969) suggested using the term bibliometrics instead of statistical bibliography. He believed that the term statistical bibliography was vague and could be confused with statistics itself or bibliographies on statistics. According to Pritchard, bibliometrics is defined as "the application of mathematics and statistical methods to books and other media of communication."

SCOPE OF BIBLIOMETRIC STUDIES

The huge body of knowledge existing today is the result of research publications in the form of journal articles, conference proceedings, books, etc. According to Ravichandra Rao (1983, p.216), bibliometric techniques are extensively used in the identification of trends in subjects such as the identification of core journals and the patterns of library use. They are also used to build models of the study of scientific communication.

Most of these models are tested and used primarily at the local level (institutional level) to:

1. Describe scientific productivity;
2. Describe the growth of publications;
3. Identify core journals;
4. Weed out documents;
5. Identify the patterns of library use.

Of the characteristics of documents which have been hypothesized in library use studies, the following are of particular interest:

1. The age of documents—the number of years since they have been published, or the number of years since they have been available for use in a library;
2. The number of citations to documents;
3. Past usage of a given document—the number of times it is circulated or number of times it is used in the library.

The basic units of bibliometric studies are authors and documents (journal articles, conference proceedings and books). The trends and patterns of scientific communications can be detected by analyzing (quantitatively as well as qualitatively) the aggregated periodical data. Ravichandra Rao (1983, p.179) defines bibliometrics and libarametry as an area in which one studies: "Information process and information handling in libraries and information centers by quantitatively analyzing the characteristics and behavior of documents, library staff, and library users." The study areas of bibliometrics and libarametry include bibliometric distribution, citation analysis, library use studies, etc.

Bibliometric Distribution

One of the sub areas in bibliometric research is distribution. The study of bibliometric distribution has led to the following important laws in bibliometrics. They are Lotka's law of scientific productivity, Bradford's law of scatter, and Zipf's law of word occurrence. The term "law" used in bibliometrics is to be interpreted differently from immutable laws found in the physical sciences. According to Wolfram (2003), the term "law" is used by informetricians in its loosest sense to describe a mathematical generalization of an observed regularity in information.

Zipf's Law on Word Frequency

Philologist G. K. Zipf (1935; , 1949) found that the relationship between the frequency of a word within a document and its rank can be represented as:

r x f = C
where r is the rank of a given word,
f is its frequency, and
C is a constant.

If the words contained within a lengthy document are listed in order of decreasing frequency, the rank of a word on that list multiplied by the frequency in the document equals a constant. See (Wolfram, 2003) for detailed descriptions, extensions, and variations of this model. Potter used a simple example of this relationship in Zipf's Law using an analysis of James Joyce's Ulysses by Zipf (Potter 1988).

He showed that the tenth most frequent word occurred 2,653 times, the hundredth most frequent word occurred 265 times, the two hundredth word occurred 133 times, and so on. Zipf found, then that the rank of the word multiplied by the frequency of the word equals a constant that is approximately 26,500.

Lotka's Law on Productivity of Authors

Alfred Lotka, chemist and mathematician, analyzed the number of publications that appeared in *Chemical Abstracts* during the period of 1907 to 1916. Based on the computation of the theoretical frequencies of publications of authors using the least square method, Lotka suggested the following inverse square law of scientific productivity (Lotka, 1926). The following two equations are taken from (Ravichandra Rao, 1983).

$$y_x = 6/(\pi^2 x^\alpha)$$
$$x = 1, 2, 3. \ldots, \quad \alpha > 0$$

where y_x denotes the relative frequency of authors publishing x number of papers. This equation can be rewritten in the following form:

$$y_x = k / x^\alpha$$
$$x = 1, 2, 3. \ldots, \quad k = 6/\pi^2 \text{ for } \alpha = 2$$

k and α are constants depending on the specific field α is approximately 2.

Lotka' law says that the number of authors who wrote n papers to a discipline over time is proportional (equal) to $1 / n^2$.

- The proportion of all contributors that make a single contribution is about 60% of all authors contributed to a field ($60\%/1^2$).
- The proportion of all contributors that make two contributions to a discipline over time is about 15% of all authors contributed to a field ($60\%/2^2$).
- The proportion of all contributors that make three contributions is about 6.7% of all authors contributed to a field ($60\%/3^2$).
- The proportion of all contributors that make three contributions is about 3.75% of all authors contributed to a field ($60\%/3^2$).

Bradford's Law of Core and Scatter in Journals

Samuel C. Bradford was a mathematician and librarian at the Science Museum in London. He formulated a general relationship between the number of articles published in a given field and the distribution of the journal that published articles in that field (Bradford, 1934, 1948). It was Vickery (Vickery, 1948) who coined the term, "Bradford's Law of Scattering". Later Garfield also coined *Garfield's law of concentration,* which specifically addresses the differences in demand for scientific journals.

In layman's terms, Bradford's law says that in a well established field of study, a small number of journals publishes a sizeable portion of the total publications in that area and an increasing number of journals publish fewer and fewer articles in the area. Bradford found that based on the study of a bibliography of geophysics, 9 journals contained 429 articles, 59 contained 499 articles, and 258 contained 404 articles.

Bradford classifies all journals in a given field into three categories (Bradford zones). The first (core) zone contains a core of a few journals. The second zone contains more journals than the number of the first zone journals. The third zone contains the rest of the journals in that field. The core contents of the Bradford's law can be stated as follows.

- The ratio among journals in the three zones is found to be about 1: n: n2, where n is referred to as the Bradford multiplier.
- Each zone is found to publish approximately the same number of articles.

LIBRARY USE STUDIES

Library use studies is one of three main bibliometrics used to measure the adequacy of a library collection. Then future library programs can better serve and satisfy the needs of the users. Further, they aim to formulate mathematical models for patterns of library use in relation to different types of users and documents. These studies include the analysis of circulation statistics, obsolescence study in the use of documents over time, study of the relationships between circulation and acquisition, etc. Readers are referred to pp. 201-215 of (Ravichandra Rao, 1983) for the overview of these research areas.

Researchers build on each others' and their own previous work. Definitions, topics and concepts are shared and interesting lines of inquiry need to be continuously followed up. To facilitate the progress of an academic field, it is important to build such a cumulative research tradition. In this process of knowledge creation, it is necessary to identify, examine, and trace the intellectual linkage to each other in a given academic field as a basis of assessing the current state of its field to guide future development. The intellectual linkages are established through the process of referencing and citation. These intellectual linkages can be systematically examined by means of counting and analyzing the various facets of intellectual activity outputs in the form of written communication.

Citation Analysis

Knowledge creation and dissemination in a discipline are facilitated through the circulation of ideas among "invisible colleges" (Crane, 1972). Each individual contributes to the body of knowledge by building on what others have already accomplished. In this process, referencing and citation are important tools to link each other's writing. The majority of published research is never cited. Citation researchers are interested in identifying the patterns of how published articles are read and cited over time. According to Derek John de Solla Price (1965, p.511), an information scientist who credited as the father of scientometrics:

It seems that, in any given year, about 35 percent of all the existing papers are not cited at all, and another 49 percent are cited only once (n=1). This leaves about 16 percent of the papers to be cited an average of about 3.2 times each. About 9 percent are cited twice; 3 percent, three times; 2 percent, 4 times; 1 percent, five times; and a remaining 1 percent, six times or more.

Citation analysis can be basically classified into two types. The first type is the counting citation of a document or set of documents authored by an individual without considering intellectual linkage. The second is the co-citation analysis of authors or documents to identify intellectual linkages among authors/publications. For examples of the first type of analysis, see (Eom, 1994; Eom & Lee, 1993). Such citation analysis is often used to compare the research productivity of an individual faculty member and/or a university specific academic program measured by citation counts.

Citation analysis of specific journals created the concept of the impact factor, which is defined as the number of citations to the journal divided by the number of articles published in that journal. To be specific, the impact factor for a journal is calculated based on a three-year period. It is the average number of times published papers are cited up to two years after publication. For example, the impact factor for year 2005 for a journal is computed as A/B:

Where A = the number of times articles published in 2003-4 were cited in indexed journals during 2005;
B = the number of articles, reviews, or notes published in 2003-4.

The next type is the cocitation analysis of multiple authors or multiple documents, which was developed under the name of "co-mentions analysis" in 1968 (Karl Erik Rosengren, 1968; Karl Erik Rosengren, 1990). Systematic analysis of co-citation can be done using many different methods including bibliographic coupling, document co-citation analysis, author co-citation analysis, and co-word analysis. Since the primary purpose of the book is author cocitation analysis, we will focus on only one of the many available tools here. For a methodological review of these four different methods, see Baker (1990).

Document Co-Citation Analysis

There are two primary types of cocitation analysis to map the intellectual structure of an academic field: document cocitation analysis and author cocitation analysis (ACA). Document cocitation analysis involves the analysis of a set of selected documents (e.g., journal articles, books, proceedings, etc.) in terms of which pairs of documents are cited together. Readers are referred to (E. Garfield, 1979) for a detailed description of the process of document co-citation analysis. Small and his colleagues at the Institute for Scientific Information (ISI) conduct their research on the document co-citation clustering and mapping techniques using ISI citation databases (Griffith, Small, Stonehill, & Dey, 1974; Henry Small & Garfield, 1985; Henry Small & Griffith, 1974; H. Small & Sweeney, 1985; H. Small, Sweeney, & Greenlee, 1985). ACA and document co-citation analysis share all the technical and methodological identical procedures. The only difference is the unit of analysis. ACA's unit of analysis is the author, whereas document cocitation analysis uses the document as the unit of analysis.

Small (1973, p.265.) compared the differences between document co-citation and bibliographic coupling in this way.

To be strongly co-cited, a large number of authors must cite the two earlier works. Therefore, [document] co-citation is a relationship which is established by the citing authors. In measuring [document] co-citation strength, we measure the degree of relationship or association between papers as perceived by the population of citing authors. Furthermore, because of this dependence on citing authors, these patterns can change over time, just as vocabulary co-occurrences can change as subject fields evolve. Bibliographic coupling, on the other hand, is a fixed and permanent relationship because it depends on references contained in the coupled documents. Co-citation patterns change as the interests and intellectual patterns of the field change.

Bibliographic Coupling

Bibliographic coupling is a technique for measuring the similarity of two *source* documents by counting the number of common bibliographic references (Kessler, 1963). If documents share one or more bibliographic references (already published, of course), they are said to be *bibliographically coupled (connected).* The strength of this connection is measured by the number of shared references. The more shared references they have, the stronger their connection is. On the other hand, document co-citation analysis counts the number of times two documents are to be cited together in *later* publications (see Figure 2). According to Garfield (2001), "Bibliographic coupling is retrospective whereas co-citation is essentially a forward-looking perspective."

Co-Word Analysis

This is a technique of analyzing a set of documents to evaluate their strength of linkage by measuring the extent to which they share important key words or terms (Rip & Courtial, 1984). The co-word analysis examines co-occurrences of key words and terms extracted from publication titles or their full text. The co-occurrences of key words measure the degree of cognitive linkages among a set of documents. The co-word frequency array (matrix) can be further analyzed via cluster analysis, multi-dimensional scaling, and network analysis to construct a co-word map (Callon, Courtial, Turner, & Bauin, 1983).

There are two major differences in ACA, document co-citation, and co-word analysis. First, during a specific research period, co-citation analysis needs citing source (citing documents, authors in citing documents) and cited references (cited authors, cited documents) of citing documents. However, co-word analysis requires only a set of journal articles in a specific research area such as decision support systems, information retrieval research, etc. Second, the content of the data ma-

Figure 2. Bibliographic coupling vs. co-citation

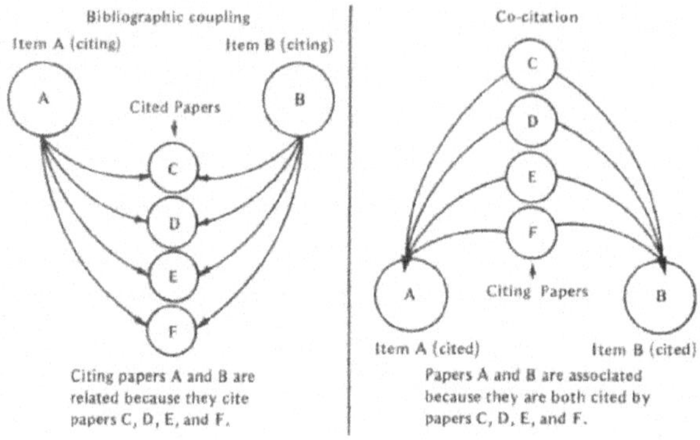

Source: Garfield (2001)

trix for each technique is different. The input to cocitation analysis is the author cocitation frequency matrix. On the other hand, document cocitation analysis and co-word analysis process the document co-citation frequency matrix and the co-word frequency matrix, respectively.

The analysis process and tools are identical. All these techniques process these matrices using hierarchical clustering, and multidimensional scaling to produce an empirical map of an academic discipline or a sub-discipline. Co-citation (document and author) analysis results do not provide the details of actual contents of all sub-specialties identified by co-citation analysis. But co-word analysis provides the content of research topics. For a good example of co-word analysis, see (Ding, Chowdhury, & Foo, 2001).

Author Cocitation Analysis

Author cocitation analysis (ACA) is a subarea of bibliometrics. ACA is a research tool whose idea originated in the late 1960s (Karl Erik Rosengren, 1968). A series of papers from researchers at the College of Information Studies at Drexel University have made ACA a popular research tool in the area of library science (Howard D. White, 1981; H. D. White, 1983; H. D. White & Griffith, 1981; H. D. White & B.C. Griffith, 1982). ACA, introduced in 1981, is a more general approach to identify, examine, and trace the intellectual structure of an academic discipline. This is done by counting the frequency with which any work of an author is cited to any work by another author in the references of citing documents (Bayer, Smart, & McLaughlin, 1990).

ACA, a major area of bibliometrics, is a technique that applies quantitative methods to various media such as books, journals, conference proceedings, and so on. ACA is "a set of data gathering, analytical, and graphical display techniques that can be used to produce empirical maps of prominent authors in various areas of scholarship" (McCain, 1990). The cocitation of authors occurs when a citing paper cites any work of authors in reference lists. Many information scientists and author cocitation analysis researchers define an author as "a body of writings by a person" or "a body of contributions by a person." The term "contributions" may be better since it can include any type of contribution that can be cited as a reference, such as speeches delivered at professional meetings, personal communications including conversation and letters, and other media. The term "person" refers to a single author or one of multiple authors. These different uses of terms are related to the citation databases used in the study.

Most commercial citation databases and software access only the first author, regardless of the number of multiple authors, when retrieving author cocitation counts. This has been the critical weakness of using the commercial citation databases and software. However, this book is based on the bibliographic database I have created, which includes all contributions such as speeches delivered at various meetings and software we have developed that can access all multiple authors. With custom-built bibliographic databases, and the bottom-up approach of the selection of author sets, ACA becomes an exploratory tool for digging up the roots (reference disciplines), locating the trunk (foundations), and sifting through the branches (subspecialties) of a tree (an academic discipline). The critical element that makes ACA an exploratory tool is the custom bibliographic databases and the author selection method of screening the entire databases to finalize the author set for ACA analysis. For an overview and discussion of the continuing relevance of ACA to the study of the intellectual structure of literatures, see a special issue of Journal of the American Society for Information Science, vol. 41, no. 6, 1990. The issue contains a brief introduction by Howard D. White (Guest Editor) and a technical overview of the steps in ACA (McCain, 1990).

Assumptions of Author Cocitation Analysis

Author cocitation analysis is based on the assumptions that "bibliographic citations are an acceptable surrogate for the actual influence of various information sources" (McCain, 1986) and that the cocitation analysis of a field yields a valid representation of the intellectual structure of the field (Bellardo, 1980; McCain, 1984, 1990; Smith, 1981). According to Bellardo (1980), the fundamental premise of cocitation analysis is that the greater the frequency a pair of documents/authors are cited together, the more likely it is that they are related in content. The cocitation frequency of authors represents relationships among authors. Authors whose works are cited together frequently are interpreted as having close relationships with one another. ACA is based on the assumptions that "cocitation is a measure of the perceived similarity, conceptual linkage, or cognitive relationship between two cocited items (documents or authors)" and "cocitation studies of specialties and fields yield valid representations of intellectual structure" (McCain, 1986).

Purposes and Benefits of Author Cocitation Analysis

Citation analysis is often used to determine the most influential scholars, publications, or universities in a particular discipline by counting the frequency of citations received by *individual* units of analysis (authors, publications, etc.) over a period of time from a particular set of citing documents. However, citation analysis cannot establish relationships among units of analysis. ACA is the principal bibliometric tool to establish *relationships* among authors in an academic field. It can thus identify subspecialties of a field and how closely each subgroup is related to each of the other subgroups. By establishing relationships among authors, ACA provides a basis of revealing the intellectual structure of literature and defining the principal subject (major area of subspecialties in an academic discipline and their contributing disciplines) through the empirical consensus of numerous authors in an academic discipline.

In her landmark ACA based research, which examined the intellectual evolution and development of the MIS area, Culnan (1986, p.156) discusses the importance of the study of the intellectual development of a field of study:

Researchers in any academic discipline tend to cluster into informal networks, or "invisible colleges," which focus on common problems in common ways (Price 1963). Within these networks, one researcher's concepts and findings are soon picked up by another to be extended, tested and refined, and in this way, each person's work builds on that of another. The history of exchanges between members of these subgroups in a discipline describes the intellectual history of the field. .....

Researchers can benefit by understanding this process and its outcomes because it reveals the vitality and the evolution of thought in a discipline and because it gives a sense of its future. In a relatively new field such as MIS, this understanding is even more beneficial because it identifies the basic commitments that will serve as the foundations of the field as it matures....

Limitations/Criticisms of Author Cocitation Analysis

ACA is a quantitative tool that cannot be used by itself to determine the intellectual structure of academic disciplines. This is a supporting quantitative tool that must be used with further qualitative analysis of bibliographic data. In regard to citation behavior of authors, Smith (1981, p. 84) enumerated fifteen reasons for citation based on the work of Garfield (1965).

1. Paying homage to pioneers
2. Giving credit for related works (homage to peers)
3. Identifying methodology, equipment, etc.
4. Providing background reading
5. Correcting one's own work
6. Correcting the work of others
7. Criticizing previous work
8. Substantiating claims
9. Alerting to forthcoming work
10. Providing leads to poorly disseminated, poorly indexed, or uncited work
11. Authenticating data and classes of fact— physical constants, etc.
12. Identifying original publications in which an idea or concept was discussed
13. Identifying original publications or other works describing an eponymic concept or term
14. Disclaiming work or ideas of others (negative claims)
15. Disputing priority claims of others (negative homage).

While citation analysis can be a useful research tool due to its unobtrusive, precise, and objective characteristics, there are limitations of ACA stemming from the citation behavior of authors and bibliographic databases. Many problems can also arise in relation to the sources of citation data and mechanics of deriving citations from existing ISI citation indexes. Table 1 summarizes the problems of citation analysis. The table is taken from (MacRoberts & MacRoberts, 1989).

The technical problems consist of multiple authorship, self-citations, homographs, synonyms, unification problems, etc. (Lindsey, 1980; Long, McGinnis, & Allison, 1980; Smith, 1981). The use of Social Science Citation Index (SSCI) and the Science Citation Index (SCI) can raise a potential problem since these sources can exhibit English language bias (Baker, 1990). The use of custom databases and the cocitation matrix generation system we have developed can eliminate many of the problems discussed above.

Multiple Authorship: SSCI lists records by first author only. All authors except first authors will not be counted when compiling the cocitation frequency matrix.

Table 1. Event-data problem of citation analysis

1.	Formal influences not cited.
2.	Biased citing.
3.	Informal influences not cited.
4.	Self-citing
5.	Different types of citations.
6.	Variations in citation rate related to type of publication, nationality, time period, and size and type of specialty.
7.	Technical limitations of citation indices and bibliographies
	a. Multiple authorship
	b. Synonyms
	c. Homonyms
	d. Clerical errors
	e. Coverage of literature

Source: MacRoberts (1989, 343.).

This has been a fundamental issue in ACA. Chapter IV fully addresses this issue and concludes that the ISI convention of relying on only the name of the first author in assembling the cocitation matrix on the investigation of the intellectual structure of academic disciplines may often fail to identify all possible underlying factors.

Name-Homographs: SSCI indexes only the author's last name and initials. Consequently, citation records by an author of the same last name and initials may not be authored by the same person. In the case of common English surnames, such as Smith, Davis, and Williams, indexing only initials creates significant problems of name-homographs such as Smith G., Smith GA., Smith GN., Smith GD., Smith GR.

Synonyms: To further complicate matters, the same author's initials can be recorded in many different ways. Some examples of synonym are Keen, P. or Keen, PGW., Lee, S. or Lee, SM. Furthermore, many individuals have the same names. Another problem is when the name changes, there is no easy way to handle this situation. The name of women authors can change when they marry or some change their names for many different reasons.

Unification Problems: This problem of unification is concerned with the way each author name and the journal title in each cited record is entered into the citation index. In other words, journal titles are entered in many un-standardized ways. For example, MIS Quarterly can be entered in to Citation Index as MIS Q, MISQ, MIS Quart., etc. due to the fact that some journals (e.g., Omega) use their own abbreviated journal name in the references.

Coverage of Literature: Due to the simple fact that SCI and SSCI do not cover all science and social science literatures, the use of these citation indices are undoubtedly problematic. See (MacRoberts & MacRoberts, 1989) for a more detailed discussion of this topic.

Recent Development in Informetrics

Since the late 1990's, a new subset of informetrics called webometrics/cybermetrics, has become part of the main stream library and information science research area. In 2004, *Journal of the American Society of for Information Science and Technology* published a special issue to discuss the emerging area (Mike Thelwall & Vaughn, 2004)

Björneborn and Ingwersen introduced a basic framework for webmatrics and provided a broad picture of relationships between the library and information science fields of informetrics, bibliometrics, cybermetrics, webometrics, and scientometrics as shown in Figure 1 (Lennart Björneborn & Ingwersen, 2004). The conceptual and terminological confusions of the emerging phenomena seems to be settling down. According to Björneborn and Ingwersen (2004, pp. 1216-1217),

A range of new terms for the emerging research field were rapidly proposed from mid-1990s, for example, netometrics (Bossy, 1995); webometry (Abraham, 1996); internetometrics (Almind & Ingwersen, 1996); webometrics (Almind & Ingwersen, 1997); cybermetrics (journal started 1997 by Isidro Aguillo); Web bibliometry (Chakrabarti, Joshi, Punera, & Pennock, 2002). This and similar more specific conceptual diversity and development often made (and make) it difficult to understand what actually is analyzed in the contributions. The transformation over a year from internetometrics to webmetrics by the same authors, Almind and Ingwersen (1996, 1997), is typical of conceptual confusion.

Cybermetrics

The term cybermetrics refers to the quantitative studies of the nature of scientific communication over the Internet and its impact on diffusion of ideas and formation, whereas bibliometrics aims to understand the communication process of authors using the analysis of journal articles to infer the intellectual structure of an academic discipline, and to assess the journal impact factor. Often webometrics and cybermetrics are used as synonyms. But Informatricians tend to agree that cybermetrics is a broader area that encompasses webometrics (Lennart Björneborn & Ingwersen, 2004).

Cybemetrics is proposed as a generic term for the quantitative study of all internet applications.

The study of the quantitative aspects of the construction and use of information resources, structures, and technologies on the whole Internet drawing on bibliometric and informetric approaches (L. Björneborn, 2004).

The coverage of cybermetrics includes the following:

- Statistical study of the World Wide Web and computer-mediated communication on the Internet (Herring, 2002) such as discussion groups(Matzat, 1998), mailing lists(Hernández-Borges, Pareras, & Jiménez, 1997), usenet newsgroup(Bar-Ilan, 1997), etc.
- Quantitative measure and analysis of the Internet backbone technology, topology, and traffic(Molyneux & Williams, 1999).
- Analysis of Web contents, link structure, web-usage in information systems or computer science, etc. such as:
  - Cybergeography and cyber cartography (Dodge, 1999; Dodge & Kitchin, 2001, 2002).
  - Web ecology (Chi et al., 1998).
  - Web mining (Cooley, Mobasher, & Srivastava, 1997; Etzioni, 1996).
  - Web graph analysis(Broader et al., 2000).
  - Web dynamics (Levene & Poulovassilis, 2001), and
  - Web intelligence (Yao, Zhong, Liu, & Ohsuga, 2001).

Webometrics

Webometrics is proposed as a generic term for the quantitative study of the World Wide Web phenomena.

The study of the quantitative aspects of the construction and use of information resources, structures, and technologies on the Web drawing on bibliometric and informetric approaches (L. Björneborn, 2004)

The coverage of webometrics includes the following four main areas (Lennart Björneborn & Ingwersen, 2004):

- Web link structure analysis/web colink analysis
- Web page content analysis

- Web usage analysis
- Web technology analysis

As Figure 1 shows, cybermetrics refers to the quantitative studies of the Internet related phenomena including discussion groups, mailing lists, computer-mediated communication, etc. Webometics is a subarea of cybermetrics which focuses on only the World Wide Web-related phenomena. Especially, the hyper-link is the core of webomerics. *Journal of American Society of for Information Science and Technology* published a special issue on Webometrics. Most of the articles in that issue investigated the issues surrounding the hyperlinks (Mike Thelwall & Vaughn, 2004).

Web colink analysis (WCA) is an emerging field in webometrics. In webometrics, Colink is established when two web pages "both have inlinks from a third pages"(M. Thelwall, 2004, p.5). The link analysis is concerned with the analysis of inlinks and outlinks. Inlinks are defined as follows (anonymous, 2008):

Backlinks (or back-links (UK)) are incoming links to a website or web page. In the search engine optimization (SEO) world, the number of backlinks is one indication of the popularity or importance of that website or page (though other measures, such as PageRank, are likely to be more important). Outside of SEO, the backlinks of a webpage may be of significant personal, cultural or semantic interest: they indicate who is paying attention to that page.

In basic link terminology, a backlink is any link received by a web node (web page, directory, website, or top level domain) from another web node (Björneborn and Ingwersen, 2004). Backlinks are also known as incoming links, inbound links, inlinks, and inward links.

The inlinks are classified into two types: internal inlinks and external inlinks. These two types of inlinks make up total inlinks. In ACA, the inclusion of self-citation has been an issue in analyzing and interpreting the ACA results. WCA may include links within the site itself or examine only external links. For the comparison of ACA and WCA, authors are referred to Zuccala (2006). Detailed comparisons are made between ACA and WCA in term of the following:

- Selecting author names and web pages
- Retrieving cocitation frequency matrix and Web colink frequency matrix
- Mapping and Interpretation of ACA and WCA

Table 2 is constructed based on Zuccala (2006) and it highlights some differences between ACA and WCA. As the table indicates, these two tools are different each other and do not share many things in common. One exception is that the data matrix is constructed by similar procedures and processed by the same multivariate statistical techniques such as factor analysis, cluster analysis, and multidimensional scaling techniques. Although differences in data sources, data currency, data selection, and data stability exist, the most critical difference occurs when interpreting the results of ACA and WCA. To interpret what each factor/cluster/dimension means, ACA and WCA need interpretation based on citation theory and link theory, respectively.

Recent Development in Author Cocitation Analysis

Although the original idea of ACA came from Rosengren (1968), ACA has become an important field of study since White and his colleagues from Drexel's College of Information Studies published a series of papers (C. E. White & Christy, 1987; Howard D. White, 1981; H. D. White, 1983; H. D. White & Griffith, 1981; H. D. White & B.C. Griffith, 1982; Howard D. White & B.C. Griffith, 1982).

There are two important recent developments in author cocitation analysis: The

Table 2. Comparison of ACA and WCA

	Author Co-citation Analysis (ACA)	**Web Colink Analysis (WCA)**
Data Source	SciSearch or Social ScieSearch Custom Databases	World Wide Web data
Data Selection	Highly cocited authors in a discipline or research area	Well-linked web pages in the area of common themes such as business or academic pages
Currency of data	Historical	Up-to-the minutes
Stability of Cocitation/ Colink	Stable and reliable	Fluctuate daily
Retrieval of homonymous data	Homonymous data	No homonymous data
Inputs	Cocitation frequency matrix	Colink frequency matrix
Interpretation of Outcomes (Maps)	Intellectual structures (cognitive linkages) of an academic field)	Some types of Web structure (geography, mission, subject-area orientation, etc.)
Theory to interpret the results	Citation theory	Hyperlink theory

use of Pearson correlations coefficients, r, as a similarity measure and several new developments in ACA visualization tools such as Pathfinder networks ((Howard D. White, 2003b), AuthorLink (Lin, White, & Buzydlowski, 2003), and VxInsight (Boyack, Wylie, & Davidson, 2002). Although there are some developments in applying common bibliometric methods to web colink analysis (Zuccala, 2006), this section discusses only two streams of developments.

Debates on Normalizing Cocitation Counts (Using Pearson's Correlations Coefficients (r) as a Similarity Measure)

The debate was ignited by Ahlgren, Jarneving, and Rousseau [AJR] in 2003. They questioned and criticized the use of Pearson's Correlations Coefficients (r) as the similarity measure, which has been the de facto standard since the publication of a technical overview of the ACA procedure article (McCain, 1990). In ACA, the raw cocitation matrix compiled is normalized to a similarity measure to perform factor analysis, cluster analysis, and multi-dimensional scaling. AJR asserted that Pearson's Correlations Coefficients has a number of drawbacks and therefore is not an optimal choice for normalizing the cocitation frequency matrix.

The drawback of Pearson's correlations coefficient, r, according to AJR, is that it does not satisfy two natural requirements AJR set forth. The focal point of their criticism of r is its sensitivity to zero-vector element values. This deals with the situation of expanding the author set to analyze by adding several more authors who were never cocited with the original author set and therefore adding zeros to the cocitation matrix. AJR requirement 1 states that the association measure between the original author set and new author to be added does/should not decrease in order for Pearson's correlations coefficients to be used as a similarity measure. In mathematical terms, AJR specified the following conditions to hold when expanding author set (Ahlgren, Jarneving, & Rousseau, 2003, p. 553).

If (X$n) denotes an original vector X where n zeros are added, and s denotes a similarity measure then s(A,B) <= s(An, Bn) for every n.

AJR requirement 2 states the association between two sub-groups within the original author set should maintain the same degree of strength (e.g., smaller, the same, or larger) and the expansion of the author set. AJR require that "if, before the expansion, the association between A and B were smaller than that between C and D, it stays smaller after the expansion." In mathematical terms, AJR specified the following conditions to hold when expanding an author set (Ahlgren, Jarneving, & Rousseau, 2003, p. 553).

If (X$n) denotes a similarity measures, then s(A,B) < s(C,D) implies s(An, Bn) < s(Cn, Dn) for every n, where (X$n) denotes an original vector X where n zeros are added.

If the similarity measures (correlation coefficients) between authors A and B are the same before and after adding a set of authors who has "0" cocitation frequency with the original author set to the original set, the use of Pearson's r is not the best choice because it does not satisfy these two requirements.

They also suggest the chi-squared distance and Salton's cosign measure as alternative measures of similarity. AJR demonstrated that the chi-squared distance measure and Salton's cosine measure satisfy the two requirements they set forth. They concluded that there is not a single best approach to this problem. Each measure has strengths and weaknesses and therefore ACA researchers should be aware of this (Ahlgren, Jarneving, & Rousseau, 2003).

White's Response to AJR [White]

In response to Ahlgren, Jarneving, and Rousseau, White argued that the problem they stated is an unrealistic one in traditional ACA. Furthermore, White showed that Pearson's r produces similar maps and clusters to the one produced by the two possible alternative measures (i.e., a cosine similarity measure and a chi square dissimilarity measure). Consequently, r is sufficient for the purpose of ACA (Howard D. White, 2003a).

White's response to AJR can be summarized as follows. First, the cocitation matrix is fixed from the beginning and the Pearson's correlation matrix produced is fixed. "Drexel-style" ACA does not focus on nontemporal change. Second, "Drexel-style" ACA selects the author set from the outset that has a high degree of interconnections. Therefore, it is highly unrealistic for hypothetical situations to expand the original author set to include the new author set with zero cocitation counts. Third, the treatment of the diagonal values is still a problem in ACA, but it is a minor one. Fourth, the results of clustering and multi-dimensional scaling produced almost the same configurations/ordination in multidimensional spaces and clustering results, regardless of the association measures selected. The term, ordination, is used by Schneider and Borlund (2007a; , 2007b). They defined it as "the collective term for multivariate statistical techniques that arranges a set of data with respect to one or more axes in a multidimensional space." Fifth, the purpose of ACA maps is to produce a bird's eye view of the field. Therefore ACA contributes to enrich the global overview of the field to show a group of mountains and forests in the field, rather than the mathematical properties of similarity measures.

Bensman's Commentary on AJR

Bensman presents three points of the AJR controversy (Bensman, 2004). First, AJR failed to give a logical explanation for the necessity of their two axioms. Furthermore it is unrealistic to expect the invariant relationships among a group of authors when adding another group of authors with zero correlations coefficients with the original author set. Second, the measures meeting AJR requirements only measure similarity and do not measure dissimilarity. Third, a point made by AJR is that Pearson's r requires the bivariate normal distribution and needs the proper transformation to meet the normal distribution condition. Bensman indirectly refuted this point saying "many parametric statistical techniques like Pearson's r are distributionally robust and give close to the same answer whether the data is raw or transformed (p. 1257)." Bensman concluded that AJR's entire system is built in the mathematical spaces and failed to provide a practical demonstration of the strength of their proposed similarity measures (Bensman, 2004).

Ahlgren et al's Responses to Bensmann's Commentary [AJR2]

In response to Bensmann's Commentary, AJR made the following reactions. First, AJR clarified that the similarity measure should not decrease and AJR do not expect the relationships to be constant. Second, in response to the second point of Bensman (the measures meeting AJR requirements only measure similarity and do not measure dissimilarity), AJR insist that Pearson's correlation coefficient in ACA used only as a similarity measure, not dissimilarity measure. Third, AJR show a cautionary attitude on this point rather than making a "big point". AJR do not believe that Pearson's r is not the best approach (Ahlgren, Jarneving, & Rousseau, 2004b).

AJR Responses to White [AJR3]

AJR acknowledge that White's comments are generally correct. However, they believed that the use of Pearson's r is not entirely correct and it is possible to correct it (Ahlgren, Jarneving, & Rousseau, 2004a).

White's Concluding Responses to AJR3

White made the final comments to AJR3. The use of Pearson's correlation coefficient "does not invalidate all the substantive claims that have been made in article that use it." White acknowledged that the use of other alternative measure of similarities could have possibly altered the ACA results obtained by using Pearson's correlation coefficient. But this AJR approach does not alter the science, but possibly our taste. White advocated the "in-depth, comparative study of the various proximity measures" in White (2003a).

An Information Theory Approach (Loet Leydesdorff, 2005)

Leydesdorff and his colleague conducted experimentations to investigate the results of using the Pearson's r and the cosign measures and concluded that the differences between the two are minimal and marginal. Leydesdorff presented a new approach, what he calls "an information theoretical approach".

Shifting the Attention from Pearson's Correlation Coefficient to Matrix Generation (L. Leydesdorff & Vaughan, 2006)

While all of the previous debates above focused on the use of Pearson's r, Leydesdorff and Vaughn emphasized that a matrix generation is a more important issue than that of proximity measure. They believe that understanding the nature of various types of matrices is more important. As an example of symmetrical and asymmetrical matrices, they used document cocitation matrix (Figure 3) and citation matrix (Figure 4), respectively. Figures 3 and 4 are very much different from each other in terms of (1) the number of columns and rows, (2) nature of entity/object, and data in the matrix (see, Table 3). Leydesdorff and Vaughn argue that only an asymmetrical citation matrix should be processed to derive the proximity matrix such as Pearson's. Therefore a similarity measure should not be derived from the symmetrical cocitation matrix. A co-citation matrix itself is a proximity matrix. Therefore, it is not necessary to apply a similarity measure to construct a proximity matrix (L. Leydesdorff & Vaughan, 2006). Using flying mileage between 10 American cities, Multi-dimensional scaling is applied to produce the two maps of the 10 cities. The first map uses the original distance matrix and the second map uses the Pearson's r. The second map is distorted. Consequently, Leydesdorff and Vaughn concluded that similarity measure should not be derived from the symmetrical cocitation matrix.

In regard to the debates on choice of proximate measures and their impacts on ordination and clustering results, Schneider and Borlund (2007a; , 2007b) introduce two important issues (matrix generation and the composition of proximity measures). The two matrix generation approaches most likely will result in different ordination and clustering results because the two approaches will produce different proximity rankings of objects. To deal with these issues, Schneider and Borlund introduced two statistical validation tools (the Mantel test and Procrustes analysis) to help ACA researchers choose suitable proximity measures. These validation tools compare and evaluate the degree of monotonicity between different proximity measures or their ordination results.

Normalization of ACA Data Using the Jaccard Index in the Web Environment

As part of the debate on the normalization of ACA data, Leydesdorff discussed the case of using the Jaccard index in the Web environment. Pearson's r or Salton's cosine can only be used only if the author cocitation data can be retrieved from commercial databases or generated from custom databases to construct a similarity matrix (Loet Leydesdorff, 2008). According to Leydesdorff (2008, p. 77), "in this environment, often the researcher may have

Figure 3. Document co-citation matrix (symmetrical matrix)

	paper 1	paper 2	paper 3	paper 4	paper 5
paper 1		5	10	7	8
paper 2	5		15	3	25
paper 3	10	15		1	12
paper 4	7	3	1		9
paper 5	8	25	12	9	

Figure 4. Citation matrix (asymmetrical matrix)

	Cited Papers					
	A	B	C	D	E	F
Citing paper 1	1	0	1	1	0	0
Citing paper 2	1	1	0	0	1	1
Citing paper 3	0	0	1	1	1	0
Citing paper 4	1	1	0	0	0	1

Table 3. Summary of differences between symmetrical and asymmetrical matrices

	Co-citation Matrix	Citation Matrix
Entity/Object	Same (Cited Paper)	different (Citing and cited paper)
No. of Rows & Columns	Same	different
Data in the Matrix	Symetrical	Asymetrical

only the index available and searches the database with a Boolean AND to construct a co-citation or, more generally, a co-occurrence matrix without first generating an occurrence matrix." Leydesdorff argued that the Jaccard index (Jaccard, 1901) may be a better choice than Pearson's r or Salton's cosine in normalizing co-citation data.

Visualization

ACA used dendrograms, two-dimensional and three dimensional maps derived from multidimensional scaling, factor loading plots to visualize the relationships among authors. Now there are new visualization tools such as pathfinder networks(PFNET), AuthorLink, and VxInsight. Researchers in many disciplines have pursued to develop theories and methods for mapping scientific frontiers for decades. Research on knowledge visualization has been multi-disciplinary endeavors from multiple disciplines such as bibliometrics, scientometrics, data mining, knowledge discovery, etc. Readers are referred to (Chen, 2003) for an overview of the fundamental issues in visual communication and science mapping, the basic principles of cartography for visual communication, basic approach to science mapping, and groundbreaking theories, techniques, and applications in science mapping, the landmark works .

Visualizing techniques in ACA includes the following (Howard D. White, 2003b):

- Multidimensional scaling (MDS)
- Clustering
- Factor loading plots
- Kohonen self-organizing maps
- Geographic style maps
- Pathfinder Networks (PFNETs)

Recently, ACA researchers have developed new visualization tools such as AuthorLink (Lin, White, & Buzydlowski, 2003). Some ACA researchers applied visualization tools developed in other disciplines to ACA analyses including Pathfinder Networks (Howard D. White, 2003b) and VxInsight (Boyack, Wylie, & Davidson, 2002).

Pathfinder Network

White introduced Pathfinder Networks (PFNETs) in ACA, which he claims has considerable advantages over MDS maps. First, PFNETs use raw cocitation counts whereas MDS maps use a coordinate matrix, derived from the distance matrix using correlations coefficients, which is transformed from cocitation frequency counts. Consequently, PFNETs reduce the labor of ACA. White demonstrated the use of PFNETs to remap paradigmatic information scientists over the period of 1972-1995. According to White (Howard D. White, 2003b), the great advantage of PFNETs is that they create a visual representation of a field in one pass using the network models. A network consists of nodes (circles) and pathes between any pair of nodes (See Figure 4). Each node represents authors in ACA. In this hypothetical PFNET, some authors (Ackoff, March, and Huber) are linked to only one other author, while other authors such as Keen, Whinston, and Simon are linked more than once. White use the term "degree centrality" to indicate the authors with the higher co-citation frequency with many other authors. These authors are represented in the PFNET like stars connected with many other researchers.

PFNETs can be made with software such as KNOT (The Knowledge Network Organizing Tool (KNOT) Software by Interlink at http://interlinkinc.net/index.html and Pajek. The KNOT software uses the Pathfinder network generation algorithm. Pathfinder algorithms process the proximity measures between pairs of items and produces a network representation of the items (Schvaneveldt, 1990). Pajek (Slovenian word for spider) is a software for large network analysis which is free for non-commercial use. It can be downloaded from its authors' website at http://vlado.fmf.uni-lk.si/pub/networks/pajeck/. The web site also includes several documents including manuals at: http://vlado.fmf.uni-lj.si/pub/networks/pajek/doc/pajekman.pdf

AuthorLink

AuthorLink is a web-based prototype system for author searching and a visualization tool. It is a system that enhances author searching in a bibliographic database created by the Institute for Scientific Information (Lin, White, & Buzydlowski, 2003). When a user enters a single author's name, AuthorLink produces a map of 24 authors who have highest cocitation counts with the author entered by the use. The map shows an intellectual linkage of those 24 authors using two mapping procedures—a self-organizing map and pathfinder networks. AutorLink system is built on the ACA and the visualization mapping algorithms of PFNETs and Kohonen's feature. The advantages of AuthorLinks are summarized by Lin (2003, pp. 696-697) as follows:

- AuthorLink puts a selected author in his or her most relevenat neighborhood of authors.
- AuthorLink presents an overview of a field or subject area.

Figure 5. Basic structure of PFNET

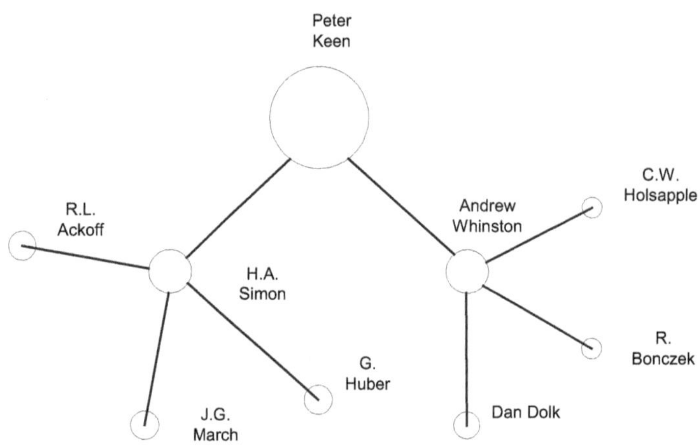

- AuthorLink lets the user observe intellectual connections of authors and discover unsuspected linkages.
- AuthorLink can distinguish similar author names that are otherwise conflated in ISI data.
- AuthorLink makes it easy for the user to explore unfamiliar territories.

Knowledge Visualization Tool

VxInsight® is a knowledge visualization tool developed by Sandia National Laboratories for presenting information as a landscape. The distinguishing feature of VxInsight® is its ability to mine and visualize the competitive nature of information for science and technology management. VxInsight® is a domain visualization tool that can be used for technology planning (Beck, Boyack, Bray, & Siemens, 1999) and knowledge mining (Davidson, Hendrickson, Johnson, Meyers, & Wylie, 1998). Readers are referred to (Boyack, Wylie, & Davidson, 2002) for a detailed description of the VxInsight® tool, its functionality, and an example of domain analyses for competitive intelligence at Sandia National Laboratories.

In the same line of research (knowledge visualization), a generic knowledge visualization approach is developed to extend traditional ACA. This approach integrates structural modeling and information visualization techniques to process author cocitation structure. This represents it in a 3D knowledge Landscape (Chen & Paul, 2001). This approach replaces multi-dimensional scaling (MDS) with the

Pathfinder network scaling technique and visualizes the intellectual grouping of authors produced by factor analysis. A new methodology is proposed to generate scientograms of major scientific domains on the basis of cocitation of Institute of Scientific Information categories (Moya-Anegón et al., 2007).

CONCLUSION

This chapter briefly introduced the concept of bibliometrics. The terms bibliometrics, scientometrics, and informetrics are frequently used interchangeably. Bibliometrics is the application of quantitative methods to communication media. Scientometrics is the application of quantitative tools to the study of scientific communications (Loet Leydesdorff, 2001). Informetrics is a subfield of information science. Nowadays, informetrics is a broader term that encompasses electronic communication media, books, and journals. We also briefly discussed the scope of bibliometric studies including bibliometric distribution, citation Analysis, library use studies, cocitation analysis, coword analysis, and bibliographic coupling.

The second part of this chapter also discussed the origin, assumptions, purposes, benefits, limitations and criticisms of ACA. Technical limitations of ACA stem from the use of ISI citation indices. They are also reviewed as a basis of section II, the fundamental issues in ACA online data retrieval. Of these various issues raised by MacRoberts & MacRoberts, most of technical limitations can be effectively managed by developing custom databases and cocitation counts generation systems.

The third part of this chapter briefly overviewed the recent developments in ACA. There are two streams of noteworthy developments. They are the controversial debates on the use of Pearson's correlations coefficient as a similarity measure and the area of visual representation tools. White acknowledged that the use of other alternative measure of similarities could have possibly altered the ACA results obtained by using Pearson's correlation coefficient. Nevertheless, he strongly defended his previous achievements in ACA, saying the use of alternative measure of similarity "does not invalidate the all substantive claims that have been made in article that use it" White advocated the "in-depth, comparative study of the various proximity measures" in White (2003a). Future research must be conducted to clarify this issue of choosing the most desirable proximity measure. Even though the alternative measures may not alter the outcomes of ACA research substantially, AJR theoretically show that the other measures could be better than Pearson's correlations coefficient when dealing with the hypothetical situations. Therefore, future research is necessary to empirically evaluate the impact of each measure on the outcome of ACA analysis. New developments in visualization tools provide

ACA researchers with a plethora of tools for accurately visualizing the intellectual structure of an academic discipline.

REFERENCES

Abraham, R. H. (1996). Webometry: Measuring the Complexity of the World

Wide Web. Retrieved July 9, 2004, from http://www.ralph-abraham.org/vita/redwood/vienna.html

Ahlgren, P., Jarneving, B., & Rousseau, R. (2003). Requirements for a Cocitation Similarity Measure, with Special Reference to Pearson's Correlation Coefficient. *Journal of the American Society for Information Science and Technology, 54*(6), 550-560.

Ahlgren, P., Jarneving, B., & Rousseau, R. (2004a). Letter to the Editor—Author Cocitation Analysis and Pearson's R. *Journal of the American Society for Information Science and Technology, 55*(9), 843.

Ahlgren, P., Jarneving, B., & Rousseau, R. (2004b). Rejoinder: In Defence of Formal Methods *Journal of the American Society for Information Science and Technology, 55*(10), 936.

Almind, T. C., & Ingwersen, P. (1996). *Informetric Analysis on the World Wide Web: A Methodological Approach to Internetometrics* (No. CIS Report 2). Centre for Informetric Studies: Royal School of Library and Information Science, Copenhagen, Denmark

Almind, T. C., & Ingwersen, P. (1997). Informetric Analyses on the World Wide Web: Methodological Approaches to "Webometrics". *Journal of Documentation, 53*(4), 404-426.

anonymous. (2008, 19 June 2008, at 19:05.). Backlink. Retrieved June 23, 2008, from http://en.wikipedia.org/wiki/Inlinks

Baker, D. R. (1990). Citation Analysis: A Methodological Review. *Social Work Research & Abstracts, 26*(3), 3-10.

Bar-Ilan, J. (1997). The "Mad Cow Disease," Usenet Newsgroups and Bibliometric Laws. *Scientometrics, 39*(1), 29-55.

Bayer, A. E., Smart, J. C., & McLaughlin, G. W. (1990). Mapping Intellectual Structure of a Scientific Subfield through Author Cocitations. *Journal of the American Society for Information Science, 41*(6), 444-452.

Beck, D. F., Boyack, K. W., Bray, O. H., & Siemens, W. D. (1999). Landscapes, Games, and Maps for Technology Planning. *CHEMTECH, 29*(6), 8-16.

Bellardo, T. (1980). The Use of Co-Citations to Study Science. *Library Research, 2*, 231-237.

Bensman, S. J. (2004). Letters to the Editor—Pearson's R and Author Cocitation Analysis: A Commentary on the Controversy. *Journal of the American Society for Information Science and Technology, 55*(10), 935.

Björneborn, L. (2004). *Small-World Link Structures across an Academic Webspace: A Library and Information Science Approach.* Unpublished Doctoral dissertation, Royal School of Library and Information Science Copenhagen, Denmark.

Björneborn, L., & Ingwersen, P. (2004). Toward a Basic Framework for Webometrics. *Journal of the American Society for Information Science and Technology, 55*(14), 1216-1227.

Bossy, M. J. (1995). The Last of the Litter: "Netometrics." Solaris, 2 (Special Issue on "Les Sciences De L'information: Bibliométrie, Scientométrie, Infométrie") Retrieved July 9, 2004, from http://biblio-fr.info.unicaen.fr/bnum/jelec/Solaris/d02/2bossy.html

Boyack, K. W., Wylie, B. N., & Davidson, G. S. (2002). Domain Visualization Using Vxinsight for Science and Technology Management. *Journal of the American Society for Information Science and Technology, 53*(9), 764-774.

Bradford, S. C. (1934). Sources of Information on Specific Subjects. *Engineering, 26*, 85-86.

Bradford, S. C. (1948). *Documentation.* London: Crosby Lockwood.

Broader, A., Kumar, R., Maghoul, F., Raghavan, P., Rajagoparan, S., & Stata, R. (2000). Graph Structure in the Web. *Computer Networks, 33*(1-6), 309-320.

Callon, M., Courtial, J.-P., Turner, W. A., & Bauin, S. (1983). From Translations to Problematic Networks: An Introduction to Co-Word Analysis. *Social Science Information, 22*, 191-235.

Chakrabarti, S., Joshi, M. M., Punera, K., & Pennock, D. M. (2002). The Structure of Broad Topics on the Web. Proceedings of the Www2002 Conference. Retrieved July 9, 2004, from http://www2002.org/CDROM/refereed/338/

Chen, C. (2003). *Mapping Scientific Frontiers: The Quest for Knowledge Visualization* London: Springer-Verlag.

Chen, C., & Paul, R. J. (2001). Visualizing a Knowledge Domain's Intellectual Structure. *IEEE Computer, 34*(3), 65-71.

Chi, E. H., Pitkow, J., Mackinlay, J., Pirolli, P., Gossweiler, R., & Card, S. K. (1998). Visualizing the Evolution of Web Ecologies. In *Proceedings of Human Factors in Computing Systems* (pp. 400-407). ACM Press.

Cooley, R., Mobasher, B., & Srivastava, J. (1997). Web Mining: Information and Pattern Discovery on the World Wide Web [Electronic Version]. *Proceedings of the ninth IEEE International Conference on Tools with Artificial Intelligence (ICTAI '97).* Retrieved July 9, 2004 from http://citeseer.nj.nec.com/cooley97web.html.

Crane, D. (1972). *Invisible Colleges: Diffusion of Knowledge in Scientific Communities.* Chicago: University of Chicago Press.

Culnan, M. J. (1986). The Intellectual Development of Management Information Systems, 1972-1982: A Co-Citation Analysis. *Management Science, 32*(2), 156-172.

Davidson, G. S., Hendrickson, B., Johnson, D. K., Meyers, C. E., & Wylie, B. N. (1998). Knowledge Mining with Vxinsight: Discovery through Interaction. *Journal of Intelligent Information Systems, 11*, 259-285.

Ding, Y., Chowdhury, G. G., & Foo, S. (2001). Bibliometric Cartography of Information Retrieval Research by Using Co-Word Analysis. *Information Processing & Management, 37*(6), 817-842.

Dodge, M. (1999). *The Geography of Cyberspace. Casa Working Paper 8.* Retrieved July 9, 2004, from http://www.casa.ucl.ac.uk/cyberspace.pdf.

Dodge, M., & Kitchin, R. (2001). *Mapping Cyberspace.* London: Routledge.

Dodge, M., & Kitchin, R. (2002). New Cartographies to Chart Cyberspace [Electronic Version]. *Geoinformatics*, 5, 38-41. Retrieved July 9, 2004 from http://www.casa.ucl.ac.uk/martin/geoinfiormatics_article.pdf.

Eom, S. B. (1994). Ranking Institutional Contributions to Decision Support Systems Research: A Citation Analysis. *Data Base, 25*(1), 35-42.

Eom, S. B., & Lee, S. M. (1993). Leading Universities and Most Influential Contributors in DSS Research: A Citation Analysis. *Decision Support Systems, 9*(3), 237-244.

Etzioni, O. (1996). The World Wide Web: Quagmire or Gold Mine? *Communications of the ACM, 39*(11), 65-68.

Garfield, E. (1965). Can Citation Indexing Be Automated? In M. E. Stevens, et al. (Ed.), *Statistical Association Methods for Mechanized Documentation (Nbs Misc. Pub. 269.)* (pp. 189). Washington, D.C.: National Bureau of Standards.

Garfield, E. (1979). Is Citation Analysis a Legitimate Evaluation Tool? *Scientometrics, 1*(4), 359-375.

Garfield, E. (2001, November 27, 2001). *From Bibliographic Coupling to Co-Citation Analysis Via Algorithmic Historio-Bibliography: A Citationist's Tribute to Belver C. Griffith.* Paper presented at the Drexel University, Philadelphia, PA.

Griffith, B. C., Small, H. G., Stonehill, J. A., & Dey, S. (1974). The Structure of Scientific Literature, Ii: Toward a Macro and Microstructure of Science. *Science Studies, 4*, 339-365.

Hernández-Borges, A. A., Pareras, L. G., & Jiménez, A. (1997). Comparative Analysis of Pediatric Mailing Lists on the Internet [Electronic Version]. *Pediatrics*, 100, e8. Retrieved July 9, 2004 from http://www.pediatrics.org/cgi/content/full/1002/2/e8.

Herring, S. C. (2002). Computer-Mediated Communication on the Internet. *Annual Review of Information Science and Technology, 36*, 109-168.

Hulme, E. W. (1923). *Statistical Bibliography in Relation to the Growth of Modern Civilization.* London: Grafton.

Jaccard, P. (1901). Distribution De La Flore Alpine Dans Le Bassin Des Drouces Et Dans Quelques Regions Voisines. *Bulletin de la Société Vaudoise des Sciences Naturelles, 37*(140), 241-272.

Kessler, M. M. (1963). Bibliographic Coupling between Scientific Papers. *American Documentation, 14*(1), 10-25.

Levene, M., & Poulovassilis, A. (2001). Web Dynamics. *Software Focus, 2*(2), 60-67.

Leydesdorff, L. (2001). *The Challenge of Scientometrics: The Development, Measurement, and Self-Organization of Scientific Communications*: Universal Publishers.

Leydesdorff, L. (2005). Similarity Measures, Author Cocitation Analysis, and Information Theory. *Journal of the American Society for Information Science and Technology, 56*(7), 769-772.

Leydesdorff, L. (2008). On the Normalization and Visualization of Author Co-Citation Data: Salton's Cocine Versus the Jaccard Index. *Journal of the American Society for Information Science and Technology, 59*(1), 77-85.

Leydesdorff, L., & Vaughan, L. (2006). Co-Occurrence Matrices and Their Applications in Information Science: Extending Aca to the Web Environment. *Journal of the American Society for Information Science and Technology, 57*(12), 1616-1628.

Lin, X., White, H. D., & Buzydlowski, J. (2003). Real-Time Author Co-Citation Mapping for Online Searching. *Information Processing & Management, 39*(5), 689-.

Lindsey, D. (1980). Production and Citation Measures in the Sociology of Science: The Problem of Multiple Authorship. *Social Studies of Science, 10*, 145-162.

Long, J. S., McGinnis, R., & Allison, P. D. (1980). The Problem of Junior-Authored Papers in Constructing Citation Counts. *Social Studies of Science, 10*, 127-143.

Lotka, A. J. (1926). The Frequency Distribution of Scientific Productivity. *Journal of the Washington Academy of Sciences, 16*(12), 317-323.

MacRoberts, M. H., & MacRoberts, B. R. (1989). Problem of Citation Analysis: A Critical Review. *Journal of the American Society for Information Science, 40*(5), 342-349.

Matzat, U. (1998). Informal Academic Communication and Scientific Usage of Internet Discussion Groups [Electronic Version]. *Proceedings IRISS '98 International Conference*. Retrieved July 9, 2004 from http://sosig.ac.uk.iriss/papers/paper19.htm.

McCain, K. W. (1984). Longitudinal Author Cocitation Mapping: The Changing Structure of Macroeconomics. *Journal of the American Society for Information Science, 35*(6), 351-369.

McCain, K. W. (1986). Cocited Author Mapping as a Valid Representation of Intellectual Structure. *Journal of the American Society for Information Science, 37*(3), 111-122.

McCain, K. W. (1990). Mapping Authors in Intellectual Space: A Technical Overview. *Journal of the American Society for Information Science, 41*(6), 433-443.

Molyneux, R. E., & Williams, R. V. (1999). Measuring the Internet. *Annual Review of Information Science and Technology, 34*, 287-339.

Moya-Anegón, F. d., Vargas-Quesda, B., Chinchilla-Rodríguez, Z., Corera-Ălvarez, E., Munoz-Fernández, F. J., & Herrero-Solana, V. (2007). Visualizing the Marrow of Science. *Journal of the American Society for Information Science and Technology, 58*(14), 2167-2179.

Price, D. J. d. S. (1965). Networks of Scientific Papers. *Science, 149*(3683), 510-515.

Prichard, A. (1969). Statistical Bibliography or Bibliometrics? *Journal of Documentation, 25*(4), 348-349.

Raisig, L. M. (1962). Statistical Bibliography in the Health Science. *Bulletin of Medical Library Association, 50*(3), 45-461.

Ravichandra Rao, I. K. (1983). *Quantitative Methods for Library and Information Science*. New York: Wiley.

Rip, A., & Courtial, J. (1984). Co-Word Maps of Biotechnology: An Example of Cognitive Scientometrics. *Scientometrics, 6*(6), 381-400.

Rosengren, K. E. (1968). *Sociological Aspects of the Literary System*. Stockholm: Natur och Kultur.

Rosengren, K. E. (1990). Who Carries the Field? Communications between Literary Scholars and Critics. In C. L. Borgman (Ed.), *Scholarly Communication and Bibliometrics* (pp. 107-128). Newbury Park, CA: Sage.

Schneider, J. W., & Borlund, P. (2007a). Matrix Comparison, Part 1: Motivation and Important Issues for Measuring the Resemblance between Proximity Measures or Ordination Results. *Journal of the American Society for Information Science and Technology, 58*(11), 1586-1595.

Schneider, J. W., & Borlund, P. (2007b). Matrix Comparison, Part 2: Measuring the Resemblance between Proximity Measures or Ordination Results by Use of the Mantel and Proclustes Statistics. *Journal of the American Society for Information Science and Technology, 58*(11), 1596-1609.

Schvaneveldt, R. (Ed.). (1990). *Pathfinder Associative Networks: Studies in Knowledge Organization*. Norwood, NJ: Ablex.

Small, H. (1973). Co-Citation in the Scientific Literature: A New Measure of the Relationship between Two Documents. *Journal of the American Society for Information Science*, 265-269.

Small, H., & Garfield, E. (1985). The Geography of Science-Disciplinary and National Mappings. *Journal of Information Science, 11*, 147-159.

Small, H., & Griffith, B. C. (1974). The Structure of Scientific Literatures: Identifying and Graphing Specialties. *Science Studies, 4*, 17-40.

Small, H., & Sweeney, E. (1985). Clustering the Science Citation Index Using Co-Citations: I. A Comparison of Methods. *Scientometrics, 7*, 391-409.

Small, H., Sweeney, E., & Greenlee, E. (1985). Clustering the Science Citation Index Using Cocitations: Ii Mapping Science. *Scientometrics, 8*, 321-340.

Smith, L. C. (1981). Citation Analysis. *Library Trends, 30*(1), 83-106.

Tague-Sutcliffe, J. (1992). An Introduction to Informetrics. *Information Processing & Management, 28*(1), 1-3.

Thelwall, M. (2004). *Link Analysis: An Information Science Approach.* San Diego, CA: Academic Press.

Thelwall, M., & Vaughn, L. (2004). Webometrics: An Introduction to the Special Issue. *Journal of the American Society for Information Science and Technology, 55*(14), 1213-1215.

Vickery, B. C. (1948). Bradford's Law of Scattering. *Journal of Documentation, 4*(3), 198-203.

White, C. E., & Christy, D. P. (1987). The Information Center Concept: A Normative Model and a Study of Six Installations. *MIS Quarterly, 11*(4), 451-458.

White, H. D. (1981). Cocited Author Retrieval Online: An Experiment with the Social Indicators Literature. *Journal of American Society for Information Science, 32*, 16-22.

White, H. D. (1983). A Cocitation Map of the Social Indicators Movement. *Journal of the American Society for Information Science, 34*(5), 307-312.

White, H. D. (2003a). Author Co-Citation Analysis and Pearson's R. *Journal of the American Society for Information Science and Technology, 54*(13), 1250-1259.

White, H. D. (2003b). Pathfinder Networks and Author Cocitation Analysis: A Remapping of Paradigmatic Information Scientists. *Journal of the American Society for Information Science and Technology, 54*(5), 423-434.

White, H. D., & Griffith, B. C. (1981). Author Cocitation: A Literature Measure of Intellectual Structure. *Journal of the American Society for Information Science, 32*(3), 163-171.

White, H. D., & Griffith, B. C. (1982). Authors as Markers of Intellectual Space: Cocitation in Studies of Science, Technology, and Society. *Journal of Documentation, 38*, 255-272.

White, H. D., & Griffith, B. C. (1982). A Cocitation Map of Authors in Judgment and Decision Research. In B. F. Anderson, D. H. Deane, K. R. Hammond, G. H. McClelland & J. C. Shanteau (Eds.), *Concepts in Judgment and Decision Research:*

Definitions, Sources, Interrelationships, and Comments (pp. 261-271). New York: Praeger.

Wolfram, D. (2003). *Applied Informetrics for Information Retrieval Research*. Westport, Conn.: Libraries Unlimited, a member of the Greenwood Publishing Group.

Wormell, I. (1998). Informetrics: An Emerging Subdiscipline in Information Science. *Asian Libraries, 7*(10), 257-268.

Yao, Y. Y., Zhong, N., Liu, J., & Ohsuga, S. (2001). Web Intelligence (Wi): Research Challenges and Trends in the New Information Age. *Lecture Notes in Artificial Intelligence, 2198*, 1-17.

Zipf, G. K. (1935). *The Psycho Biology of Language*: Houghton-Mifflin, New York.

Zipf, G. K. (1949). *Human Behaviour and the Principle of Least Effort*: Addison-Wesley, Cambridge, MA.

Zuccala, A. (2006). Author Cocitation Analysis Is to Intellectual Structure as Web Colink Analysis Is To ...? *Journal of the American Society for Information Science and Technology, 57*(11), 1487-1502.

Section II
Fundamental Issues in ACA Online Data Retrieval

Chapter II
An Introduction to Online Data Retrieval and Issues

ABSTRACT

This chapter introduces online citation data retrieval using the Web of Science and Dialog Classic. The Web of Science provides access to multidisciplinary citation index information from approximately 8,700 high impact research journals in the world. Users can navigate to electronic full-text journal articles, with complete bibliographic data, cited reference data, and direct links to the full text. A citation index, developed by The Institute for Scientific Information (ISI), is an alphabetical listing by author of all the references found in footnotes and bibliographies of journals covered in the index. This chapter overviews three search options: general search, cited reference search, and advanced search. The following section provides some useful information about the procedure to retrieve cocitaion frequency counts using Dialog Classic, using the free ONTAP® (ONline Training And Practice) site. This chapter points out several technical limitations of using ISI online citation index databases, including Multiple Authorship: all citation index files permit retrieving records only by last name and initials of first author only. Another limitation is Name-citation records by an author of the same last name and initials may not be authored by the same author. Another limitation is Synonyms: To further complicate matters, the same author's initials can be recorded in many different ways. Some examples of synonym are Keen, P. or Keen, P. G. W. Lee, S. or Lee, S. M.

INTRODUCTION

There are basically two ways of compiling cocitation frequency data. The first approach is retrieval from the Thompson's ISI databases using Dialog software. The second approach is to develop custom databases using database management systems such as Microsoft Access or Foxbase. Spreadsheets can also be used to develop databases and to process the cocitation frequency data. This chapter briefly introduces a way to retrieve the cocitation frequency counts. We will start with citation searching and later co-citation searching. The subsequent two chapters are about the design of custom databases and the development of citation counts generation system, which can overcome most of technical limitation of using the ISI citation index databases and retrieval systems.

CITATION SEARCHING THROUGH THE WEB OF SCIENCE

Citation searching is a good approach when the user wants to know how many articles are citing articles of a certain author (i.e., Albert Einstein or Geraldine DeSanctis). Citation can be searched by using the Web of Science/web of knowledge or dialog classic. The Web version of SCI appeared in 1997 is the Web of Science®. The Web of Science provides access to multidisciplinary citation index information from approximately 8,700 high impact research journals in the world. Users can navigate to electronic full-text journal articles, with complete bibliographic data, cited reference data and navigation, and direct links to the full text (anonymous, 2007). The following are the features and benefits of the Web of Science® product information retrieved on June 17, 2007 from http://scientific.thomson.com/products/wos/.

Features

- Access to the *Science Citation Index®* (1900-present), *Social Sciences Citation Index®* (1956-present), *Arts & Humanities Citation Index®* (1975-present), Index Chemicus® (1993-present), and *Current Chemical Reactions®* (1986-present), plus archives 1840 - 1985 from INPI
- Navigate backward in time using cited references to uncover the research that influenced an author's work
- Navigate forward in time using Times Cited to discover the impact a paper or other published item has had on current research
- Link to the full text of primary literature
- Export records directly to leading biblographic management programs: *EndNote®, Reference Manager® and ProCite®*

- Access groundbreaking scientific journal articles from the entire twentieth century, through the *Century of Science* initiative
- Use the **Analyze tool** to view an analysis of search results
- Set up citation alerts*
- See the number of and access shared references for each related record
- Use term lists (dictionary) for author names, cited authors, cited work, and source title
- Take advantage of enhanced personalization features
- Access all records retrieved by searches

Benefits

- Discover who is citing your research and the impact your work is having on the global research community
- Uncover the seminal research of an important theory or concept
- Measure the influence of colleagues' or competitors' work
- Follow the path and direction of today's hottest ideas and concepts
- Determine if a theory has been confirmed, changed or improved
- Find out how a basic concept is being applied
- Track a topic through years of research literature
- Verify the accuracy of references
- Locate relevant articles missed through a topic or subject search

This version of the Science Citation Index includes abstracts for many of the articles and the e-mail addresses of the authors. A powerful feature of the Web version is the capability to find citations to most of an author's journal publications even if the author was not listed as the first author on the publication. The Web of Science is the primary tool for citation searching, but it cannot be used to retrieve co-citation counts.

Science Citation Index on the Web of Knowledge platform does allow some additional citation analyses beyond citation searching. These include a Find Related Records feature, which identifies additional articles with similar citation patterns, and Citation Report, which provides data and graphs of cited reference patterns within a results set based on an author, journal or topic search. However, access to the Web of Science is generally on an institutional, not an individual, basis.

This is the entry point of ISI Web of Knowledge portal. The Web of knowledge consists of searchable database products (Web of Science, Web Citation Index) and analytical tools (Journal Citation Index), and other resources.

Figure 1. The entry point of the web of knowledge

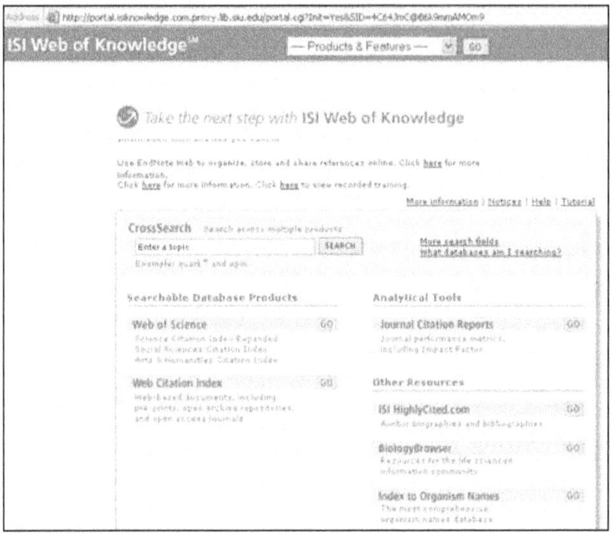

Figure 2. Web of Science: Selecting citation databases and timespan

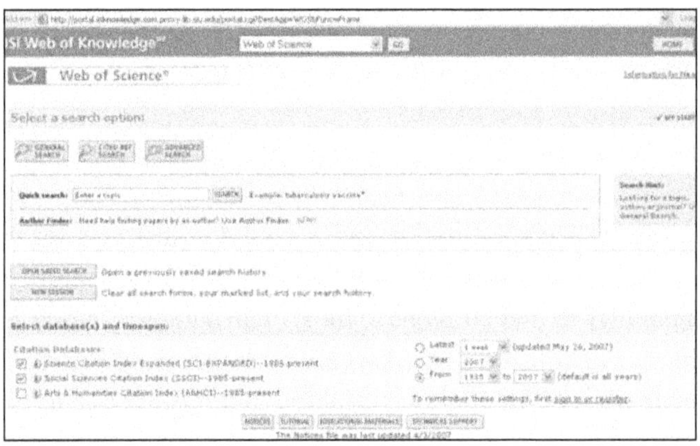

To search for the Web of Science, the user should have a reasonable level of understanding of the citation databases listed at the bottom of the first screen of the Web of Science. The user selects one or more of the citation databases. The specific time period of investigation must be specified: multi-year period (e.g., 1985-2007), or a specific year (e.g., 2007 choose a year between 1985 and 2007), or latest shorter period (e.g., 1 week, 2 week, or 4 week).

CITATION INDEX FILES

The Web of Science has the following citation index files as the bibliographic databases. A citation index, developed by The Institute for Scientific Information (ISI), is an alphabetical listing by author of all the references found in footnotes and bibliographies of journals covered in the index. The entries in the citation indexes include the author name, year of publication, abbreviated book title/abbreviated journal title, volume number, and starting page. Source indexes that provide detailed bibliographic records accompany the citation indexes. As described below, the citation indexes cover thousands of journals deemed to be "core literature" by the ISI. The ISI database does/can not index every journal published in all areas. The ISI is continuously evaluating the impact of each journal to add or drop. Non-indexed journals or books can be listed in the cited references. Citation indexes are used to study the development of research subspecialties in an academic discipline in author cocitation analysis.

Arts & Humanities Citation Index

This file covers 1,300 of the world's leading arts and humanities journals, plus relevant social and natural science journals. It also has additional records from the Current Contents series of publications.

Science Citation Index

This file contains all of the records from more than 6,100 international scientific and technical journals. The areas covered include engineering technology, physical sciences, agriculture, biology, environmental sciences, clinical medicine, and the life sciences.

Social Sciences Citation Index

This file covers the area of social, behavioral, and related sciences. Social SciSearch contains all of the records published in the Social Sciences Citation Index.

Example of cited references in Citation Index.

Cited References:

ALLEN TFH, 1982, HIERARCHY PERSPECTIV
AUGER P, 1989, HIERARCHICALLY ORGAN
AUGER P, 1980, V8, P82, INT J GEN SYS
AUGER P, 1985, V112, P595, J THEOR BIOL

AUGER P, 1983, V65, P269, MATH BIOSCI
AUGER P, 1986, V3, P41, SYST RES
EHRESMANN AC, 1987, V49, P13, B MATH BIOL
KLIR GJ, 1985, ARCHITECTURE SYSTEMS
LOTKA AJ, 1925, ELEMENTS PHYSICAL BI
MAY RM, 1976, THEORETICAL ECOLOGY
MESAROVIC MD, 1970, THEORY HIERARCHICAL
PATTEE HH, 1973, HIERARCHY THEORY CHA
SIMON HA, 1969, SCI ARTIFICIAL
WEISS P, 1971, HIERARCHICALLY ORG S
WHYTE LL, 1969, HIERARCHICAL STRUCTU

After selecting citation databases and timespan, the user chooses one of the following three search options: general search, cited reference search, and advanced search.

GENERAL SEARCH

Enter terms or phrases separated by the operators AND, OR, NOT, or SAME, and then press SEARCH. The search will be added to the search history. The General Search option searches author, article title, subject keywords, abstract, periodical title, publication year, etc. The Web of Science is a more user friendly system than Dialog Classic, which will be discussed later. The user can begin the general search by entering information on the topic or author of articles. Suppose you are interested in knowing the articles written by Eom SB or Eom HB. These two are the names of the same person. You will begin the search process by entering "EOM SB OR EOM HB into the blank area in the general search screen.

Figure 3 displays the general search option screen. The general search option of the Web of Science needs three different kinds of information from the user: Citation databases, timespan, and detailed search conditions using six fields. The user specifies the search conditions by using the six available fields (see Figure 4). The user adds search information by filling out one or more available fields. Adding six fields is interpreted as compound conditions.

As the outputs from the general search show, the general search may be used to find:

- A list of articles/book reviews by author: The output screen shows 23 results of articles, and book reviews by Eom.
- Up to three coauthors

- Citation frequencies
- Each of listed records can be fully displayed by clicking the hyperlinked title.
    - Author and coauthor
    - Cited references (total number of references in each article)
        - First author, abbreviated title of books/journals, vol., no., year.
    - Total number of citation given to each article
        - All authors up to three, hyperlinked title, journal, vol., no., pages, times cited.

Figure 3. Web of Science: General search option

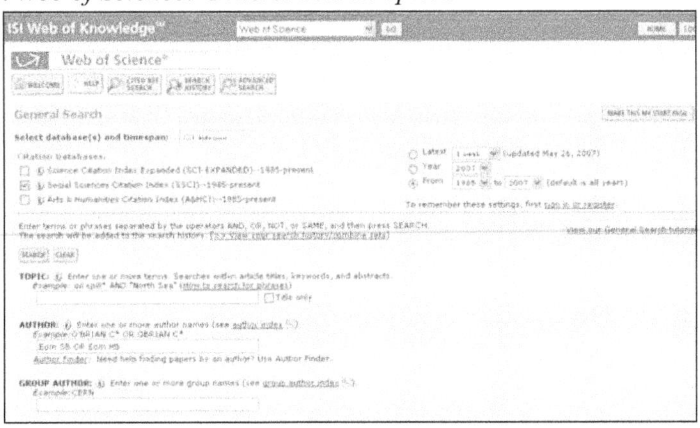

Figure 4. Web of Science: General search fields available

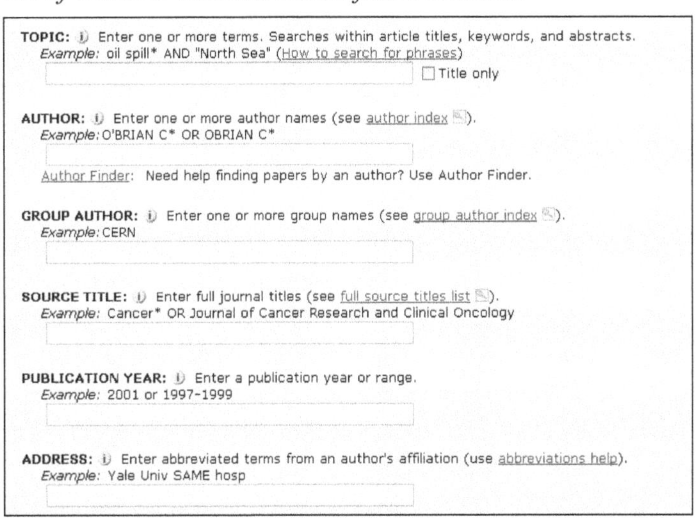

Figure 5. General search results of AU=Eom SB or Eom HB between 1985-2007

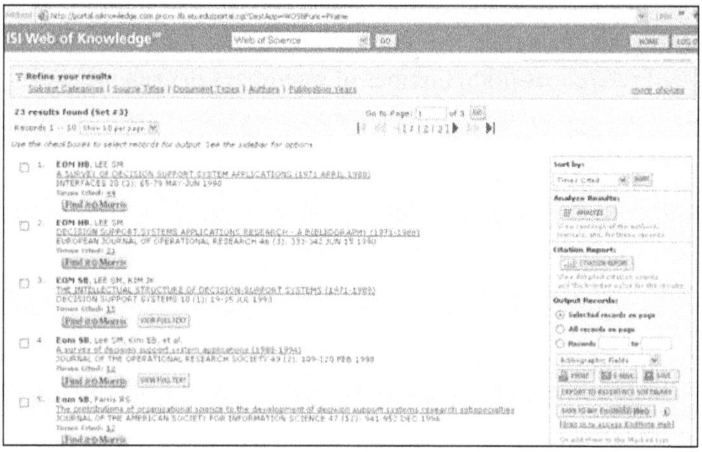

Figure 6. Full record of an item listed in the general search output

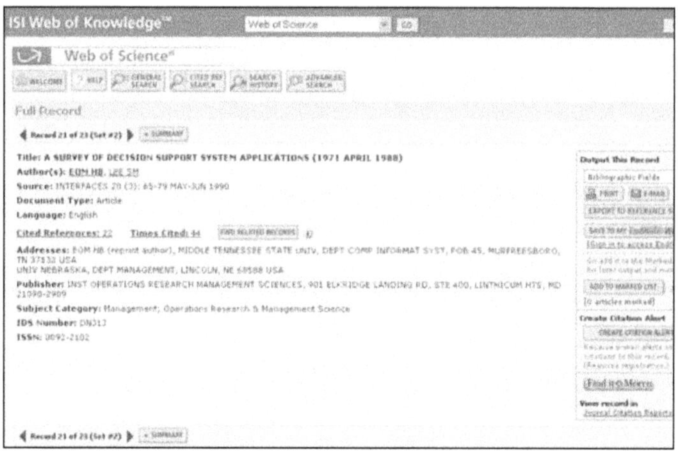

The full record of one of the articles authored by Eom shows that there are 22 cited references (Figure 7) in the article entitled "A Survey of Decision Support System Applications (1971- April 1988)". This article is cited 44 times between 1985 and 2007. The user may be interested in knowing the cited references of Eom's survey article. Clicking the hyperlinked page <u>Cited References: 22</u> in Figure 6 results in the list of 22 cited references of Eom's Survey article (Figure 7). Figure 7 displays the first 8 of 22 cited references.

Figure 7. Cited references of Eom's DSS survey article

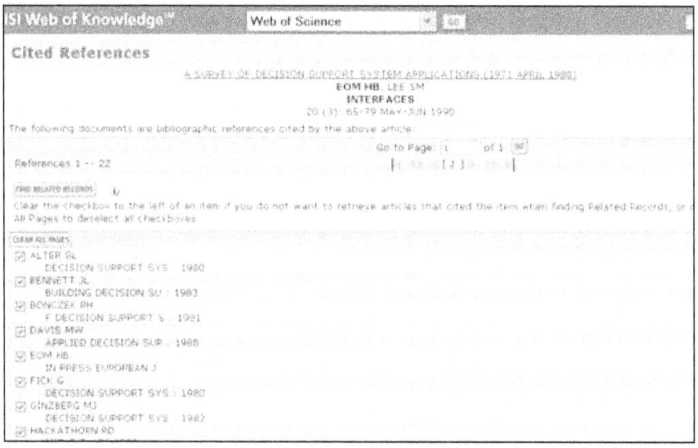

Number of Times Cited

The Web of Science provides the user with the information on which article is citing Eom's survey article. Figure 8 lists part of 44 articles that cites Eom's DSS survey article.

Citation Report for Each Author

Another feature of the Web of Science is a citation report for each author searched. Figure 9 shows two histograms: Published items and citations in each year. Specifically, there are 23 articles in the databases. The sum of the times cited is 190. Average citations per article (8.26) is computed by dividing 190 total cited times by the total number of articles (23).

Figure 10 reflects citations to source items indexed within the Web of Science. This report lists 23 articles in each row. The column displays the total number of citations by each year and the average citation per year for entire contributions by each author and yearly citations.

Cited Reference Search

The general search finds the published articles of a given author's name. The output of citation search/cited reference search is the list of journal articles which include the contributions (articles/books/proceedings, etc.) of authors as references. Any

Figure 8. Forty four documents in the database that cite Eom's DSS survey article

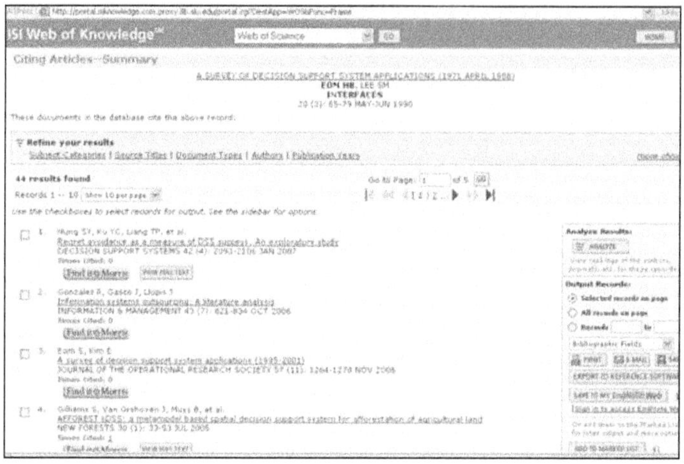

Figure 9. Citation report for author

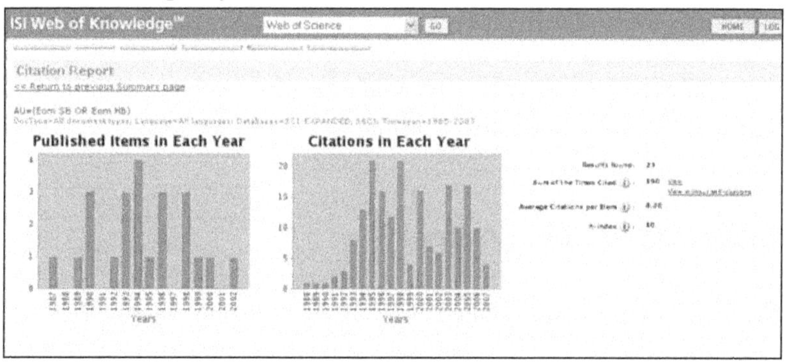

Figure 10. Citation to source items within the Web of Science

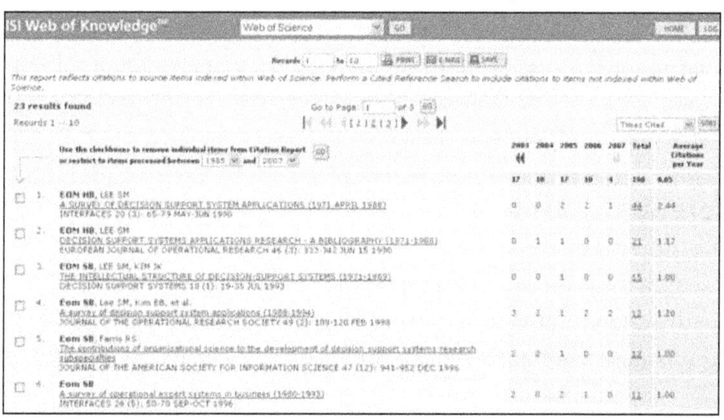

Figure 11. Selection of database(s) and timespan

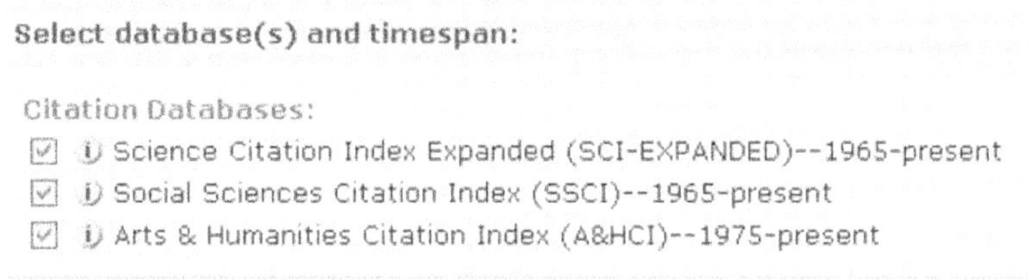

type of search begins with the selection of a database and the time span on the lower portion of the screen span as shown below.

Choose a time span for your search on the right side of the screen.

Click on **Cited Reference Search** near the middle of the screen.

Type in the author's last name, followed by the first and second initials. Use the asterisk if the second initial is unknown or uncertaint. Cited Reference Search searches bibliographies and footnotes of articles. After entering cited author EOM SB or EOM HB, it produced the result of cited reference search—cited reference index. The index lists the following:

- Times cited for each article
- Cited author
- Cited works – Journal name and article tile with only expanded title options
- Year
- Volume
- Page
- Article id
- View record

Related Record Search & Combining Searches

Related Record Searches: Each record carries a button labeled <Find Related Records>. Hitting this button will retrieve articles which have citations in their bibliographies in common with the record you have. These are displayed by articles with the most overlap in their bibliographies at the top. Presumably, articles that share many bibliographic citations are similar. The cited reference and related reference searches are especially useful in cases where you have no useful keywords but do have a relevant article.

Combining Searches: The Advanced Search feature allows Boolean combination of result sets using AND, OR, NOT, etc. The results of General and Cited Reference searches may also be combined.

TECHNICAL LIMITATIONS OF USING CITATION INDEX

Multiple Authorship

All citation index files permit the user to retrieve records only by last name and initials of first author only. All authors except first authors will not be counted when compiling cocitation frequency matrix. The index file contains the author's first name in many different variations. Due to the variations in the first name field such as Keen, P., Keen, PGW., Eom, S. or Eom, SB., it is often suggested to use a wildcard character (*), if you are not certain of middle initials.

Name-Homographs

SSCI only indexes the author's last name and initials. Consequently, citation records by an author of the same last name and initials may not be authored by the same author. The user should be aware the possible consequences of inexact result of query due to name-homographs (e.g., Smith, G., Smith, GA, Smith, GN, Smith, GD, Smith, GR, etc.). It is almost impossible for the indexers to differentiate among authors bearing the same first and middle initials.

Figure 12. Selection of a time span

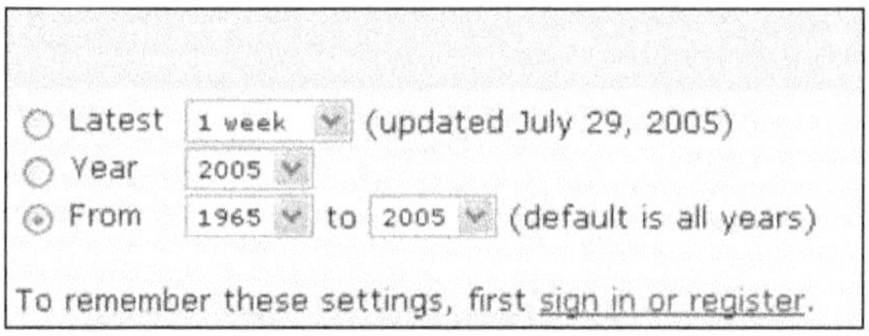

Figure 13. Selection of a search option

Figure 14. Cited reference search results

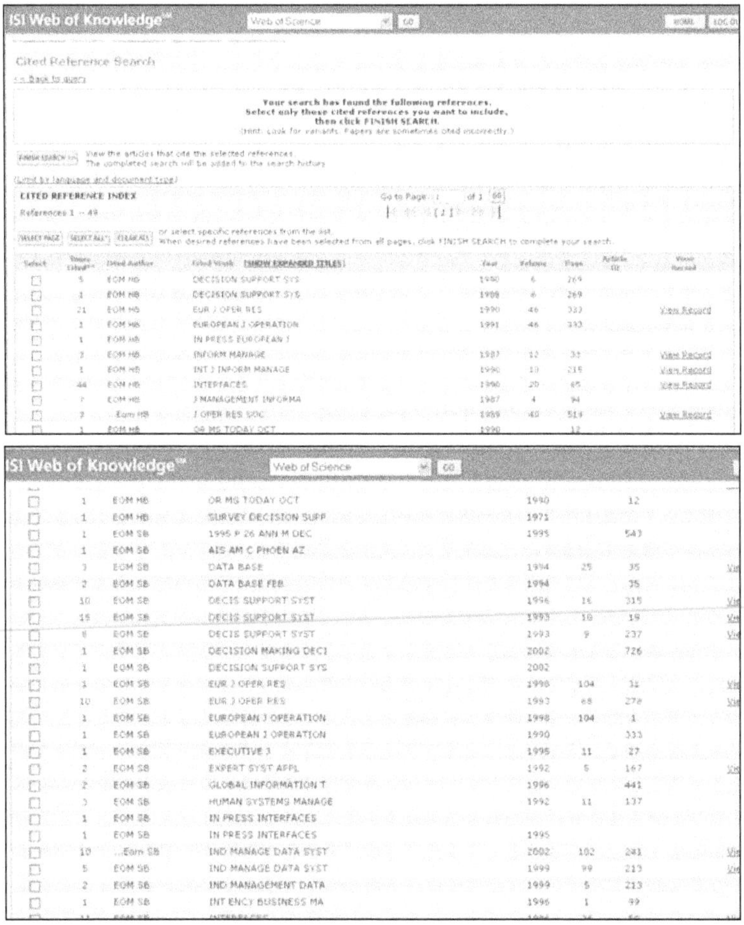

Synonyms

To further complicate the matter, the same author's initials are recorded in many different ways. Some examples of synonyms are Keen, P. or Keen, PGW., Lee, S. or Lee, SM. Furthermore, there are many individuals who have same names. Another problem is that there are no easy ways to handle name changes. The name of women authors can change when they marry or some change their names for other reasons.

CO-CITATION SEARCH USING DIALOG

ACA research is based on author cocitation frequency counts data. The cocitation data can be retrieved using the dialogue classic. Using SciSearch or Social Science Citation Index for the data may not be the best way. SciSearch is a very expensive database, costing $318/hour for access and over $7 for every record displayed or downloaded. The user will quickly run up several hundred dollars in costs just searching and getting numbers without even getting any records. The other alternatives are Research Services and Search Services group at Thomson ISI company. They do custom research for a fee. ACA researchers can purchase data (source articles and referenced articles in a relational database format from Thomson). Additional data mining or analytical programs are generally used to manipulate the data further for more detailed and /or graphical results.

The following section provides some useful information to retrieve cocitation frequency counts using Dialog Classic. In order to conduct the co-citation counts retrieval, the user must have the user name and passwords from the dialog company of Thompson. The following example demonstrates the entire procedure for first time users using the free ONTAP® (**ON**line **T**raining And **P**ractice). You may sign up for your own Dialog account by contacting the Dialog Help Desk.

The company posts a set of user ids and passwords on the web site at: http://training.dialog.com/sem_info/ontap_pw.html.

The ONTAP databases are created for practice purposes. They consist of 38 databases that cover a wide range of subjects content including ONTAP® SciSearch®- a Cited Reference Science Databases [294]. The ONTAP® databases usually contain only 6 months to one year of coverage in the past period, not the most current data.

Co-Citation Search Procedures

Step 1: Go to DialogWeb (http://www.dialogweb.com/) or DialogClassic on the Web http://www.dialogclassic.com)

Step 2: Logging on to the system by entering the ONTAP user number in the box labeled "User ID" and the ONTAP password in the box labeled "Password".

This is the entry point for accessing the dialog classic service at http://www.dialogclassic.com/. Accessing the free ONTAP databases requires the user to use an ONTAP user number and password. This is NOT the one you receive when you sign-up as a subscriber to Dialog.

The company offers many training opportunities for inexperienced users at http://training.dialog.com/

Additionally, the company allows the user "free practice" without incurring charges through ONTAP® (**ON**line **T**raining **A**nd **P**ractice) databases. The practice databases are similar to the real databases except they have only limited small number of records. Practice databases are available for DialogWeb and DialogClassic and Dialog DatastarWeb.

Step 3: Select and Open a database.

Type in "BEGIN 294" on the Command pane. The system accepts "begin" "BEGIN" "B" or "b".

After logging on to the system, the first command is to select and open a database file from available databases. There are 38 databases as of May 31, 2007.

Begin 294 or B 294 and press the enter button.

Step 4: Cited author retrieval by issuing Select (S) command

You will notice that the session pane on the right lists two previous activities up to this point and the database's pane show the database selected. The next step is to retrieve the co-citation frequency between a pair of authors. The syntax of ISI query language is simpler than other query languages such as Oracle SQL.

To retrieve the cocitation counts between Simon HA and Keen P, we will follow the following steps.

Enter the command: Select CA=Simon HA

The system command is not case-sensitive. You could have entered select ca=simmon ha, or SELECT CA=SIMON HA, or Select Ca=Simmon Ha.

After the command, the system displays the output of the query

? s ca= simon ha
S1 7 CA=SIMON HA

The first line beginning with ? is the command entered. The second line beginning with S1 is the output from the system. "S1" is output set 1 and 7 is the number of total records of cited author "SIMON HA".

The following figure shows that three search commands are issued to retrieve the total number of cited records of cited author = ackoff rl, cited author=keen p, cited author = simon ha, and cited author=keen p,. The output shows that the number of cited references for each of them are 0, 7, and 1, respectively.

To find the cocitation frequency between Simon HA and Keen P, a new select command is issued to combine the results of two previous searches (select s2 and s3).

The output of the system indicates that the total number of cocitation frequency between Simon and Keen is 1. That result is stored as set 4 (S4), a intersection of S2 and S3.

Figure 15. Log in screen

Figure 16. Selecting and opening a database

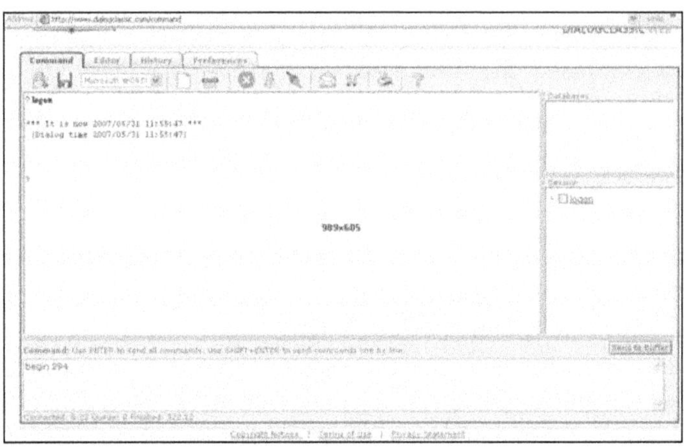

Figure 17. Dialogue classic screen before query

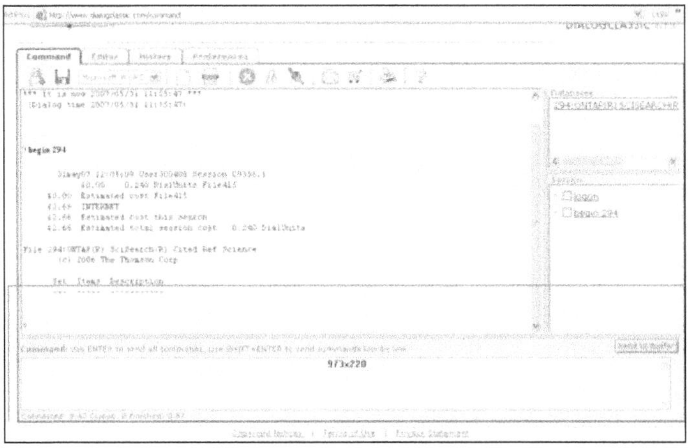

? s s2 and s3

```
        7  S2
        1  S3
S4      1  S2 AND S3
```

Retrieval Using CR (Cited References)

Most of the cocitation frequency counts are retrieved by using CR command. The CR command is used to find references to a paper. For example, SELECT CR=SIMON HA 1977 retrieves papers that include HA SIMON's 1977 article in their references. In ACA, the researchers are interested in the number of papers which cites the works of HA SIMON. Therefore, the following command is used to retrieve the number of papers which include HA SIMON's paper in the references.

Select CR= SIMON HA

Figure 19 displays the results of CR commands. In principal, CA= SIMON H? and CR=SIMON H? produces the same number of outputs. However, CR=SIMON HA did not find any records at all. CA= SIMON HA retrieves 7 records whose references include the work of SIMON H. Therefore, it is suggested that the author's name should be entered as SIMON H?, rather than SIMON HA. This sample ONTAP database contains several records whose references contain SIMON H. The truncation symbol (?) is a wildcard character typically used in many database commands to retrieve any records of SIMON H, SIMON HA, or SIMON HB, etc. Therefore, the use of "?" can be a source of inaccuracy.

Figure 18. Co-citation output screen

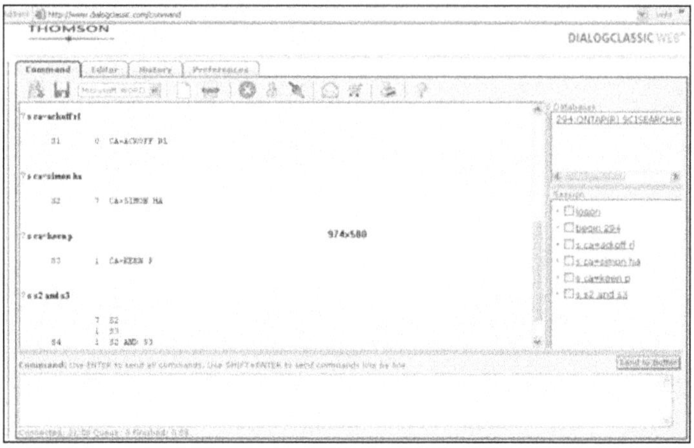

Figure 19. Use of CR and CA

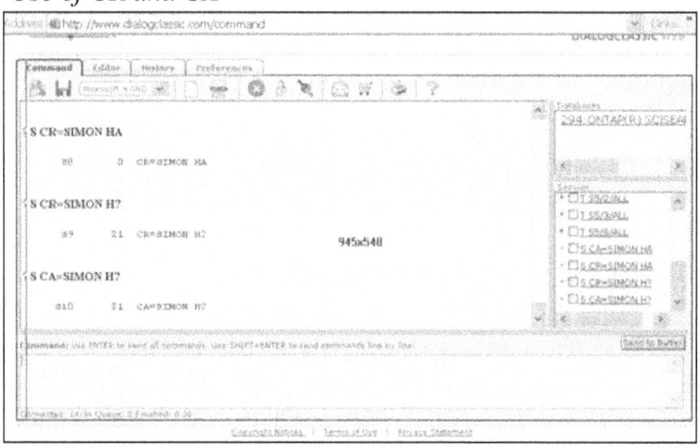

Figure 20. Co-citation retrieval

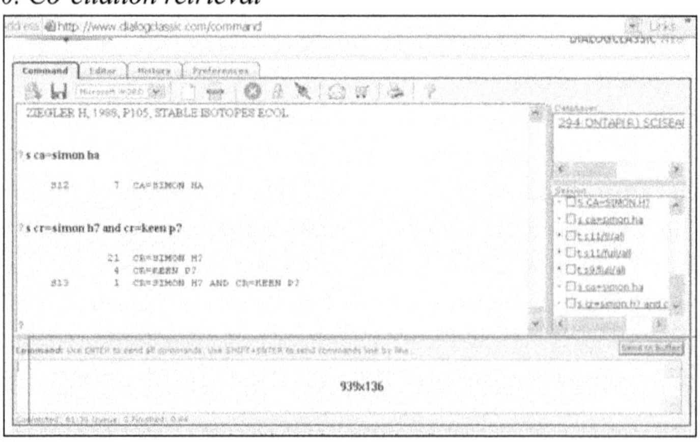

Figure 20 shows a single command that retrieves a count of all papers whose references contains works of SIMON H and KEEN P., where as Figure 18 shows the three different commands to achieve the same result.

Displaying the Outputs

Type commands display outputs from the system. For example, the command (t s2/9/all) will display the following information.

*s2: All information of output set #2, which is cited references that contains HA Simon's articles and the corresponding source articles.

*9: indicates that full contents of records will be displayed.

*all: all records in set 2.

The following is only one of the 7 entire records saved as word documents by pressing diskette icon at Top left.

2/9/1
DIALOG(R)File 294: ONTAP(R) SciSearch(R) Cited Ref Science
(c) 2006 The Thomson Corp. All rights reserved.
10594902 **Genuine Article#:** EQ469 **Number of References:** 15
Title: DYNAMICS IN A HIERARCHICALLY ORGANIZED SYSTEM - BIOLOGICAL EXAMPLES
Author(s): AUGER P
Corporate Source: FAC PHARM DIJON,BIOPHYS LAB,7 BD JEANNE DARC/F-21000 DIJON//FRANCE/
Journal: MATHEMATICAL AND COMPUTER MODELLING , 1990 , V 14 , P 680-685
Language: ENGLISH **Document Type:** ARTICLE
Geographic Location: FRANCE
Subfile: SciSearch; CC ENGI--Current Contents, Engineering, Technology & Applied Sciences
Journal Subject Category: MATHEMATICS, APPLIED
Abstract: We consider a dynamical system containing a large number of elements which can be in different states. In order to define a hierarchical partition of the system, strongly coupled

states are put together creating groups of states. To each group is associated a single global variable. For such a two level system, we study dynamical aspects, and particularly the jumping from a microscopic level towards a more macroscopic level. In this way, we obtain dynamical equations governing the time behavior of the global variables which are shown to be slow-time varying with respect to microscopic variables. Each group reaches an intra-group equilibrium and the fast-varying microscopic variables are replaced by their time averages. This process leads to an important reduction in the number of variables. We also study interactions between the different levels of organization, which can be of two types, top-bottom and bottom-top couplings. In general, intra-group equilibrium depend upon the values of the global variables. This is top-bottom coupling. In many cases, the parameters of the global dynamical equations also depend upon the intra-group equilibrium. This is bottom-top coupling. Then, we present an ecological application by studying a model of coupled individual and population levels.

Descriptors--Author Keywords: HIERARCHICALLY ORGANIZED SYSTEMS; INDIVIDUAL AND GROUP DYNAMICS; TOP-BOTTOM AND BOTTOM-TOP COUPLINGS; COUPLED INDIVIDUAL AND POPULATION LEVELS

Identifiers-- KeyWords Plus: POPULATIONS; MODEL

Research Fronts: 89-0176 001 (DECISION SUPPORT SYSTEMS; MODEL MANAGEMENT; COGNITIVE SCIENCE; COMPUTATIONAL PARADIGM; DESIGN OF AN INTELLIGENT DSS)
89-0906 001 (LARGE-SCALE STEADY-STATE SYSTEMS; VECTOR-VALUED FUNCTIONS; PARAMETRIZED MULTICRITERIA OPTIMIZATION)
89-1368 001 (FUNCTIONAL-DIFFERENTIAL EQUATIONS; INFINITE DELAY SYSTEMS; UNIFORM STABILITY)
89-3082 001 (QUALITATIVE SIMULATION; MATHEMATICAL FOUNDATIONS OF BOND GRAPHS; LOAD BALANCING STRATEGY; STOCHASTIC GENERATION)

Cited References:

ALLEN TFH, 1982, HIERARCHY PERSPECTIV
AUGER P, 1989, HIERARCHICALLY ORGAN
AUGER P, 1980, V8, P82, INT J GEN SYS
AUGER P, 1985, V112, P595, J THEOR BIOL
AUGER P, 1983, V65, P269, MATH BIOSCI
AUGER P, 1986, V3, P41, SYST RES

EHRESMANN AC, 1987, V49, P13, B MATH BIOL
KLIR GJ, 1985, ARCHITECTURE SYSTEMS
LOTKA AJ, 1925, ELEMENTS PHYSICAL BI
MAY RM, 1976, THEORETICAL ECOLOGY
MESAROVIC MD, 1970, THEORY HIERARCHICAL
PATTEE HH, 1973, HIERARCHY THEORY CHA
SIMON HA, 1969, SCI ARTIFICIAL
WEISS P, 1971, HIERARCHICALLY ORG S
WHYTE LL, 1969, HIERARCHICAL STRUCTU

This example of a record from the citation index file clearly shows that cited author, as well as cited reference, only records the first author of each reference. All authors except first authors will not be counted when compiling a cocitation frequency matrix. This has been a fundamental issue in ACA. As this example illustrates, SSCI indexes only the author's last name and initials. The multiple ways of entering an author's initials into the database further aggravates the problem of name homonym, homophone, or homograph. All of these refer to authors that are identical in spelling, but are actually different authors. Consequently, citation records of Lee, S. may include the cited references of Lee, Sang M., or Lee, Sang C., or Lee, Soon, etc. Furthermore, Keen, Peter G.W. cab be entered as Keen, P. or Keen, P.G. or Keen, P.G.W..

Name Synonyms refer to identical authors even though their names are spelled differently. The name of women authors can change when they marry and some change their names for many different reasons. When a name changes, there is no easy way to handle this situation with the citation index files.

Another problem with the ISI citation index file is unification problems. This problem of unification is concerned with the way each author's name and journal title in each cited record is entered into the citation index. In other words, journal titles are entered in many un-standardized ways. For example, MIS Quarterly can be entered in to Citation Index as MIS Q, MISQ, MIS Quart., etc. due to the fact that some journals (e.g., Omega) use their own abbreviated journal name in the references.

Using Basic Commands

This basic commands is an abbreviated version of section 1 of basic skill review (Judith, Mary, Raymond, & Beverly, 2002). See Exhibit A.

SOURCE

Anonymous. (2007). Web of Science Product Information. Retrieved June 17, 2007, 2007, from http://scientific.thomson.com/products/wos/

Judith, L., Mary, W., Raymond, N., & Beverly, C. (2002). Basic Skills Review for Dialog ® Command Language Searchers (Dialog classic & Dialogweb). Retrieved May 31, 2007, 2007, from http://training.dialog.com/sem_info/courses/pdf_sem/basic_skills_review.pdf

Exhibit A.

Basic Commands

COMMAND	EXAMPLE	DESCRIPTION
BEGIN (B)	**b990** **b 990,640** **b papersmj** **b 9 current**	Used to open a database or to move from one database to another. Indicate the database with the appropriate file number(s) or OneSearch® category. More than one file number can be entered, separated by commas, to access multiple databases. File numbers for all Dialog databases and OneSearch categories are listed in the *Dialog Database Catalog.* They are also available on DialogWeb at http://www.dial.Qgweb.com .
SELECT (S)	**s Olympics** **s glaucoma and therap?**	Retrieves records containing the search terms or groups of terms specified and stores them in a numbered set. The truncation symbol on Dialog is a question mark (?). Use it at the end of a word stem to retrieve variant forms of the word, *e.g.,* **s therap?** retrieves therapy, therapies, and therapeutic, etc.

Section 1: Using Basic Commands Basic Skills Review

COMMAND	EXAMPLE	DESCRIPTION
TYPE (T)	**t sl/6, k/l-10** **t sl/9/2,6**	Displays information continuously. Command has three parts: set number*(e.g.,* sl which specifies the set of records to display); format *(e.g.,* 6,k which specifies the portions of the record to display); and records *(e.g.,* l-l0 which specifies what records to display).

		Note: The DISPLAY (D) command, which displays information one screen at a time, may also be used. For example, **d s1/8/all** allows you to see all records from set 1 in format 8, one screen at a time (press <enter> to see the next screen of output.
DISPLAY SETS (DS)	**ds**	Displays the search history since the last BEGIN command.
LOGOFF	**logoff**	Ends the online session and provides estimated cost information. Other commands that can be used to disconnect include STOP, BYE, LOG, and OFF. Note: To logoff temporarily (to take a break or attend to other business) for up to 30 minutes use the LOGOFF HOLD command. When you log on again through the same telecommunication network, you will be able to resume your search where you left off. The system will indicate that you have been reconnected in the file(s) where you were searching prior to the LOGOFF HOLD. Use the DISPLAY SETS (DS) command to review your set history if you have forgotten what sets you created in the last session.

Logical Operators

OPERATOR	FUNCTION
AND	Retrieves records that contain all specified search terms.
OR	Retrieves records that contain at least one of the search terms.
NOT	Eliminates search terms from a set of records.

Formats for Viewing Records

CLASSIC DIALOG	DIALOGWEB	OUTPUT DISPLAYED
Format 6	FREE	Displays titles and publication date.
Format 8	SHORT	Displays titles, publication date and subject indexing (descriptors).
Format 3	MEDIUM	Displays bibliographic citations.
Format 9	FULL	Displays complete records.
Format K	KWIC	Displays keywords (search terms) in the context of an article. May be combined with other formats, *e.g.,* t sl/6,k/l-10.

Note: Formats can vary across databases. Check Dialog Bluesheets for database-specific formats (See Section 4).

Chapter III
The Impact of the ISI Convention of Relying on only the Name of the First Author on ACA Results:
An Empirical Investigation

ABSTRACT

Virtually all ACA studies using Thomson's ISI citation indexes used only the first author to retrieve the cocitation counts. Therefore, this has been a methodological issue in ACA study. First, this chapter introduces two classifications of author cocitations (Rousseau & Zuccala, 2004; Zhao, 2006). Second, literature survey of three studies is conducted to review what has been done to deal with this issue. Third, we discuss the data and cocitation matrix generation system used in our study. Fourth, using our data produced by the cocitation matrix generation system, principal component analysis and non-metric multi-dimensional scaling (MDS) are applied to compare the differences in the process and outcomes of using different cocitation matrices. Finally, we discussed the results of comparative analyses of our study and three other studies in this area. In doing so, the treatment of the diagonal values in a cocitation matrix, the number of authors in all author ACA and first author ACA are critical determinants of ACA outcomes such as number of factors, total percent of variance explained, goodness of fit, and so forth. In future research, it is imperative to use the same assumption in regard to the treatment of the diagonal values, the selection of authors using the same criterion, and so forth. If these are not uniformly standardized, the results are always incomparable and have less meaning. Three conclusions can be reached based on our study. First, all author-based ACA is better to capture all influential researchers in a field than first author-based ACA. It identifies more subspecialties. Finally, all author-based ACA and first author based ACA produce little differences in stress values.

INTRODUCTION

The majority of author cocitation analysis (ACA) have relied on the Institute for Scientific Information (ISI) citation databases. ISI convention allows only the retrieval of papers citing works of which the author is the first or sole author. As McCain states (1990, p. 435), "Authors who consistently publish as coauthors, in second or later position, will not be visible in the retrivals based on these contributions."

The purpose of this chapter is to empirically examine the impact of the ISI convention of relying on only the name of the first author in assembling the cocitation matrix on the investigation of the intellectual structure of academic disciplines. Virtually all ACA studies using Thomson's ISI citation indexes only used the first author to retrieve the cocitation frequency matrix. Therefore, this has been a methodological issue in ACA study. There are many full-text research articles and citation indexes that index all authors of cited works on the web (Zhao, 2006). Further, it is possible to use popular software such as Microsoft Excel to develop co-citation generation system (McIntire, 2007).

First, we introduce two classifications of author cocitations (Rousseau & Zuccala, 2004; Zhao, 2006). Second, literature survey of three studies is conducted to review what has been done to deal with this issue. Third, we discuss the data and cocitation matrix generation system used in our study. Our system is based on Foxbase database management systems and computes author co-citation frequencies between any pair of all (primary and non-primary) authors under study(Sean B. Eom, 2003; Sean B. Eom, 2006). Fourth, using our data produced by the cocitation matrix generation system, principal component analysis and non-metric multi-dimensional scaling (MDS) are applied to compare the differences in the process and outcomes of using different cocitation matrices. Finally, we discussed the results of comparative analyses of our study and three other studies in this area. In doing so, we found that the treatment of the diagonal values in a cocitation matrix, and the number of authors in all author ACA and first author ACA are critical determinants of ACA outcomes such as number of factors, total percent of variance explained, goodness of fit, etc.

CLASSIFICATION OF AUTHOR CO-CITATIONS BY ROUSSEAU AND ZUCCALA

Author co-citation analysis analyzes the author cocitation frequency matrix to infer the intellectual structure of an academic discipline. However, the term author co-citation has not been fully defined. Because of imprecise definitions and restrictions of ISI databases, many ACA studies may have used various cocitation counts. This imprecise definition of author co-citation can lead to imprecise co-citation counts on the diagonal values as well as non-diagonal values.

Rousseau and Zuccala (2004) present a classification of author co-citation forms.

1. Forms Pure first-author cocitations.
2. Pure cocitations.
3. General cocitations.
4. Co-author/cociations.

Pure First-Author Cocitations

The pure first-author cocitation occurs when the reference list of a publication includes at minimum one publication with A as the first or single author and at minimum one publication with B as the first or single author, with one exception that all articles with these two authors (A and B) as co-authors are excluded in the process of computing co-citation frequency. The examples used here come from the reference list of (Sean B. Eom & Farris, 1996). Example A shows that there is one occurrence of cocitation between Eom and Farhoomand. Example B does NOT yield pure first- author cocitation between Smith and Thomas.

Example A
Eom, S. B. (1996). A Survey of Operational Expert Systems in Business (1980-1993). *Interfaces, 26*(5), 50-70.
Eom, S. B., Lee, S. M., & Kim, J. K. (1993). The Intellectual Structure of Decision Support Systems (1971-1989). *Decision Support Systems, 10*(1), 19-35.
Farhoomand, A. F. (1987). Scientific Progress of Management Information Systems. *Data Base, 18*(4), 48-56.

Example B taken from Rousseau and Zuccala
Smith, K. & Thomas, T. (1992) The k-procedure, The Gamma Journal.
Thomas, T. & Smith, K. (1992) More details about The k-procedure, The Gamma Journal.

Pure Co-Citations

The pure first-author cocitation occurs when the reference list of a publication includes at minimum one publication with A as a coauthor and at minimum one publication with B as a coauthor, with one exception that all articles with these two authors (A and B) as co-authors are excluded in the process of computing co-citation frequency. The term "pure" means all contributions coauthored by A and B will not be counted in the counting process. Example C below shows that the cocitations occur between Eom and Keen, Eom and Scott Morton, Lee and Keen, Lee and Scott Morton, Kim and Keen, and Kim and Scott Morton. However, under the pure first author co-citation counting method, co-citation occurs only between Eom and Keen. Example D gives one pure cocitation count between Smith and Thomas. Example E yields no cocitation counts between Smith and Thomas. By definition, example E can be interpreted as having only one reference of Rao and Thomas.

Example C
Eom, S. B., Lee, S. M., & Kim, J. K. (1993). The Intellectual Structure of Decision Support Systems (1971-1989). *Decision Support Systems, 10*(1), 19-35.
Keen, P. G. W., & Scott Morton, M. S. (1978). *Decision Support Systems: An Organizational Perspective.* Reading, MA: Addison-Wesley.

Example D from Rousseau and Zuccala
Janssens, J. & Smith. K., The breakthrough. New World: Delta Publishers.
Peters, P., Zhang, W., & Thomas, T. The Breakthrough revisited. The Alpha Journal.

Example E from Rousseau and Zuccala
Smith, K. & Thomas, T. (1992) The k-procedure, The Gamma Journal.
Thomas, T. & Smith, K. (1992) More details about The k-procedure, The Gamma Journal.
Rao, S. & Thomas, T. (1994) Problems with the k-procedure, The Gamma Journal.

General Co-Citations

The general cocitation occurs when the reference list of a publication includes at minimum one publication with A as a coauthor and at minimum one other publication with B and another coauthor (C). Using example E, the pure cocitation counting method yields NO cocitation counts between Smith and Thomas. However, the general co-citation method yields one cocitation count between them, because there is one reference of Rao & Thomas.

Co-Author/Co-Citation Scores

As shown in Figure 1, co-citation/co-author counting occurs when the reference list of a publication includes at minimum one publication coauthored by A and B, or cited together using any of the previous methods (pure first-author co-citations, pure cocitations, and general co-citation). Figure 1 implies that when any co-citation counting methods (outer circles) that embed other methods (inner circle), there are more cocitation counting methods available in addition to the methods used the inner circles.

Figure 1. Classification hierarchy of author co-citation forms

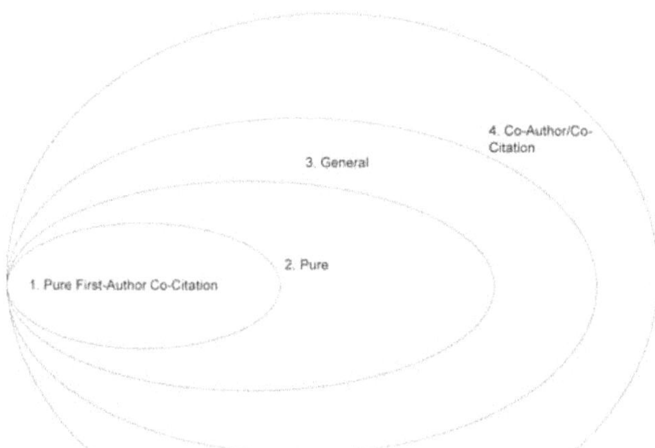

Source: Rousseau and Zuccala (2004, p. 516.)

CLASSIFICATION OF AUTHOR CO-CITATIONS BY ZHAO

As Table 1 illustrates, Zhao used three different ways of counting cociation counts in his research. The traditional first author only method in the first column in Table 1 recognizes the occurance of the first co-citation between Lee and Hair in only citing paper 2. This is because paper 2 contains a reference article by Lee and another reference article by Hair and Lee. The second and third columns show all-author co-citation counts. The difference between inclusive and exclusive all-author co-citations is that the exclusive approach excludes any co-citation counts within a citing paper. Paper 3 cites only one paper co-authored by Lee and Hair. This paper will not be used as a co-citation count.

When comparing the two classification schemes by Rousseau and Zuccala (2004) and Zhao (2006), the pure first author cocitation method and co-author/cocitation method of Rousseau and Zuccala are equivalent to Zhao's first author cocitation method and inclusive of the all author cocitation method, respectively. The exclusive all author cocitaion method by Zhao is equivalent to the general cocitation method of Rousseau and Zuccala.

Table 1. Illustration of the difference between three co-citation definitions (Authors Lee and Hair being co-cited?)

Reference list		Counting method		
Citing Paper		First-author co- citation	Inclusive all-author co-citation	Exclusive all-author co-citation
paper 1: Lee,, K. (2000) Lee, K., & Hair, S. (1998)	CSLT and XML. *XML Journal* S. RDF and OWL. *The Semantic Web*	No	Yes	Yes
paper 2: Lee, K. (2000) Hair, S., & Lee, K. (2003)	CSLT and XML. *XML Journal.* RDF-Schcma. *Ontologia*	Yes	Yes	Yes
paper 3: Lee, K.., & Hair, S. (1998).	RDF and OWL. *The Semantic Web*	No	Yes	No
Total number of co-citations between Lee and Hair		I	3	2

Source: Zhao (2006, p. 1579).

LITERATURE SURVEY

There are only a handful of studies that deal with the issue of the impact of different cocitation counting method s on the intellectual structure of a discipline.

The Persson Study

Persson (2001) used 7001 citing papers and their 106, 876 citations in the library and information science area in the CD-ROM version of Social Science Citation Index. This covered 1986-1996 to investigate the impacts of first author only vs. all author co-citation method on visualizing the intellectual structure of a field. Using the data, the two co-citation MDS maps of 57 authors in information science are constructed using the two different counting methods. Persson also complied a citation counts table that show the citation counts of each author under two different counting methods (all author vs. first author only). The table identified that a large number of well-known prolific authors did not get many citation counts under the first author only method.

Persson demonstrates that "all author" citation counts should be preferred when visualizing the structure of research fields. "First author" only citation studies distort the picture in terms of most influential researchers, while the subfield structure tends to be same for both methods. Persson disregarded more than 90% of references to papers not indexed by the ISI databases as sources papers. Persson concludes that first-author based ACA may leave out several influential researchers compared to all author based ACA. Nevertheless, the subspecialties identified by both approaches are almost identical.

The Zhao Study

Zhao (2006) compared the results of two different types of cocitation counting: first author only cocitation vs all author cocitation. He distinguished three different definitions of counting co-citations. He counted only the first five authors instead of all of the authors and called it a simplified approach. As Zhao admits, due to the limited number of authors, the results of her investigation may not tell the whole story as to the impact of the current prevailing author counting method on the assessment of the intellectual structure of an academic field. The primary objectives of her investigation are to find how and why results under the two different approaches might be different from each other.

Zhao's data include 312 publications and their 4578 citations, from the NEC Corporation Research Institute's *Research Index,* available at http:citseer.ist.psu.edu/. She developed a Java program to search all documents indexed by ResearchIndex.

Using another Java program developed by Zhao that generates cocitation counts, a raw cocitation frequency counts matrix is compiled. Later, authors whose frequency matrix contains more than 95% of blank cells (zero cocitation frequency with others) are deleted from the matrix. The final author set consists of 100 authors.

Table 2 summarizes major findings from factor analysis. Table 2 shows that when the same number of authors is selected, first author only ACA produces more factors that all author based ACA. The MDS results indicate that "the picture produced through this simplified all-author co-citation counting contains author groups that are more coherent, and is therefore considerably clearer."

The Schneider and Others Study

To extend the Zhao study, Schneider et al (Jesper Wilborg Schneider, Larsen, & Ingwersen, 2007) used the citation index which is extracted from a corpus of full text XML documents. The XML version of the IEEE Computer Society journal includes 16,819 articles over the period of 1995 to 2004. The citing 16, 819 articles contains a total of 212, 657 cited references. Each article on the average contains 12.6 references. An arbitrary citation trash hold of 75 cited authors is used for both all-author ands firs author ACA studies. Their study comparatively investigates the impacts of all author vs. first author cocitation analyses on various dimensions of ACA outcomes. In this study, they also try to tackle an issue of matrix generation and transformation raised by (Leydesdorff & Vaughan, 2006; Jesper W. Schneider & Borlund, 2007a, 2007b).

In regard to the debates on choice of proximate measures and their impacts on ordination and clustering results, Schneider and Borlund (2007a; , 2007b) introduce

Table 2. Results of factor analysis by Zhao

	First author analysis	All author analysis	Remarks
Number of factors	11	5	The eigen value criterion
Total variance explained	96%	97%	Authors with Factor loadings greater than .4 are included.
Specialties found	4 major areas and 7 minor other areas	4	

two important issues (matrix generation and the composition of proximity measures). The two matrix generation approaches most likely will result in different ordination and clustering results because the two approaches will produce different proximity rankings of objects. To deal with these issues, Schneider and Borlund introduced two statistical validation tools (the Mantel test and Procrustes analysis) to help ACA researchers choose suitable proximity measures. These validation tools compare and evaluate the degree of monotonicity between different proximity measures or their ordination results.

Their research aims to answer the following two questions: Do a different data set and matrix generation techniques result in different outcomes from the previous studies? Unlike Zhao (2006), who used only the first five authors, Schneider and others used all authors in their in the counting.

Co-Citation Matrix Generation

As we introduced in Chapter I under the section heading Recent Developments in ACA, many methodological issues were raised recently. One of them is the issue of cocitation counts matrix generation and transformation. Leydesdorff and Vaughn argue that only the asymmetrical citation matrix should be processed to derive the proximity matrix such as Pearson's, and therefore similarity measure should not be derived from the symmetrical cocitation matrix. A co-citation matrix itself is a proximity matrix. Therefore, it is not necessary to apply a similarity measure to construct a proximity matrix (Leydesdorff & Vaughan, 2006).

In their study, they compiled two different cocitation matrices – symmetrical and asymmetrical. Symmetrical cocitation matrix (the Drexel approach) is the one we used throughout this book. For details in creating asymmetrical cocitation matrix, see (Leydesdorff & Vaughan, 2006; Jesper Wilborg Schneider, Larsen, & Ingwersen, 2007). Unlike Zhao who counted only the first five authors, they counted all of the authors in a cited reference. An arbitrary citation threshold of 75 is used to compile the cocitation data.

Diagonal Values

In Schneider and others' study, diagonal values are treated as missing data due to possible implications. As discussed in Chapter I, the diagonal value is an important issue in ACA. In Chapter IV, we fully investigated the impact of treating six different approaches on the outcome of ACA analysis. If the pure cocitation counts are not used, the next best alternatives are as follows. They are the missing value approaches, mean cocitation value approach, and the highest off-diagonal value approaches in the order of the highest total variance explained.

Results

The results of factor analysis and MDS are summarized in Tables 3 and 4. This information indicates that the inclusion of all authors results in producing 2 dimensional MDS visualization that better fit the data with lower stress values. It also indicates that all author ACA may lead to a stronger concentration in the MDS map. As far as the differences between the two matrix generation approaches, the Drexel approach produced the results that have much lower stress values and are more concentrated into groups. Therefore their study supports part of Zhao's conclusion but did not support the idea that first author ACA produces more factors.

DATA AND THE COCITATION GENERATION SYSTEM

The data for this research is collected as part of on-going research to investigate the intellectual structure of the decision support systems research area. The database consists of 692 citing articles in the DSS area over the period of 1971-1990. A citing

Table 3. Results of factor analysis by Schneider and others

	First author analysis	All author analysis	Remarks
Number of factors with the conventional approach	29	25	The Kaiser-Guttman Rule (eigen value criterion)
Total variance explained with the conventional approach	59%	63%	Authors with Factor loadings greater than .4 are included.
Number of factors with the Drexel approach	16	15	
Total variance explained with the Drexel approach	80%	86%	

Table 4. Results of MDS analysis (Stress 1 values) by Schneider and others

Configuration	First author/ conventional approach	First author/ Drexel approach	All author/ conventional approach	All author/ Drexel approach
2 dimensions	.251	.174	.219	.132
3 dimensions	.186	.102	.181	.096

article is selected if: (1) it discussed the development, implementation, operation, use, impact of DSS, or DSS components; or (2) for DSS articles related to contributing disciplines, they were explicitly related to the development, implementation, operation, use, impact of DSS, or DSS components. Subsequent steps were taken to create a database file that consists of a total of 15,030 *cited* references.

We have developed a cocitation generation system using Fox-base database management systems (Sean B. Eom, 2003; Sean B. Eom, 2006). The system computes author co-citation frequencies between any pair of all (primary and non-primary) authors under study. The system enables the users to overcome a restriction of using ISI databases by giving access to cited coauthors as well as the first authors.

RESEARCH METHODOLOGY

This study is based on the assumption that "bibliographic citations are an acceptable surrogate for the actual influence of various information sources" (McCain, 1984). It also assumes that the cocitation analysis of a field yields a valid representation of the intellectual structure of the field (Bellardo, 1980; McCain, 1984, 1986; Smith, 1981). The focus of this study is the investigation of the impact of the ISI convention of relying on only the name of the first author in assembling the cocitation matrix on the intellectual structure of academic disciplines. Therefore, the author cocitation generator is modified to count only the first author. Alternatively, we may delete all author names in the middle name field to generate first author only cocitaion counts.

Selection of Author and Compilation of Cocitation Frequency Matrix

Authors for this study were selected by an objective approach. The majority of ACA research uses a subjective approach selecting authors based on the researchers' judgment. For example, Schneider and others chose an arbitrary citation threshold of 75 cited authors. Zhao selects 100 authors from a database based on 312 publications. Authors for this study were selected by a two-stage process.

The first step is the loose screening based on the minimum citation of 60 or more times during the period of 1969-1990, based on the prior work of Culnan (1986). Rather than selecting the author set based on purely researchers' judgment, the first step of our approach uses citation frequency (60) of each candidate author from 15,030 *cited* references. This stage yields a list of about 100 names, whose citation frequency exceeded the threshold value of 60. For any ACA research using commercial database such as ISI social sciences citation index, it is a very difficult, if not impossible, task due to the time and cost of database search. However, it is a reasonable approach that can be adopted by any ACA researcher using custom databases. There are no predefined strict rules for determining the citation threshold itself.

In the second step, the final author set was determined. While the first step involves the use of citation counts of each author, the second step is based on co-citation counts for further screening. Since the primary focus of ACA is to establish the relationship among selected authors, the second stage is to measure the *connectivity* among the authors selected in the previous step. Due to the possible instability of small cocitation counts, author cocitation analysis researchers introduced several ad hoc criteria for further screening a large pool of candidate authors to finalize a list of authors. The criteria include a *mean cocitation rate* above a certain lower limit per author in each time period (e.g., nine for 10 years of Social Scisearch data), cocitation with at least one-third of the entire author set, or restricting the final author set to the 20% receiving the highest number of citations and cocitations in initial retrieval trials. For further details on several different approaches to compiling a predetermined list of authors, see McCain (1990). However, all these criteria were suggested for application to the commercial on-line databases such as SCISEARCH and SOCIAL SCISEARCH.

Our databases are significantly different from those commercial databases in terms of size of records. Besides, the cocitation matrix generation system we developed gives access to cited coauthors as well as first authors. Due to these differences, we could not follow the suggested criteria e.g., (McCain, 1990) such as "a mean cocitation rate of 'x' or more cocitations in each time period." Rather, we had to invent a new criterion through the method of trial and error. We experimented with the sensitivity of changing the cocitation threshold on the final outcomes (number of meaningful factors to accurately represent DSS research subspecialties). With our databases, we conclude that the number of cocitations of an author with himself/herself can be a better criterion to determine the final author set due to the simplicity. Applying the mean cocitation criteria to our Excel worksheet file (the output from the cocitation matrix generation system) involves too many computations. Whenever we delete/add an author to the final author set, we need to compute the mean cocitation rate of each author again.

Using the cocitation rate of 25 with himself/herself in the period (1969-1990), the initial set of 100 authors was reduced to 67 as the final author set for further analysis. Counting the cocition rate with himself/herself is equivalent to "inclusive all-author cocitaion" counting method (Zhao, 2006). Later, we experimented and lowered the threshold from the 25 cocitation rate to a 20 cocitation rate. Although the number of authors increased from 67 to 80, the number of meaningful factors remained the same, and the major conclusions reported here did not change. Also, it is important to point out that cocitation thresholds themselves, as sole connection criteria, are suspect in a highly multidisciplinary area. One should look at the overall connectedness and the focused cocitation counts as well.

Compilation of Cocitation Frequency Matrix and Diagonal Values

To overcome a standard problem with the Institute for Scientific Information (ISI) databases which code only the first author of a cited work, a Fox-Base based matrix generation system was developed to compute a cocitation frequency between any pair of authors. The cocitation matrix generation system we developed gives access to cited coauthors as well as first authors.

In order to investigate the impact of the ISI convention of relying on only the name of the first author in assembling the cocitation matrix on the identification of the intellectual structure of DSS field, we have assembled two cocitation matrices using two different co-citation counting methods (First author only counting and inclusive all author counting). The co-citation matrix using the inclusive all author counting method consists of 67 author variables and 67 observations, while the first author only counting method produced a matrix of 47 variables and 47 observations. The different number of authors, 67 vs. 47, under each counting method must be interpreted as an important factor that will produce different intellectual structures of an academic discipline. Recall that Zhao selected the same number (100 authors) of authors under the two different cocitation counting methods. This is an important issue for further clarification. As said earlier, we used two different stages of author selection process to be included in the final author set. The first step is based on citation frequencies. The second step is co-citation frequencies. Applying the same threshold of citation and cocitation frequencies to the two co-citation frequency matrices generated by the first author and all author counting methods should produce different numbers of the final author set. This is because the non-primary (first) authors usually do not receive any credit at all. Consequently non-primary authors co-citation counts substantially decrease under the first author only counting method.

Several ad hoc criteria for finalizing a list of authors include the following.

- a *mean cocitation rate* above a certain lower limit per author in each time period (e.g., nine for 10 years of Social Scisearch data),
- cocitation with at least one-third of the entire author set, or
- restricting the final author set to the 20% receiving the highest number of citations and cocitations in initial retrieval trials.

These are ad hoc criteria for those who are using commercial databases. Although ACA researchers using other custom databases can not blindly apply these criteria, it is still important to consider the possible instability of small cocitation counts. What it means is that the first author only co-citation counting method should produce a smaller number of the final author set. To produce comparable results to Schneider and others, diagonal values are treated as missing data. It is unknown how the diagonal values are treated in Zhao's study.

Results of the Factor Analysis

The two cocitation matrices of 67 authors (all author ACA) and 47 authors (first author ACA) are processed by the SAS factor procedures. Originally, the diagonal values of cocitation frequency counts matrix are computed by adding the three highest off-diagonal values and dividing by two, as suggested by (McCain, 1990). To make a meaningful comparison with Schneider et al. study, another co-citation matrix with the diagonal values treated as missing values is prepared. Therefore, four different cocitation matrices are processed to see the magnitude of the diagonal value impacts on the ACA outcomes. First, I will introduce the results of using the cocitation counts matrix with the missing diagonal value treatment, so that direct and meaning comparisons can be made with the results of Schneider et al. Inclusive all author cocitation matrix and first author only matrix are analyzed by the principal component analysis (promax rotation) with the latent root criterion (eigenvalue 1 criterion) to obtain the initial solution of 6 and 5 factors, respectively. Total cumulative percent of variances are 89.53% and 92.62% of each, respectively.

The study of Schneider and others (Jesper Wilborg Schneider, Larsen, & Ingwersen, 2007) produced a 15-factor model for the Drexel approach to all author co-citation analysis, which explains 86% of the variance; a 16-factor model for the Drexel approach to first author cocitation analysis, which explains 80% of the variance. The summary of the two previous studies indicated that the first author cocitation study produced more numbers of factors while explaining a smaller percentage of variances. This result is most contrasting when comparing with this study of Eom. Contrary to the conventional wisdom (increasing number of factors increases the percentage of variance explained), increasing number of factors decreases the percentage of variance explained. The message is clear. While both authors (Zhao and Schneider et al.) keep the same number of authors in the cocitation matrices, the instability of small cocitation counts in the co-citation matrices increased.

The most important implication of these three studies is that first author only cocitation analysis undeniably decreases the percentage of variance explained. By omitting non-primary authors contributions, we may possibly misrepresent the intellectual structure of an academic discipline.

Secondly, by keeping the same number of authors the researchers are producing the results of an increasing number of factors that may poorly represent the research subspecialties. Thirdly, the result of Eom's study clearly shows that not counting non-primary authors contribution in ACA analysis results in the loss of several meaningful factors.

As Table 3 shows, the treatment of the diagonal values is a powerful factor that determines variances of the factor solution. This indicates that ACA scholars need to pay more attention and further investigate the diagonal values impacts on ACA outcomes. In that regard, our discussions in Chapter III can be a useful starting point.

Results of Multidimensional Scaling Analysis

A study by Zhao claimed that "the picture produced through this simplified all-author co-citation counting contains author groups that are more coherent, and is

Table 5. Eigenvalues of all author ACA

Eigenvalues of the Correlation Matrix: Total = 67 Average = 1

	Eigenvalue	Difference	Proportion	Cumulative
1	31.4492008	18.2153433	0.4694	0.4694
2	13.2338575	4.5277371	0.1975	0.6669
3	8.7061204	5.7179785	0.1299	0.7969
4	2.9881420	0.9669530	0.0446	0.8415
5	2.0211889	0.4335541	0.0302	0.8716
6	1.5876348	0.6851982	0.0237	0.8953
7	0.9024366	0.0995275	0.0135	0.9088
8	0.8029091	0.1653553	0.0120	0.9208
9	0.6375538	0.1644479	0.0095	0.9303

Table 6. Eigenvalues of first author ACA

Eigenvalues of the Correlation Matrix: Total = 47 Average = 1

	Eigenvalue	Difference	Proportion	Cumulative
1	24.3816517	14.2841490	0.5188	0.5188
2	10.0975027	4.9948716	0.2148	0.7336
3	5.1026311	2.8298053	0.1086	0.8422
4	2.2728258	1.0682554	0.0484	0.8905
5	1.2045704	0.5813621	0.0256	0.9162
6	0.6232083	0.1963470	0.0133	0.9294
7	0.4268614	0.0558913	0.0091	0.9385
8	0.3709701	0.0693831	0.0079	0.9464
9	0.3015870	0.0168864	0.0064	0.9528
10	0.2847006	0.0399047	0.0061	0.9589
11	0.2447958	0.0048418	0.0052	0.9641

Table 7. Results of factor analysis by Eom

	First author analysis (total number of authors 47)	All author analysis (total number of authors 67)	Remarks
Number of factors with the diagonal values treated as missing values	5	6	The Kaiser-Guttman Rule (eigen value criterion)
Total variance explained with the diagonal values treated as missing values	92.62%	89.53%	
Number of factors with the diagonal values computed by adding three highest off-diagonal values and dividing it by 2.	7	9	
Total variance explained with the diagonal values computed by adding three highest off-diagonal values and dividing it by 2.	81.94%	84.66%	

therefore considerably clearer." Schneider et al. (Jesper Wilborg Schneider, Larsen, & Ingwersen, 2007) made a similar claim that "the all-author ACA maps result in stronger concentration of the co-cited authors into clusters, regardless of the approach to matrix generation."

The purposes of MDS are to help researchers identify the "hidden structures" in the data and visualize relationships among/within the hidden structures to give clearer explanations of these relationships to others (Hair, Anderson, & Tatham, 1987; Kruskal & Wish, 1990). Three SAS procedures (MDS, PLOT, and G3D) are necessary to convert the author cocitation frequency matrix to two or more dimensional pictures of data. The distance matrix produced earlier by using the PROC distance procedure should be converted to a coordinate matrix. The coordinate matrix is used to produce two-dimensional plots and annotated three-dimensional scatter diagrams. The PLOT and G3D procedures process the coordinate matrix to visualize the similarity and dissimilarity within each group of an academic discipline, as well as the similarity and dissimilarity among the various subspecialties within an academic discipline. In ACA study, 3D scatter plots without labels on data points provide little information for the ACA researchers. The annotate facility in the SAS system produces figures with the name of the author on each data point. The PROC MDS procedure includes many of the features of the ALSCAL procedure (Young, Lewyckyj, & Takane, 1986) and some features of the

Tables 8 through 11 show stress 1 values of four MDS solutions based on the method of counting cocitation counts (all author vs. first author) and the number of dimensions (two vs. three). Table 14 compares all three studies to reach conclusions in regard to the impact of cocitation counting methods on Stress 1 (badness-of-fit) values.

The three dimensional MDS map as shown in Figures 2 and 3 (the badness of fit values are .0670 and .0572, respectively) enabled us to map the overall relationships between the DSS subspecialties and contributing disciplines, as well as the interrelationships within both the DSS subspecialties and the contributing disciplines to DSS. Multidimensional scaling maps show the big picture of inter-cluster relationships. The placement of authors on the center of the MDS map means that those authors are linked with a substantial portion of the author set, with relatively high correlations. Placement near the periphery represents a more focused linkage. Figures 4 through 7 present the two dimensional MDS map, which is the projection of the first and second planes presented in the three-dimensional MDS maps.

DISCUSSIONS

To make a meaningful comparison, there are several important factors, assumptions and criteria to be applied to every study. Here are some core requirements.

Table 8. All author ACA MDS solution in three dimensions – Iteration history

The SAS System 14:04 Saturday, January

Multidimensional Scaling: Data=WORK.DISTANCEM.DATA
Shape=TRIANGLE Condition=MATRIX Level=ORDINAL
Coef=IDENTITY Dimension=3 Formula=1 Fit=1

Mconverge=0.01 Gconverge=0.01 Maxiter=100 Over=2 Ridge=0.0001 Alternate=ROW=34

		Badness-of-Fit	Change in	Convergence Measures	
Iteration	Type	Criterion	Criterion	Monotone	Gradient
0	Initial	0.1845	.	.	.
1	Monotone	0.1539	0.0307	0.0995	0.5902
2	Gau-New	0.1036	0.0503	.	.
3	Monotone	0.0981	0.005446	0.0297	0.5079
4	Gau-New	0.0890	0.009147	.	.
5	Monotone	0.0800	0.009021	0.0291	0.4624
6	Gau-New	0.0722	0.007763	.	.
7	Monotone	0.0705	0.001718	0.0135	0.2525
8	Gau-New	0.0690	0.001523	.	.
9	Monotone	0.0684	0.000534	0.007593	0.1988
10	Gau-New	0.0679	0.000523	.	0.1725
11	Gau-New	0.0671	0.000842	.	0.0607
12	Gau-New	0.0670	0.0000428	.	0.0292
13	Gau-New	0.0670	8.3729E-6	.	0.0152
14	Gau-New	0.0670	2.1831E-6	.	0.008216

Convergence criteria are satisfied.

Table 9. First author ACA MDS solution in three dimensions – Iteration history

non-mteric mds with level=ordinal option 14:46 Saturda

Multidimensional Scaling: Data=WORK.DIST.DATA
Shape=TRIANGLE Condition=UN Level=ORDINAL
Coef=IDENTITY Dimension=3 Formula=1 Fit=1

Mconverge=0.01 Gconverge=0.01 Maxiter=100 Over=2 Ridge=0.0001 Alternate=ROW=24

		Badness-of-Fit	Change in	Convergence Measures	
Iteration	Type	Criterion	Criterion	Monotone	Gradient
0	Initial	0.1737	.	.	.
1	Monotone	0.1086	0.0650	0.1336	0.5824
2	Gau-New	0.0730	0.0357	.	.
3	Monotone	0.0698	0.003162	0.0190	0.3239
4	Gau-New	0.0670	0.002838	.	.
5	Monotone	0.0627	0.004238	0.0225	0.2913
6	Gau-New	0.0621	0.000611	.	.
7	Monotone	0.0600	0.002175	0.0158	0.2652
8	Gau-New	0.0596	0.000316	.	.
9	Monotone	0.0589	0.000760	0.008721	0.2454
10	Gau-New	0.0586	0.000272	.	0.2390
11	Gau-New	0.0574	0.001197	.	0.1132
12	Gau-New	0.0573	0.000135	.	0.0818
13	Gau-New	0.0573	0.0000203	.	0.0785
14	Gau-New	0.0572	0.0000157	.	0.0740
15	Lev-Mar	0.0572	0.0000140	.	0.0630
16	Gau-New	0.0572	0.0000101	.	0.0602
17	Lev-Mar	0.0572	4.4083E-6	.	0.0544
18	Gau-New	0.0572	5.5702E-6	.	0.0532
19	Lev-Mar	0.0572	2.9199E-6	.	0.0483
20	Gau-New	0.0572	3.0724E-6	.	0.0481
21	Lev-Mar	0.0572	2.3201E-6	.	0.0436
22	Gau-New	0.0572	4.7546E-7	.	0.0456
23	Lev-Mar	0.0572	9.4071E-8	.	0.0434
24	Lev-Mar	0.0572	2.5484E-6	.	0.0423
25	Lev-Mar	0.0572	3.7349E-7	.	0.0401
26	Lev-Mar	0.0572	1.8155E-6	.	0.0395

Table 10. All author ACA MDS solution in two dimensions – Iteration history

The SAS System 15:22 Saturday,

Multidimensional Scaling: Data=WORK.DISTANCEM.DATA
Shape=TRIANGLE Condition=MATRIX Level=ORDINAL
Coef=IDENTITY Dimension=2 Formula=1 Fit=1

Mconverge=0.01 Gconverge=0.01 Maxiter=100 Over=2 Ridge=0.0001 Alternate=ROW=34

		Badness-of-Fit	Change in	Convergence Measures	
Iteration	Type	Criterion	Criterion	Monotone	Gradient
0	Initial	0.2286	.	.	.
1	Monotone	0.1605	0.0680	0.1618	0.5138
2	Gau-New	0.1242	0.0363	.	.
3	Monotone	0.1216	0.002609	0.0242	0.3459
4	Gau-New	0.1179	0.003768	.	.
5	Monotone	0.1139	0.003978	0.0249	0.2508
6	Gau-New	0.1128	0.001082	.	.
7	Monotone	0.1123	0.000541	0.0104	0.1867
8	Gau-New	0.1117	0.000586	.	.
9	Monotone	0.1112	0.000448	0.008483	0.1480
10	Gau-New	0.1108	0.000452	.	0.1307
11	Gau-New	0.1095	0.001245	.	0.0178
12	Gau-New	0.1095	0.0000105	.	0.008184

Convergence criteria are satisfied.

Table 11. First author ACA MDS solution in two dimensions – Iteration history

non-mteric mds with level=ordinal option 14:46 Saturday,

Multidimensional Scaling: Data=WORK.DIST.DATA
Shape=TRIANGLE Condition=UN Level=ORDINAL
Coef=IDENTITY Dimension=2 Formula=1 Fit=1

Mconverge=0.01 Gconverge=0.01 Maxiter=100 Over=2 Ridge=0.0001

Iteration	Type	Badness-of-Fit Criterion	Change in Criterion	Convergence Measures: Monotone	Convergence Measures: Gradient
0	Initial	0.2085	.	.	.
1	Monotone	0.1199	0.0886	0.1691	0.5665
2	Gau-New	0.0967	0.0231	.	.
3	Monotone	0.0933	0.003385	0.0244	0.2459
4	Gau-New	0.0924	0.000974	.	.
5	Monotone	0.0889	0.003462	0.0246	0.1266
6	Gau-New	0.0886	0.000273	.	.
7	Monotone	0.0884	0.000253	0.006624	0.0899
8	Gau-New	0.0880	0.000380	.	0.0180
9	Gau-New	0.0880	0.0000228	.	0.0146
10	Gau-New	0.0880	0.0000190	.	0.0152
11	Gau-New	0.0879	0.0000207	.	0.0157
12	Gau-New	0.0879	0.0000219	.	0.0158
13	Gau-New	0.0879	0.0000219	.	0.0155
14	Gau-New	0.0879	0.0000207	.	0.0148
15	Gau-New	0.0878	0.0000186	.	0.0137
16	Gau-New	0.0878	0.0000159	.	0.0125
17	Gau-New	0.0878	0.0000130	.	0.0111
18	Gau-New	0.0878	0.0000102	.	0.009701

Convergence criteria are satisfied.

Figure 2. All author ACA G3D plot (67 authors)

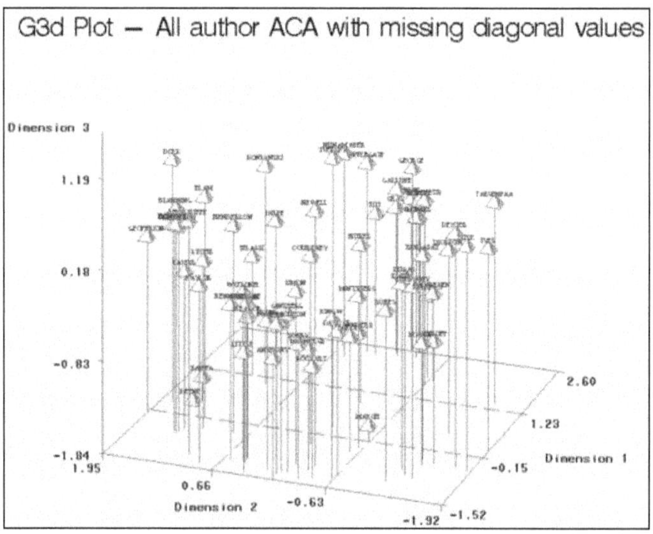

Figure 3. First author ACA 3D plot (47 authors)

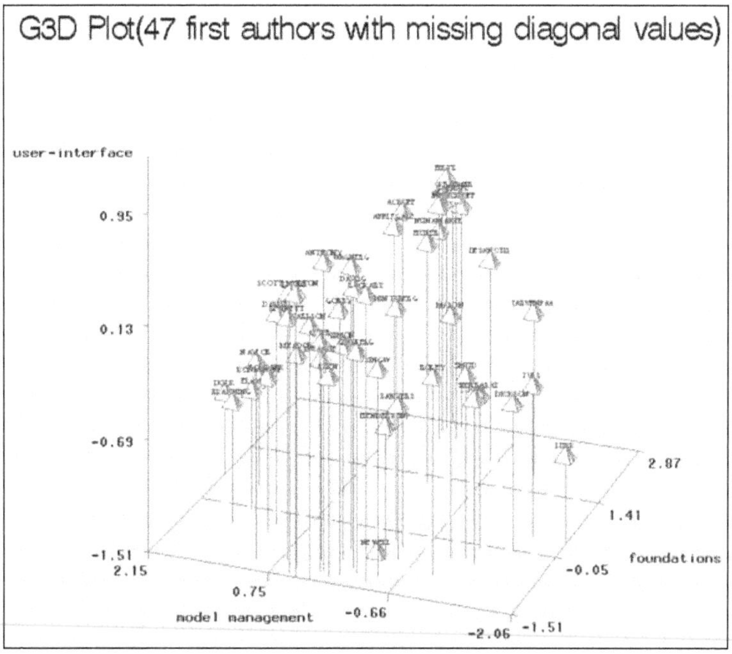

*Figure 4. All author ACA – Plot of dim1*dim2 (67 variables)*

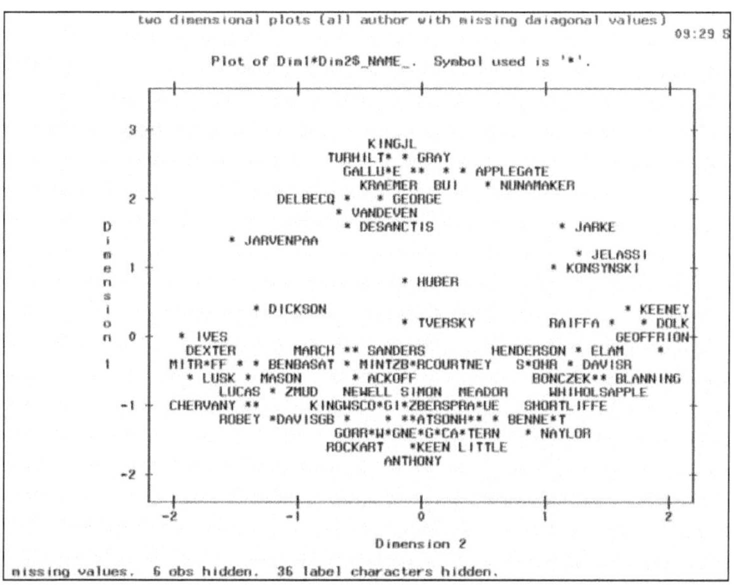

*Figure 5. First author ACA – Plot of dim1*dim2 (47 authors)*

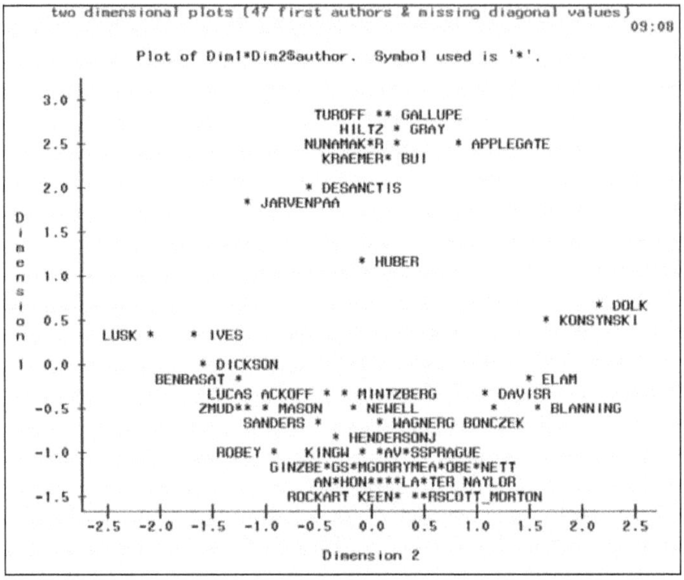

*Figure 6. All author ACA – Plot of dim2*dim3 (67 authors)*

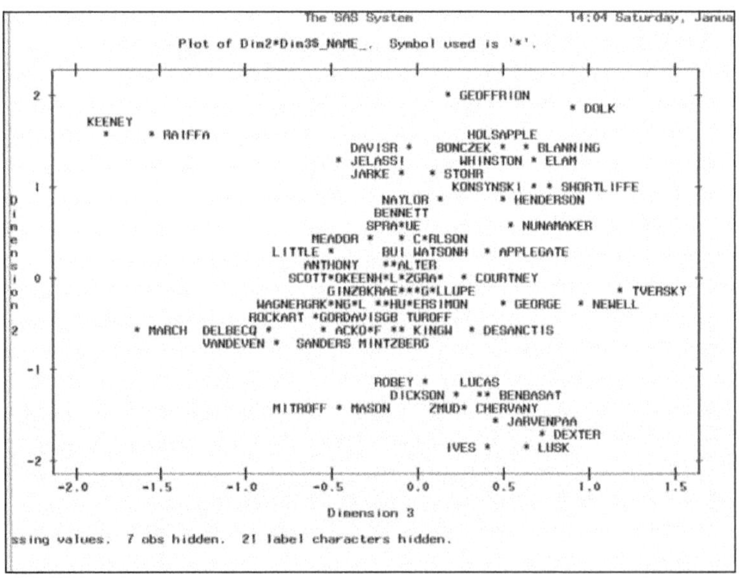

*Figure 7. First author ACA – Plot of dim2*dim3 (47 authors)*

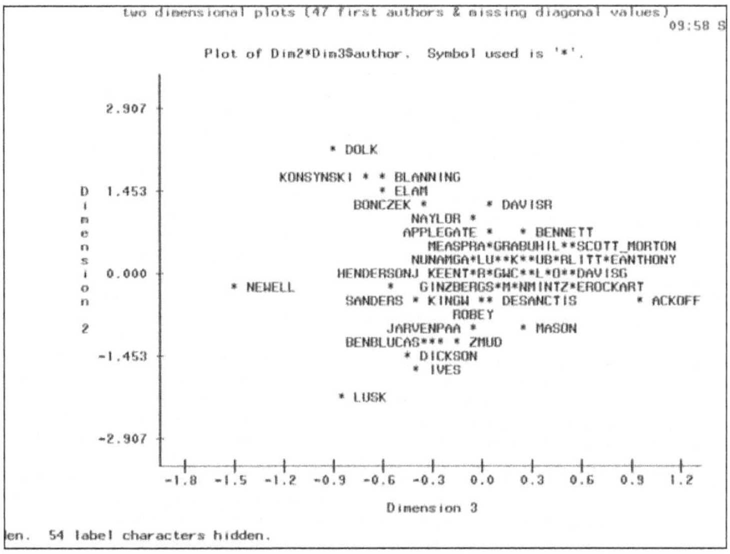

The Treatment of the Diagonal Values

This chapter and Chapter V undoubtedly show that the treatment of the diagonal values will trigger substantially different results that are not comparable at all. For that reason, I prepared two different cocitation matrices with two different diagonal values (missing value treatment and the value derived by adding three highest off-diagonal values and dividing it by two).

Selection Criteria of Author Set

Zhao chose 100 authors from the database of 312 citing publications and their 4578 citations.

The current study chose 67 authors that come from 692 citing articles and 15,030 *cited* references. If the current study applied the similar criteria used by Zhao, the final author set could have been at least 200. Generally speaking, the lower threshold value increases the number of author set, which in turn increases the instability of ACA outcomes. An example of unstable outcomes includes the fact that author counts with an excessive number of zeros tend to decrease the percentage of variance explained with possibly less number of meaning factors. Consequently, the comparison of results of ACA studies with different selection criteria becomes less meaningful.

Certainly, there are differences among many researchers in selecting the criteria. To make it worse, the criteria selected subjectively by each researcher are not applied correctly. In the studies of (Jesper Wilborg Schneider, Larsen, & Ingwersen, 2007; Zhao, 2006), a fixed number of authors was chosen to conduct first author and all authors ACA. My argument is that under all author ACA, cocitation frequency counts for each author should increase. Therefore, more authors will meet the selection criteria. Consequently, the total number of authors under different approaches should be different as in (Persson, 2001) and Eom in this chapter.

Matrix Generation

The majority of ACA studies use the Drexel approach to ACA based on symmetric cocitation matrix (n x n). Traditionally, multivariate analyses use asymmetric data matrix of variables by cases to derive proximity matrices. However, in ACA, a symmetric cocitation counts matrix is transformed to a proximity matrix. A symmetric matrix consists of two identical cociation data of the lower half and the upper half. The identical co-citation counts between author A and B appear twice in the lower and upper half of the matrix. According to (Jesper Wilborg Schneider, Larsen, & Ingwersen, 2007), the double counts lead to a magnification problem. In addition, the transformation of a cocitation matrix into a proximity matrix creates another problem in regard to the treatment of the diagonal values discussed earlier.

Comparison of Factor Analysis Results

From the reviewed literature, one study (Jesper Wilborg Schneider, Larsen, & Ingwersen, 2007) treated the diagonal value as missing values and the other study (Zhao, 2006) did not reveal that information. We will compare the results of our study with missing diagonal value treatment with that of Schneider et al. first.

Tables 12 and 13 suggest that all author analysis appears to produce factor solutions with a larger percentage of cumulative total variance explained. In terms of the number of factors, these results are incomparable because of the different number of authors (one varying and the other two fixed) in each study. The summary of the two previous studies (Jesper Wilborg Schneider, Larsen, & Ingwersen, 2007 ; Zhao, 2006) with the fixed number of authors indicated that the first author cocitation study produced more numbers of factors while explaining a smaller percentage of variances.

All three results seem incomparable due to the different approaches used. If any further investigation on this subject deems necessary, ACA researchers recognize that the total number of authors under all author and first author ACA should be different. This is because all author ACA counts all non-primary authors in addition to first authors. This is an important fact either ignored or overlooked by previous research. However, Persson agreed with this point in page 343 by saying "One obvious difference between the maps will be that author names included will differ. When going from first author citations to all author citations new names will enter the map while others will have to leave." The only thing I disagree with is that no authors will have to leave from the author set. Nevertheless, Persson himself produced MDS map of same number of authors under two different approaches (Persson, 2001). I suspect their ways of collecting cocitation counts make it extremely difficult to select the final author set based on the total number of cocitation counts.

Comparison of MDS Results

There is a clear difference between Eom and other two (Zhao and Schneider et al.). The two studies of Zhao and Schneider et al show that all author ACA maps fit better than first author ACA, although the statistical differences in Stress 1 values may not be significant. It is reasonable to conclude that all author ACA produced better fitting 2 dimensional MDS maps. Remember that these two studies used the fixed number of authors under two different approaches. We pointed out that this was a significant drawback of these studies repeatedly. On the other hand, Eom's study tells a different story that that first author ACA maps fit better than all author ACA. The reasonable justification for this difference is the varying number of authors in these studies.

Table 12. Comparative results of factor analysis by Eom and Schneider et al. (Diagonal values are treated as missing values, using the Drexel approach)

	First author analysis (total number of authors 47)	All author analysis (total number of authors 67)	Remarks
Number of factors	5	6	Eom
Total variance explained	81.94%	84.92%	Eom
Total number of authors	47	67	Eom
Number of factors	16	15	Schneider et al.
Total variance explained	80%	86%	Schneider et al.
Total number of authors	75	75	Schneider et al.

Table 13. Comparative results of factor analysis by Eom and Zhao (the Drexel approach)

	First author analysis	All author analysis	Remarks
Number of factors with the diagonal values computed by adding three highest off-diagonal values and dividing it by 2.	7	9	Eom
Total variance explained with the diagonal values computed by adding three highest off-diagonal values and dividing it by 2.	81.94%	84.66%	Eom
Total number of authors	47	67	Eom
Number of factors with unknown diagonal value treatment	11	5	Zhao.
Total variance explained with unknown diagonal value treatment	96%	97%	Zhao
Total number of authors	100	100	Zhao.

The comparative results of MDS analysis (Table 14) do not seem to support the major conclusion of Zhao: "the picture produced through this simplified all-author co-citation counting contains author groups that are more coherent, and is therefore considerably clearer." When comparing Zhao's Figures 1 and Figure 2, I agree with her conclusion that all-author cocitation based ACA produced a two dimensional map showing a more focused linkage at the right top section as well as at the bottom middle section.

Visual inspection of the two MDS 2 dimensional maps of Zhao (Figure 1 and Figure 2) does not show such changing configuration on the map. Stress 1 values of the configurations (Table 14) show the badness of fit. The badness of fit is equivalent to Young's s-stress formula 1.

CONCLUSION

Three conclusions can be reached based on our study.

Table 14. Stress -1 values for MDS configurations

Study	Configuration	First Author ACA	All author ACA	Differences in Stress Values
Schneider et al.	2 dimensions	.174	.132	.042
	3 dimensions	.102	.096	.008
Zhao	2 dimensions	.07	.05	.02
	3 dimensions	n.a.	n.a.	
Eom	2 dimensions	.0878	.1095	-.0217
	3 dimensions	.0572	.0670	-.0098

All Author Based ACA is Better to Capture All Influential Researchers in a Field than First Author Based ACA

There are only three studies that investigate the impact of different cocitation counting methods on the intellectual structure of an academic discipline. In terms of data and methodology, all three studies suffer from flaws in either data or methodology. A critical deficiency of Persson's study comes from the fact that more than 90% of references to papers not indexed by the ISI databases were left out of consideration. Despite the deficiency of the data, Persson concludes that all author based ACA better captures all influential researchers in a field. Eom's study supports Persson's conclusion. All author based ACA identified 67 authors while first author based ACA captured only 46 authors.

All Author Based ACA Identifies More Subspecialties

We have shown that the ISI convention of relying on only the name of the first author in assembling the cocitation matrix may often fail to identify all possible underlying factors. This is the most significant conclusion. The most important implication of these three studies is that first author only cocitation analysis undeniably decreases the percentage of variance explained. By omitting non-primary authors contributions, we may possibly misrepresent the intellectual structure of an academic discipline. Disregarding non-primary authors contribution in ACA analysis results in loss of several meaningful factors. When comparing the results of all author ACA and first author ACA, counting only the first author produced only 6 factors, while all author ACA resulted in a nine factor solution.

This study, therefore, disconfirms the results of the two previous studies by Zhao and Persson in regard to the number of the research subspecialties. Persson claims that the subspecialties identified are almost identical regardless of cocitation counting methods employed. On the other hand, Zhao concludes that all author based ACA produced fewer research subspecialties than produced through the traditional first-author based ACA analysis. When interpreting the result of any multivariate statistical studies including ACA studies, the important question is the meaning of each factor. Determining the optimal number of factors in ACA analysis critically depends on the interpretability of each factor, not the eigen value criterion nor the scree test result. The interpretability of each factor should be the most important factor to determine the final factor solution.

The first author only ACA result of Zhao's study vividly shows that many factors appeared in the final solution (Table 4 of Zhao's study) and may possibly have a problem of interpretability. According to McCain (1990, p. 440), "Only authors with loadings greater than ±0.7 are likely to be useful in interpreting the factor, and only loadings above ±0.4 or ±0.5 are likely to be reported." Applying this criterion to Zhao's Table 4, many factors may not be good candidates for the final solution. This is the result of increased instability of the cocitation matrix by including authors with a low number of cocitation counts.

What is the underlying reason for this result of Zhao? As said earlier, counting only the first author should decrease the cocitation frequencies in a given bibliographic database. If so, the final author set for ACA study should be decreased in number if there is a selection criteria to be applied whether it is a *mean cocitation rate,* or a cocitation with at least one-third of the entire author set, or restricting the final author set to the 20% receiving the highest number of citations, or any other criterion.

All Author Based ACA and First Author Based ACA Produce Little Differences in Stress Values

The results of MDS analysis in this study do not seem to support the major conclusion of Zhao: "the picture produced through this simplified all-author co-citation counting contains author groups that are more coherent, and is therefore considerably clearer." When comparing Zhao's Figures 1 and Figure 2, we agree with her conclusion that all-author cocitation based ACA produced a two dimensional map showing a more focused linkage at the right top section as well as at the bottom middle section.

Visual inspection of the two MDS 2 dimensional maps of Zhao (Figure 1 and Figure 2) does not show such changing configuration on the map. As shown in Table 14, all author based ACA and first author ACA produce little differences in stress values. For example, our study results show that all author based ACA produced larger stress-1 values, which means poor fit. The real question is how statistically significant the difference between the two is. In the case of two dimensional configuration, the difference in stress value is .02. We do not believe that this is a significant difference that can reach different conclusion. In conclusion, all author ACA and first author ACA produce little differences in stress values.

In future research, it is imperative to use the same assumptions and criteria in regard to the treatment of the diagonal values, the selection of authors, etc. If these are not uniformly standardized, the results are always incomparable and have less meaning at all.

REFERENCES

Culnan, M. J. (1986). The Intellectual Development of Management Information Systems, 1972-1982: A Co-Citation Analysis. *Management Science, 32*(2), 156-172.

Eom, S. B. (2003). *Author Cocitation Analysis Using Custom Bibliographic Databases—an Introduction to the SAS Approach*. Lewiston, New York: The Edwin Mellen Press.

Eom, S. B. (2006). A Decision Support System for Delineating the Intellectual Structure of Academic Disciplines: Development and Implementation. In F. Adam, P. Brézillon, S. Carlsson & P. Humhreys (Eds.), *Creativity and Innovation in Decision Making and Decision Support: Papers from the 2006 IFIP International Conference on Creativity and Innovation in Decision Making and Decision Support, 28 June - July 2006, London, Uk* (Vol. 2, pp. 1029-1043). London, England: Decision Support Press.

Eom, S. B., & Farris, R. (1996). The Contributions of Organizational Science to the Development of Decision Support Systems Research Subspecialties. *Journal of the American Society for Information Science, 47*(12), 941-952.

Hai, Jr., J. F. Anderson, R. E., & Tatham, R. L. (1987). *Multivariate Data Analysis with Readings* (2nd ed.). New York: Macmillan Publishing Company.

Kruskal, J. B., & Wish, M. (1990). *Multidimensional Scaling*. Newbury Park, CA: Sage Publications.

Leydesdorff, L., & Vaughan, L. (2006). Co-Occurrence Matrices and Their Applications in Information Science: Extending Aca to the Web Environment. *Journal of the American Society for Information Science and Technology, 57*(12), 1616-1628.

McCain, K. W. (1990). Mapping Authors in Intellectual Space: A Technical Overview. *Journal of the American Society for Information Science, 41*(6), 433-443.

McIntire, J. S. (2007). *The Clothing and Textile Research Base: An Author Cocitation Study.* Unpublished Master's Thesis, University of Missouri, Columbia, Columbia, Missouri.

Persson, O. (2001). All Author Citations Versus First Author Citations. *Scientometrics, 50*(2), 339-344.

Rousseau, R., & Zuccala, A. (2004). A Classification of Author Co-Citations: Definitions and Search Strategies. *Journal of the American Society for Information Science and Technology, 55*(6), 513-529.

Schneider, J. W., & Borlund, P. (2007a). Matrix Comparison, Part 1: Motivation and Important Issues for Measuring the Resemblance between Proximity Measures or Ordination Results. *Journal of the American Society for Information Science and Technology, 58*(11), 1586-1595.

Schneider, J. W., & Borlund, P. (2007b). Matrix Comparison, Part 2: Measuring the Resemblance between Proximity Measures or Ordination Results by Use of the Mantel and Proclustes Statistics. *Journal of the American Society for Information Science and Technology, 58*(11), 1596-1609.

Schneider, J. W., Larsen, B., & Ingwersen, P. (2007 June 25-27). *Comparative Study between First and All-Author Co-Citation Analysis Based on Citation Indexes Generated from Xml Data* Paper presented at the 11th International Conference on Scientometrics and Informetrics, Madrid, Spain.

Young, F. W., Lewyckyj, R., & Takane, Y. (1986). *The Alscal Procedure.* Cary, NC: SAS Institute.

Zhao, D. (2006). Towards All-Author Co-Citation Analysis. *Information Processing & Management, 42*(6), 1578-1591.

Chapter IV
Diagonal Values in ACA

ABSTRACT

Diagonal values in the cocitation frequency counts matrix are a fundamental issue in ACA study. Diagonal values are the co-citation frequency counts between the author himself/herself excluding self-citation. Retrieving exact values of diagonal values in the co-citation matrix requires a manual and time consuming procedure. For that reasons, ACA researchers suggested many different approaches to create, not retrieving the real values, the diagonal cells in the cocitation matrix. They include the mean cocitation count, missing values, zeroes, highest off-diagonal counts, adjusted off-diagonal values, and the number of times cocited with himself/herself. The majority of ACA researchers seem to prefer to use either the adjusted value approach by adding three highest off-diagonal values and divided by two or the missing value approach. This chapter empirically examines the impact of these different approaches on the ACA outcomes. Based on the results of this study, if the pure cocitation counts are not used, the next best alternatives are as follows. They are the missing value approaches, mean cocitation value approach, and the highest off-diagonal value approaches in the order of the highest total variance explained.

INTRODUCTION

Diagonal values in the cocitation frequency counts matrix are a fundamental issue in ACA study. Diagonal values are the co-citation frequency counts between the author himself/herself, not including self-citations. Diagonal values are the total number of citing papers that include at least two contributions by the author in the references, excluding self-citations. The retrieval of the true values of diagonal values in the co-citation matrix requires a time consuming procedure. For that reason, ACA researchers suggested many different approaches to artificially create the value of the diagonal cells in the cocitation matrix, rather than pure cocitation between the author himself/herself. The alternative approaches suggested by ACA researchers essentially create the diagonal values by arithmetic computation using non-diagonal values of each author.

The purpose of this chapter is to empirically examine the impact of many alternative ways of handling diagonal values in lieu of co-citation between the author himself/herself. Raw cocitation matrix is the prime input to the ACA process. In addition to the issue of how to compute co-citation frequencies, another issue occurring during the compilation of the raw cocitation matrix is the value of diagonal values. The off-diagonal $value_{ij}$ is the cocitation frequency between $author_i$ and $author_j$. either retrieved from ISI databases or generated from custom databases. It has been suggested that the number of times that an author has been cocited with himself (excluding self-citations) would yield a mathematically complete matrix (Ahlgren, Jarneving, & Rousseau, 2003). The problem is that retrieving pure cocitation counts is an extremely time consuming process that can be difficult to be programmed, whether we use the Thompson ISI citation index and Dialog Classic to retrieve the cocitation counts or custom databases and cocitation matrix generation systems such as developed by Eom (Eom & Farris, 1996) or McIntyre (McIntire, 2007), ACA researchers, therefore, have suggested several different possible ways of treating the diagonal values, in lieu of retrieving the true value. No prior study in this area exists that systematically compares the results of ACA using cocitation counts matrix with different diagonal values.

NUMBER OF TIMES COCITED WITH HIMSELF/HERSELF

A possible solution for the diagonal value problem is suggested by Ahlgren, Jarneving, and Rousseau (2003, p.551.) and White (Howard D. White, 2003). The diagonal values should be the number of articles in the bibliographic database that cite at least two (different) works authored or coauthored by the person. Ahlgren, Jarneving, and Rousseau suggest excluding self-citations. This number is very difficult to get when using ISI databases. The custom database and cocitation matrix generation system we have developed allowed us to retrieve the numbers so we can yield a co-citation frequency matrix, which Ahlgren, Jarneving, and Rousseau describe as "a mathematically complete matrix." One solution for the diagonal value problem is suggested by Ahlgren, Jarneving, and Rousseau (2003, p.551.)

We think that perhaps the best solution for the "diagonal" problem (if such solution is deemed necessary) is to use the number of times for an author, say AU, has been cocited with himself (excluding self-citations). This is the number of articles in the pool under study that cite at least two (different) works (co)authored by AU. Such a method would yield a mathematically complete matrix, which is always easier to study (statistically) than an incomplete one (Fienberg, 1980). Our suggestion is inspired by Eom and Faris (Eom & Farris, 1996), who, however, use the number of times an author is cocited with him/herself as a way to determine the author set to be analyzed.

White (2003, p.1253) also discussed the issue of diagonal values in the cocitation matrix (Howard D. White, 2003). He believes that "the most meaningful entry would be a count of the publications in which two or more of the author's works are jointly cited."

As Ahlgren, Jarneving, and Rousseau point out, following Eom and Farris (1996), the most meaningful entry would be not a raw cocitation count, but a count of the publications in which two or more of the author's works are jointly cited. This number, however, is not easy to get in standard data gathering from ISI databases (Eom creates his own data sets). The raw citation count is easy to get, but it is not a cocitation count. It may also be falsely high when different authors with the same name (homonyms) contribute to it, or falsely low when counts associated with all variants of an author's name (allonyms) have not been found and combined.

SIX ALTERNATIVE APPROACHES

The best way to count the diagonal value of $author_i$ is manually examining the references of each citing article to see whether each citing article contains at least two contributions of $author_{i,}$ excluding self-citations. Needless to say, this is an extremely time-consuming procedure because all citing articles' references should be examined one by one. Furthermore, it is virtually impossible to identify all citing articles during the investigation period if we use ISI databases. In the case of using custom databases, it is doable but extremely time-consuming tasks. The cocitation matrix generation system we developed can easily identify the total number of citing articles that contain at least one contribution by $author_i$. For example, our system process the citing database of N (e.g.,1000) citing articles to identify n (e.g., 300) citing articles with at least one contribution by $author_i$. Next, each of n articles must be processed manually to see whether the cocitation between author himself or herself occurs. It is difficult to develop a system that is intelligent enough to distinguish each citation as self-citations or not. Furthermore, it may be considered a simple task for the cocitation generation system to identify the two contributions of $author_i$ in a citing article, but it turns out to be a complex task. Because of this reason, ACA researchers invented the following 6 possible approaches to fill the diagonal cells in cocitation counts matrix.

1. Missing values (McCain, 1990, 1991; Ramos-Rodríguez & Ruíz-Navarro, 2004; H.D. White & McCain, 1998)
2. The mean cocitation count for each author (McIntire, 2007)
3. Zero
4. Highest off-diagonal cocitation counts
5. Three highest off-diagonal values divided by two (M. J. Culnan, 1986; M.J. Culnan, O'Reilly, & Chatman, 1990; Eom & Farris, 1996; H. D. White & Griffith, 1981)
6. Raw Cocitation

Missing Values

Diagonal values are often treated as missing data (McCain, 1990; Ramos-Rodríguez & Ruíz-Navarro, 2004; H.D. White & McCain, 1998). To process the dataset with missing numerical value treatment, the diagonal values must be replaced with a single period (.) as shown below. The factor procedure cannot be applied to the dataset with a single period. Therefore, the dataset with missing numerical values must be processed to create a correlation matrix first. The correlation matrix can be used to do factor analysis as the following SAS code demonstrates.

Mean Cocitation Counts

The diagonal values are computed from the off-diagonal cell values. From the SAS dataset used in the case of the missing value approach, the diagonal value of ALTER is computed as follows.

> ALTER's Diagonal Value = (16+25+66+8+15+34+30+22+25+23+92+16+14+15+12+ 83+38+67+27)/19
> The total number of authors is 20. The total number of co-citation between Alter and other 19 authors is 19.

Zero

All diagonal cell values are zeros. All values of single dot (.) in Table 5.1 will be replaced with zeros.

Table 1. The cocitation matrix with SAS code

```
/* BEGIN SAS CODE */

* input data;
data sample;
length author $ 12;
input author $ ALTER BENBASAT BONCZEK CARLSON DESANCTIS DICKSON
GINZBERG GORRY HENDERSON HOLSAPPLE HUBER KEEN KONSYNSKI LUCAS MA-
SON_R MITROFF SCOTT_MORTON SIMON SPRAGUE WHINSTON;

cards;
ALTER          . 16 25 66 8 15 34 30 22 25 23 92 16 14 15 12 83 38 67 27
BENBASAT       16 . 13 23 33 43 13 13 16 13 22 44 12 28 24 22 38 31 31 13
BONCZEK        25 13 . 53 18 13 15 11 34 84 23 46 34 11 8 8 42 33 59 93
CARLSON        66 23 53 . 18 17 32 34 33 49 34 112 24 18 22 18 101 54 133 54
DESANCTIS      8 33 18 18 . 43 8 7 16 16 49 33 33 18 16 17 27 28 27 19
DICKSON        15 43 13 17 43 . 13 9 13 13 35 45 19 29 23 23 39 34 26 14
GINZBERG       34 13 15 32 8 13 . 15 14 14 21 53 8 14 13 10 42 17 38 15
GORRY          30 13 11 34 7 9 15 . 8 12 12 46 4 11 16 15 62 39 32 13
HENDERSON      22 16 34 33 16 13 14 8 . 33 22 45 37 7 16 17 34 26 44 34
HOLSAPPLE      25 13 84 49 16 13 14 12 33 . 23 46 32 11 8 8 41 33 56 87
HUBER          23 22 23 34 49 35 21 12 22 23 . 47 35 16 20 17 39 37 49 24
KEEN           92 44 46 112 33 45 53 46 45 46 47 . 29 39 40 36 174 82 126 50
KONSYNSKI      16 12 34 24 33 19 8 4 37 32 35 29 . 6 16 16 23 28 35 34
LUCAS          14 28 11 18 18 29 14 11 7 11 16 39 6 . 18 16 27 22 20 11
R_MASON        15 24 8 22 16 23 13 16 16 8 20 40 16 18 . 56 37 24 25 8
MITROFF        12 22 8 18 17 23 10 15 17 8 17 36 16 16 56 . 33 27 22 8
SCOTT_MORTON   83 38 42 101 27 39 42 62 34 41 39 174 23 27 37 33 . 82 105 45
SIMON          38 31 33 54 28 34 17 39 26 33 37 82 28 22 24 27 82 . 58 37
SPRAGUE        67 31 59 133 27 26 38 32 44 56 49 126 35 20 25 22 105 58 . 61
WHINSTON       27 13 93 54 19 14 15 13 34 87 24 50 34 11 8 8 45 37 61 .
;

* print input data;
proc print data=sample;
run;
* create correlation matrix;
proc corr data=sample out=test;
run;

* print correlation matrix;
proc print data=test;
run;

* input correlation matrix into FACTOR;
proc factor data=test;
run;
/* END SAS CODE */
```

Highest Off-Diagonal Cocitation Counts

All values of single dot (.) in Table 1 will be replaced with the highest off-diagonal counts. For example, Alter's diagonal cell value is 92. Table 1 lists 19 co-citation counts at the ALTER row. Among them, 92 is the highest number.

ALTER . 16 25 66 8 15 34 30 22 25 23 92 16 14 15 12 83 38 67 27

Adjusted Off-Diagonal Values

Some ACA researchers recommend adjusted off-diagonal values by diving the sum of an author's highest three cocitation counts by two (H. D. White & Griffith, 1981). White and Griffith (1981, p. 165) explains their rationale for their approach this way:

In our first analysis, we entered total citations to each author (including self-citations); these scores were sometimes disproportionately large and a few contained homonym errors. (By contrast, the co-citations of oeurvres, the author intersections, generate in our tests less than 2% error.) After some thoughts on the manner in which values in such a matrix might be distributed, we decided that taking the three highest intersections and dividing by two would generate diagonals which would approximate the next highest scores in the distribution, thus indicating in a general way the relative importance of a particular author within the field.

In our example, the ALTER's diagonal value, 121, is computed as summing the highest off-diagonal values (92+83+67) and dividing by 2.

The Raw Citation Counts

White also discussed a possible use of the raw citation count of the author as the diagonal value. Our custom database and retrieval systems can eliminate the problem of homonyms and allonyms. We decided to use the raw citation counts of the authors as the diagonal value. The raw citation count of author we are experimenting with is the number of cited references that include at least one contribution (authored or co-authored) of the author. In the following example of two references, the raw citation count is 2 (total number of reference articles that contains one or more of the author's contributions), not 3 (total number of times cited). This number can be possibly used to substitute the number of articles in the bibliographic database that cite at least two (different) works authored or coauthored by the author. This number is certainly larger than the cocitation number with himself or herself. In the example below, article 1 contributes co-citaition counts of 1 since it is co-citing two contributions of the author, while article 2 does not have any cocitation with the author himself. Article 2 contributes to add 1 raw citation count of author EOM.

References of article 1

Eom, S. B. (1996). A Survey of Operational Expert Systems in Business (1980-1993). *Interfaces, 26*(5), 50-70.

Eom, S. B., Lee, S. M., & Kim, J. K. (1993). The Intellectual Structure of Decision Support Systems (1971-1989). *Decision Support Systems, 10*(1), 19-35.

Farhoomand, A. F. (1987). Scientific Progress of Management Information Systems. *Data Base, 18*(4), 48-56.

References of article 2

Smith, K. & Thomas, T. (1992) The k-procedure, The Gamma Journal.

Thomas, T. & Smith, K. (1992) More details about The k-procedure, The Gamma Journal.

Min, H., & Eom, S. B. (1994). An Integrated Decision Support System for Global Logistics. *International Journal of Physical Distribution and Logistics Management, 24*(1), 29-39.

COMPARATIVE RESULTS FROM FACTOR ANALYSIS

The following table summarizes the results of six alternative treatments of diagonal values on the outcomes of ACA in terms of number of factors, total variance explained, and sum of the squared residuals from pure co-citation. The correlations coefficients are very sensitive to the diagonal values since the diagonal values affect the mean values of each author. The changed correlation coefficients, in turn, affect factor loadings.

The sum of the squared residuals (errors) (the last column in Table 1) is computed as follows.

$$SSE = \sum e_i^2 = \sum (y_i - \text{pure cocitaion count})^2$$

Where y_i = Cocitation count under each approach

The factor solutions in Table 2 are categorized into either the six factor solution or the nine factor solution. All of the nine factor solutions include three factors that contain small number of authors with factor loadings less than .7 (see appendix at the end of this chapter). The careful analysis of Table 2 leads to the following relationships among the number of factors, total variance explained, and sum of

Table 2. Comparative results of all alternatives

Approaches	Number of factors produced	Total variance explained	Sum of the squared residuals from co-citation with himself/herself
1. Cocitation with himself/ herself	6	88.23%	0
1. Missing values	6	.8953%	
1. Mean values	6	86.05%	10,142
1. Zero	6	80.86%	34,128
1. Highest off-diagonal value	9	90.04%	85,139
1. Three highest off-diagonal values divided by two	9	84.82%	166,317
1. Raw co-citation	9	83.69%	175,558

the squared residuals from co-citation with himself/herself. Within each of the two groups of alternative approaches with the same number of produced factors, there exists an inverse relationship between the sum of the squared residuals from co-citation with himself/herself research and total variance explained. As the sum of the squared residual increases, the total variance explained decreases.

Table 3 illustrates that the missing value and mean values are acceptable surrogate for the true cocitation frequency counts, when the true cocitaion frequency counts with himself/herself are not readily available. The percent of variance explained (Columns #2, #4, and #6) are almost identical and the number of authors/variables (Columns #3, #5, and #7) in each factor is almost identical. Table 3 shows that the factor loadings of each author under each approach is the only noticeable difference among these approaches. For example, the first six authors' factor loadings of factor 1 under three different approaches which we conclude as acceptable surrogate is shown in Table 4. The top three authors (Ginzberg, Alter, and Gorry) in terms of the highest factor loadings in the factor structure output with diagonal values of cocitation with himself/herself are appearing in the output of two alternative approaches with different factor loadings. The correlation coefficients are sensitive to the diagonal values since the diagonal values affect the mean values of each author. The changed correlation coefficients, in turn, affect factor loadings. Therefore, ACA researchers should not give too much meaning of slightly different factor loadings of each variable.

Table 3. Further comparative results of three approaches in terms of % of variance explained and the number of variables/authors loaded in each factor

	True		Missing Values		Mean Values	
Factor	% of Variance explained	Number of variables with loadings >.70 and >.40	% of Variance explained	Number of variables with loadings >.70 and >.40	% of Variance explained	Number of variables with loadings >.70 and >.40
1	39.62	24/45	40.23	24/46	39.19	24/46
2	21.77	15/19	21.95	15/19	21.21	15/19
3	27.01	13/34	27.38	13/35	26.74	14/34
4	25.7	12/31	26.15	12/34	25.64	12/34
5	15.2	5/22	16.86	7/24	15.53	4/22
6	7.8	2/6	9.15	3/7	8.09	2/7

Table 4. The first six authors' factor loadings of factor 1 under three different approaches

True values		Missing Values		Mean values	
GINZBERG	0.95819	ALTER	0.98174	GINZBERG	0.95905
ALTER	0.94989	SCOTTMORTON	0.96767	ALTER	0.95379
GORRY	0.93886	GINZBERG	0.96425	GORRY	0.94266
KEEN	0.93777	KEEN	0.96198	ANTHONY	0.93542
SCOTTMORTON	0.93226	GORRY	0.94648	WAGNERGR	0.93214
ANTHONY	0.9289	CARLSON	0.93699	LITTLE	0.92847

CONCLUSION

Diagonal values in the cocitation frequency counts matrix are a fundamental issue in ACA study. Due to the difficulties of obtaining the true cocitation frequency counts with the author himself or herself, at least six alternative approaches have been suggested. Each of them has been used in many ACA studies. Especially, a large number of ACA studies used the adjusted cell values as the diagonal cell values, following the suggestion of White and Griffith (H. D. White & Griffith, 1981). The comparative results illustrate that the adjusted cell approach produced poorer results than that of other 3 treatments (2, 3, 5) in terms of total variance explained. There are only two alternative methods that produced worse results: putting zeros on the diagonal cells (#4) and raw cocitation counts(#7).

A major conclusion from this study is that the true value of cocitation frequency counts with the author himself or herself, excluding self-citation, may be lower than the one suggested by some ACA researchers. The alternative approaches that suggested the higher diagonal values (the highest off-diagonal value approach, the adjusted diagonal value approach, the raw cocitation approach) produced the larger number of factor solutions than the one produced by using the true diagonal values.

This study identified an important relationship among the number of factors, total variance explained, and sum of the squared residuals from co-citation with himself/herself. Within each of the two groups of alternative approaches with the same number of produced factors, there exists a inverse relationship between sum of the squared residuals from co-citation with himself/herself research and total variance explained. As the sum of the squared residual increases, the total variance explained decreases. The results of this study suggest that if the cocitation counts with the author himself/herself are not available, the next best alternatives are as follows in the order of the highest total variance explained.

- Missing values
- Mean co-citation values
- The highest off-diagonal values

Although the highest off-diagonal value approach resulted in the factor solution that explains 90.04% of total variance, it produces a nine factor solution. This conclusion is based on the result of only a set of data that contains 57 authors. Nonetheless, this is the first study that examines the impact of different diagonal values on the results of ACA study. More replicated future research is needed to solidify the findings of this research.

REFERENCES

Ahlgren, P., Jarneving, B., & Rousseau, R. (2003). Requirements for a Cocitation Similarity Measure, with Special Reference to Pearson's Correlation Coefficient. *Journal of the American Society for Information Science and Technology, 54*(6), 550-560.

Culnan, M. J. (1986). The Intellectual Development of Management Information Systems, 1972-1982: A Co-Citation Analysis. *Management Science, 32*(2), 156-172.

Culnan, M. J., O'Reilly, C. A., & Chatman, J. A. (1990). Intellectual Structure of Research in Organizational Behavior, 1972-1984: A Co-Citation Analysis. *Journal of the American Society for Information Science, 41*, 453-458.

Eom, S. B., & Farris, R. (1996). The Contributions of Organizational Science to the Development of Decision Support Systems Research Subspecialties. *Journal of the American Society for Information Science, 47*(12), 941-952.

Fienberg, S. E. (1980). *The Analysis of Cross-Classified Categorical Data*. Cambridge, MA: The MIT Press.

McCain, K. W. (1990). Mapping Authors in Intellectual Space: A Technical Overview. *Journal of the American Society for Information Science, 41*(6), 433-443.

McCain, K. W. (1991). Mapping Economics through the Journal Literature: An Experiment in Journal Cocitation Analysis. *Journal of the American Society for Information Science, 42*(4), 290-296.

McIntire, J. S. (2007). *The Clothing and Textile Research Base: An Author Cocitation Study*. Unpublished Master's Thesis, University of Missouri, Columbia, Columbia, Missouri.

Ramos-Rodríguez, A.-R., & Ruíz-Navarro, J. (2004). Changes in the Intellectual Structure of Strategic Management Research: A Bibliometric Study of the Strategic Management Journal, 1980-2000. *Strategic Management Journal, 25*, 981-1004.

White, H. D. (2003). Author Co-Citation Analysis and Pearson's R. *Journal of the American Society for Information Science and Technology, 54*(13), 1250-1259.

White, H. D., & Griffith, B. C. (1981). Author Cocitation: A Literature Measure of Intellectual Structure. *Journal of the American Society for Information Science, 32*(3), 163-171.

White, H. D., & McCain, K. (1998). Visualizing a Discipline: An Author Cocitation Analysis of Information Science, 1972-1995 *Journal of the American Society for Information Science, 49*, 327-335.

APPENDIX

Table 5.

Rotation: Promax; Diagonal values: Cocitaion with him/herself					
Factor1		Factor2		Factor3	
Foundation		GSS		Model Management	
GINZBERG	0.95819	GRAY	0.96897	BLANNING	0.96696
ALTER	0.94989	GALLUPE	0.96337	ELAM	0.95186
GORRY	0.93886	TUROFF	0.95504	DOLK	0.92709
KEEN	0.93777	HILTZ	0.95202	BONCZEK	0.91941
SCOTTMORTON	0.93226	BUI	0.94376	HOLSAPPLE	0.9163
ANTHONY	0.9289	KRAEMER	0.94276	WHINSTON	0.90834
KINGW	0.92671	APPLEGATE	0.91926	GEOFFRION	0.89159
LITTLE	0.92593	KINGJL	0.91008	STOHR	0.88798
WAGNERGR	0.92592	GEORGE	0.89151	HENDERSON	0.88571
DAVISGB	0.92079	DESANCTIS	0.88228	DAVISR	0.83132
SIMON	0.91352	NUNAMAKER	0.8757	KONSYNSKI	0.78805
ROCKART	0.91232	DELBECQ	0.80718	SHORTLIFFE	0.7346
BENNETT	0.912	VANDEVEN	0.80148	NAYLOR	0.7204
MEADOR	0.90859	HUBER	0.79738	WATSONH	0.69436
WATSONH	0.90783	JARKE	0.74973	BENNETT	0.67987
SPRAGUE	0.9024	JARVENPAA	0.66026	MEADOR	0.66237
CARLSON	0.89788	KONSYNSKI	0.65955	SPRAGUE	0.65459
NAYLOR	0.82663	JELASSI	0.61599	COURTNEY	0.64153
COURTNEY	0.78834	DICKSON	0.51694	CARLSON	0.63475
ROBEY	0.78217			JARKE	0.57775
ACKOFF	0.75853			SIMON	0.56215
MINTZBERG	0.75103			JELASSI	0.55277
SANDERS	0.74677			ALTER	0.53211
ZMUD	0.72886			TVERSKY	0.5268
LUCAS	0.69515			KEEN	0.50659
CHERVANY	0.68459			NEWELL	0.50215
STOHR	0.672			SCOTTMORTON	0.50059
HENDERSON	0.66857			LITTLE	0.4951

continued on following page

Table 5. continued

NEWELL	0.66148			GINZBERG	0.48908
MASON	0.65231			HUBER	0.47588
SHORTLIFFE	0.63825			RAIFFA	0.46289
DAVISR	0.61562			MINTZBERG	0.43801
BENBASAT	0.608			NUNAMAKER	0.4295
MITROFF	0.60222			ANTHONY	0.41444
WHINSTON	0.58459				
HOLSAPPLE	0.57543				
BONCZEK	0.56382				
HUBER	0.52925				
BLANNING	0.52726				
ELAM	0.48325				
DICKSON	0.46443				
DEXTER	0.46416				
MARCH	0.45693				
TVERSKY	0.45488				
LUSK	0.42692				
Variance	26.546		14.584		18.097
% Variance	39.62%		21.77%		27.01%
Factor4		Factor5		Factor6	
User Interface		Organization Science		MCDM	
BENBASAT	0.9673	MARCH	0.87898	KEENEY	0.82337
LUSK	0.96079	TVERSKY	0.7911	JELASSI	0.7098
DEXTER	0.94799	MINTZBERG	0.77836	RAIFFA	0.682
LUCAS	0.93127	NEWELL	0.74369	GEOFFRION	0.55638
IVES	0.93086	ACKOFF	0.72563	JARKE	0.55602
CHERVANY	0.91544	MITROFF	0.67404	STOHR	0.51572
ZMUD	0.91302	MASON	0.60427		
DICKSON	0.89774	SIMON	0.52718		
ROBEY	0.8223	ROCKART	0.50634		
JARVENPAA	0.74767	DAVISGB	0.49542		
MITROFF	0.7446	SANDERS	0.49297		
MASON	0.73921	ZMUD	0.4664		
KINGW	0.67713	ANTHONY	0.45789		

continued on following page

Table 5. continued

MINTZBERG	0.67631	GORRY	0.4519		
DAVISGB	0.66961	COURTNEY	0.45118		
SANDERS	0.64468	ROBEY	0.44512		
COURTNEY	0.6342	BENBASAT	0.44022		
SIMON	0.59604	DICKSON	0.43986		
ACKOFF	0.5957	LUCAS	0.43141		
NEWELL	0.5726	CHERVANY	0.42199		
GINZBERG	0.55093	IVES	0.41919		
DESANCTIS	0.55027	KINGW	0.41318		
WAGNERGR	0.52758				
GORRY	0.52643				
HUBER	0.52074				
KEEN	0.5156				
SCOTTMORTON	0.51412				
TVERSKY	0.50604				
ROCKART	0.47632				
ALTER	0.46005				
SPRAGUE	0.41899				
ANTHONY	0.41815				
Variance	17.221		10.182		5.223
% Variance	25.70%		15.20%		7.80%

Table 6. Factor structure correlations (1970-90)

Rotation method: promax; Diagonal values: missing values					
Factor1		Factor2		Factor3	
Foundations		GSS		Model Management	
ALTER	0.98174	GRAY	0.97698	BLANNING	0.97168
SCOTTMORTON	0.96767	GALLUPE	0.96274	ELAM	0.95172
GINZBERG	0.96425	HILTZ	0.96081	HOLSAPPLE	0.93467
KEEN	0.96198	TUROFF	0.95816	BONCZEK	0.93258
GORRY	0.94648	KRAEMER	0.95685	WHINSTON	0.92933
CARLSON	0.93699	BUI	0.95616	DOLK	0.92527
ANTHONY	0.93284	KINGJL	0.94534	HENDERSON	0.90348

continued on following page

Table 6. continued

LITTLE	0.93224	APPLEGATE	0.91641	STOHR	0.89679
WAGNERGR	0.93131	GEORGE	0.8846	GEOFFRION	0.89389
SPRAGUE	0.92816	NUNAMAKER	0.86791	DAVISR	0.83876
KINGW	0.92478	DESANCTIS	0.86131	KONSYNSKI	0.77087
BENNETT	0.92001	DELBECQ	0.81933	SHORTLIFFE	0.75166
ROCKART	0.91622	VANDEVEN	0.8133	NAYLOR	0.72549
DAVISGB	0.91555	HUBER	0.78926	WATSONH	0.69588
MEADOR	0.91519	JARKE	0.75889	BENNETT	0.68443
WATSONH	0.9138	JARVENPAA	0.65695	MEADOR	0.66182
SIMON	0.91214	KONSYNSKI	0.63956	SPRAGUE	0.65566
NAYLOR	0.8237	JELASSI	0.62622	CARLSON	0.6475
COURTNEY	0.79494	DICKSON	0.50152	COURTNEY	0.64078
ROBEY	0.7715			JARKE	0.57551
ACKOFF	0.76286			JELASSI	0.55722
SANDERS	0.74379			SIMON	0.55391
MINTZBERG	0.74075			ALTER	0.53503
ZMUD	0.71664			TVERSKY	0.51748
LUCAS	0.68905			NEWELL	0.4976
STOHR	0.67899			LITTLE	0.49569
CHERVANY	0.67401			GINZBERG	0.49311
HENDERSON	0.66874			SCOTTMORTON	0.48772
NEWELL	0.66033			KEEN	0.48733
SHORTLIFFE	0.64786			HUBER	0.47246
MASON	0.63718			RAIFFA	0.45198
DAVISR	0.6214			MINTZBERG	0.43625
BENBASAT	0.60635			NUNAMAKER	0.4256
MITROFF	0.58905			ANTHONY	0.41186
WHINSTON	0.58843			ACKOFF	0.41174
HOLSAPPLE	0.57959				
BONCZEK	0.57009				
BLANNING	0.53608				
HUBER	0.52316				
MARCH	0.50625				
ELAM	0.49224				
DICKSON	0.4692				

continued on following page

Table 6. continued

DEXTER	0.4591				
TVERSKY	0.45573				
LUSK	0.42535				
GEOFFRION	0.41475				
Variance	26.9536		14.7047		18.345
% Variance	40.23%		21.95%		27.38%

Factor4		Factor5		Factor6	
User Interface		Organization Science		MCDM	
BENBASAT	0.9675	MARCH	0.86841	KEENEY	0.90778
LUSK	0.96337	MINTZBERG	0.79586	RAIFFA	0.86112
DEXTER	0.95414	ACKOFF	0.77567	JELASSI	0.70821
IVES	0.93455	TVERSKY	0.77342	GEOFFRION	0.57622
LUCAS	0.9327	NEWELL	0.75548	JARKE	0.5723
ZMUD	0.92394	MITROFF	0.74334	STOHR	0.53897
CHERVANY	0.91957	MASON	0.70006	DAVISR	0.44026
DICKSON	0.89212	SIMON	0.57479		
ROBEY	0.82991	ROCKART	0.53534		
MASON	0.75733	DAVISGB	0.53009		
MITROFF	0.75249	SANDERS	0.52492		
JARVENPAA	0.74629	ZMUD	0.51238		
KINGW	0.69148	COURTNEY	0.49171		
DAVISGB	0.6784	GORRY	0.48472		
MINTZBERG	0.67541	DICKSON	0.48447		
SANDERS	0.65486	ROBEY	0.48198		
COURTNEY	0.64386	ANTHONY	0.47913		
ACKOFF	0.61591	BENBASAT	0.47044		
SIMON	0.58598	LUCAS	0.46828		
NEWELL	0.56792	CHERVANY	0.46717		
GINZBERG	0.56361	KINGW	0.44683		
WAGNERGR	0.55333	IVES	0.43826		
GORRY	0.53347	LUSK	0.42258		
HUBER	0.5162	MEADOR	0.40902		
DESANCTIS	0.51348	HUBER	0.4031		

continued on following page

Table 6. continued

ROCKART	0.50633	SCOTTMORTON	0.40205		
SCOTTMORTON	0.50326	HENDERSON	0.40074		
TVERSKY	0.49625				
KEEN	0.49155				
ALTER	0.47061				
ANTHONY	0.42357				
SPRAGUE	0.41847				
WATSONH	0.4069				
CARLSON	0.40473				
Variance	17.5173		11.2989		6.132681
% Variance	26.15%		16.86%		9.15%

Table 7. Factor structure correlations (1970-1990)

Rotation: promax; Diagonal value: mean values					
Factor1		Factor2		Factor3	
Foundations		GSS		Model Management	
GINZBERG	0.95905	GRAY	0.96419	BLANNING	0.95875
ALTER	0.95379	KRAEMER	0.9457	ELAM	0.93596
GORRY	0.94266	BUI	0.9389	DOLK	0.91984
ANTHONY	0.93542	HILTZ	0.93771	STOHR	0.89717
WAGNERGR	0.93214	TUROFF	0.9356	GEOFFRION	0.89604
LITTLE	0.92847	GALLUPE	0.92628	HENDERSON	0.89014
KINGW	0.92566	KINGJL	0.92227	HOLSAPPLE	0.88315
DAVISGB	0.92059	APPLEGATE	0.90074	BONCZEK	0.875
ROCKART	0.92009	GEORGE	0.88661	WHINSTON	0.86862
BENNETT	0.91334	NUNAMAKER	0.85052	DAVISR	0.83251
MEADOR	0.91056	DESANCTIS	0.82296	KONSYNSKI	0.7513
WATSONH	0.90521	DELBECQ	0.81089	SHORTLIFFE	0.74232
SIMON	0.89344	VANDEVEN	0.80471	NAYLOR	0.72368
CARLSON	0.8865	HUBER	0.75929	WATSONH	0.70153
SCOTTMORTON	0.87359	JARKE	0.74733	BENNETT	0.68586

continued on following page

Table 7. continued

SPRAGUE	0.86424	JARVENPAA	0.65943	MEADOR	0.66959
KEEN	0.84722	KONSYNSKI	0.62381	COURTNEY	0.64913
NAYLOR	0.82556	JELASSI	0.61759	CARLSON	0.6349
COURTNEY	0.79079	DICKSON	0.5005	SPRAGUE	0.63245
ROBEY	0.77966			JARKE	0.57981
ACKOFF	0.75705			JELASSI	0.55665
MINTZBERG	0.74567			SIMON	0.55348
SANDERS	0.74511			ALTER	0.53988
ZMUD	0.72497			TVERSKY	0.52713
LUCAS	0.69744			NEWELL	0.50354
CHERVANY	0.6836			LITTLE	0.49921
STOHR	0.67029			GINZBERG	0.49719
HENDERSON	0.66732			SCOTTMORTON	0.49399
NEWELL	0.65879			KEEN	0.48314
MASON	0.6502			HUBER	0.47706
SHORTLIFFE	0.64551			RAIFFA	0.46562
BENBASAT	0.61633			MINTZBERG	0.44236
DAVISR	0.61404			NUNAMAKER	0.42111
MITROFF	0.59951			ANTHONY	0.41128
WHINSTON	0.58164				
HOLSAPPLE	0.57268				
BONCZEK	0.56318				
BLANNING	0.53169				
HUBER	0.5213				
MARCH	0.51365				
ELAM	0.48967				
DICKSON	0.46545				
DEXTER	0.46486				
TVERSKY	0.44997				
LUSK	0.43219				
GEOFFRION	0.40262				
Variance	26.257		14.214		17.916
% Variance	39.19%		21.21%		26.74%

continued on following page

Table 7. continued

Factor4		Factor5		Factor6	
User Interface		Organization Science		MCDM	
LUSK	0.95092	MARCH	0.88075	KEENEY	0.80618
DEXTER	0.93322	TVERSKY	0.7902	JELASSI	0.74274
IVES	0.92983	MINTZBERG	0.76943	RAIFFA	0.68229
BENBASAT	0.91958	NEWELL	0.75185	JARKE	0.57353
LUCAS	0.9175	ACKOFF	0.74091	GEOFFRION	0.56297
ZMUD	0.91632	MITROFF	0.67761	STOHR	0.53952
CHERVANY	0.90984	MASON	0.61501	DAVISR	0.44214
DICKSON	0.84893	ROCKART	0.52125		
ROBEY	0.83028	DAVISGB	0.50935		
MASON	0.74775	SANDERS	0.50662		
MITROFF	0.7446	ZMUD	0.48521		
JARVENPAA	0.74253	GORRY	0.46921		
KINGW	0.68598	DICKSON	0.46749		
MINTZBERG	0.68586	COURTNEY	0.46709		
DAVISGB	0.67902	ANTHONY	0.46592		
SANDERS	0.65923	ROBEY	0.45781		
COURTNEY	0.65162	BENBASAT	0.45671		
ACKOFF	0.60269	LUCAS	0.44662		
SIMON	0.58122	CHERVANY	0.43916		
NEWELL	0.57614	KINGW	0.43034		
GINZBERG	0.5636	IVES	0.42389		
WAGNERGR	0.55104	SIMON	0.42086		
GORRY	0.53436				
HUBER	0.51478				
TVERSKY	0.50943				
ROCKART	0.5025				
DESANCTIS	0.5018				
SCOTTMORTON	0.49826				
ALTER	0.47227				
KEEN	0.46733				
ANTHONY	0.41892				
SPRAGUE	0.41472				

continued on following page

Table 7. continued

WATSONH	0.40789				
CARLSON	0.40698				
Variance	17.177		10.407		5.421
% Variance	25.64%		15.53%		8.09%

Table 8. Factor structure correlations (1970-1990)

Rotation method: promax; diagonal values: zero					
Factor1		Factor2		Factor3	
Foundations		GSS		Model Management	
GINZBERG	0.94326	GRAY	0.93977	BLANNING	0.93371
GORRY	0.92931	KRAEMER	0.91986	GEOFFRION	0.89719
WAGNERGR	0.928	KINGJL	0.9114	ELAM	0.89636
ANTHONY	0.92752	BUI	0.90648	DOLK	0.89408
ROCKART	0.92546	HILTZ	0.90361	STOHR	0.88704
LITTLE	0.9231	TUROFF	0.90124	HENDERSON	0.85294
ALTER	0.92279	GALLUPE	0.88056	HOLSAPPLE	0.84614
KINGW	0.91719	GEORGE	0.87748	DAVISR	0.8363
DAVISGB	0.91027	APPLEGATE	0.8706	BONCZEK	0.83393
MEADOR	0.90483	NUNAMAKER	0.8173	WHINSTON	0.82642
BENNETT	0.9041	DELBECQ	0.79281	SHORTLIFFE	0.75014
WATSONH	0.88965	VANDEVEN	0.78605	NAYLOR	0.74116
SIMON	0.85596	DESANCTIS	0.75532	BENNETT	0.70654
CARLSON	0.84411	JARKE	0.7277	WATSONH	0.70456
NAYLOR	0.81773	HUBER	0.68354	KONSYNSKI	0.69175
SPRAGUE	0.81066	JARVENPAA	0.66106	MEADOR	0.67681
SCOTTMORTON	0.80875	JELASSI	0.62063	COURTNEY	0.65065
ROBEY	0.78121	KONSYNSKI	0.57854	CARLSON	0.62046
COURTNEY	0.78118	DICKSON	0.47892	SPRAGUE	0.60363
KEEN	0.77123			JARKE	0.57657
ACKOFF	0.77018			JELASSI	0.56967
SANDERS	0.74645			ALTER	0.54777
MINTZBERG	0.74014			SIMON	0.53926

continued on following page

Table 8. continued

ZMUD	0.72406			TVERSKY	0.53714
LUCAS	0.69774			LITTLE	0.51686
CHERVANY	0.68344			GINZBERG	0.5129
STOHR	0.66672			NEWELL	0.51216
HENDERSON	0.66198			SCOTTMORTON	0.49081
NEWELL	0.66174			RAIFFA	0.48276
MASON	0.65132			HUBER	0.46922
SHORTLIFFE	0.64784			KEEN	0.46741
BENBASAT	0.61959			MINTZBERG	0.45507
DAVISR	0.61433			ANTHONY	0.43918
MITROFF	0.59957			ACKOFF	0.43135
WHINSTON	0.57037			SANDERS	0.41321
HOLSAPPLE	0.56081			NUNAMAKER	0.40616
BONCZEK	0.55267			WAGNERGR	0.40499
BLANNING	0.53441			GORRY	0.40401
MARCH	0.52499				
HUBER	0.50305				
ELAM	0.49702				
DICKSON	0.48144				
DEXTER	0.47593				
TVERSKY	0.47047				
LUSK	0.44629				
GEOFFRION	0.41315				
IVES	0.4043				
Variance	25.651		13.519		17.732
% Variance	38.29%		20.18%		26.47%

Factor4		Factor5		Factor6	
User Interface		Organization Science		MCDM	
LUSK	0.92894	MARCH	0.88594	KEENEY	0.81232
IVES	0.91405	TVERSKY	0.80774	JELASSI	0.71815
DEXTER	0.90138	NEWELL	0.7614	RAIFFA	0.5597
CHERVANY	0.89045	MINTZBERG	0.71283	GEOFFRION	0.52724

continued on following page

Table 8. continued

LUCAS	0.88648	ACKOFF	0.69654	JARKE	0.51085
ZMUD	0.88236	MITROFF	0.60449	STOHR	0.49114
BENBASAT	0.85582	MASON	0.51971	DAVISR	0.41262
ROBEY	0.82516	ROCKART	0.51881		
DICKSON	0.77425	SANDERS	0.51197		
MASON	0.73128	DAVISGB	0.50928		
JARVENPAA	0.72992	DICKSON	0.49508		
MITROFF	0.72825	ANTHONY	0.48877		
KINGW	0.68805	COURTNEY	0.48266		
MINTZBERG	0.68787	GORRY	0.48182		
DAVISGB	0.68123	ZMUD	0.46026		
SANDERS	0.66317	BENBASAT	0.43899		
COURTNEY	0.64211	ROBEY	0.43394		
ACKOFF	0.63698	LUCAS	0.42072		
NEWELL	0.58366	MEADOR	0.41998		
WAGNERGR	0.56095	IVES	0.41688		
GINZBERG	0.56092	RAIFFA	0.41579		
SIMON	0.55312	KINGW	0.41452		
GORRY	0.5411	CHERVANY	0.40891		
TVERSKY	0.52933				
ROCKART	0.5213				
HUBER	0.48489				
ALTER	0.47737				
SCOTTMORTON	0.47683				
DESANCTIS	0.4466				
ANTHONY	0.43752				
KEEN	0.42535				
WATSONH	0.41805				
MARCH	0.41228				
CARLSON	0.40412				
Variance	16.736		10.459		4.891
% Variance	24.98%		15.61%		7.30%

Table 9.

Rotation method: promax; diagonal values: the highest off diagonal values					
Factor1		Factor2		Factor3	
Foundations		GSS		Model Management	
KEEN	0.97386	GALLUPE	0.96884	BLANNING	0.96269
SCOTTMORTON	0.97162	GRAY	0.95183	WHINSTON	0.95801
ALTER	0.9611	BUI	0.946	BONCZEK	0.95795
CARLSON	0.92477	HILTZ	0.94322	HOLSAPPLE	0.95563
SPRAGUE	0.92077	TUROFF	0.93878	ELAM	0.93636
GINZBERG	0.88382	KRAEMER	0.93843	HENDERSON	0.90712
GORRY	0.8766	APPLEGATE	0.93562	STOHR	0.90574
LITTLE	0.87572	KINGJL	0.91535	DOLK	0.86827
SIMON	0.87533	NUNAMAKER	0.88716	GEOFFRION	0.84586
ANTHONY	0.87329	DESANCTIS	0.87308	DAVISR	0.81076
BENNETT	0.86828	GEORGE	0.86538	KONSYNSKI	0.7479
KINGW	0.84416	HUBER	0.7804	SHORTLIFFE	0.71503
WATSONH	0.83559	JARKE	0.73699	SPRAGUE	0.71464
WAGNERGR	0.83486	DELBECQ	0.69867	NAYLOR	0.69231
DAVISGB	0.82994	VANDEVEN	0.69431	CARLSON	0.68715
MEADOR	0.82175	KONSYNSKI	0.68046	WATSONH	0.66484
ROCKART	0.8127	JARVENPAA	0.60319	BENNETT	0.63859
NAYLOR	0.77107	JELASSI	0.54066	COURTNEY	0.61302
COURTNEY	0.7026	DICKSON	0.47854	MEADOR	0.60336
ROBEY	0.6806			SIMON	0.57033
ACKOFF	0.64846			JARKE	0.55539
MINTZBERG	0.63026			KEEN	0.5421
ZMUD	0.62767			ALTER	0.52261
STOHR	0.62392			SCOTTMORTON	0.5123
SANDERS	0.61765			JELASSI	0.49769
HENDERSON	0.61205			GINZBERG	0.46267
LUCAS	0.60251			HUBER	0.45829
CHERVANY	0.5926			LITTLE	0.43642
SHORTLIFFE	0.58614			TVERSKY	0.41933
DAVISR	0.5672				
BENBASAT	0.5414				

continued on following page

Table 9. continued

WHINSTON	0.53956				
HOLSAPPLE	0.52897				
MASON	0.52622				
BONCZEK	0.51627				
NEWELL	0.50275				
HUBER	0.48376				
BLANNING	0.4771				
MITROFF	0.47039				
ELAM	0.42835				
Variance	22.262		13.497		16.897
% Variance	33.23%		20.14%		25.22%
Factor 4		Factor 5		Factor 6	
User Interface		Organization Science		MCDM	
BENBASAT	0.97101	MARCH	0.86954	KEENEY	0.96186
DEXTER	0.93975	NEWELL	0.8363	RAIFFA	0.94822
LUSK	0.93566	SIMON	0.79417	GEOFFRION	0.63681
LUCAS	0.92775	MINTZBERG	0.78929	JELASSI	0.59086
IVES	0.9054	TVERSKY	0.7889	JARKE	0.49475
ZMUD	0.90039	ACKOFF	0.67849	STOHR	0.42119
DICKSON	0.89939	ANTHONY	0.57388		
CHERVANY	0.88894	DAVISGB	0.54244		
ROBEY	0.78126	KEEN	0.52429		
JARVENPAA	0.69428	COURTNEY	0.52428		
MASON	0.6533		0.52364		
MITROFF	0.63772	MITROFF	0.50367		
DAVISGB	0.61727	SANDERS	0.4957		
KINGW	0.61591	GORRY	0.49305		
SIMON	0.59063	HUBER	0.48605		
COURTNEY	0.57577	DAVISR	0.48558		
MINTZBERG	0.57315	MASON	0.48485		
KEEN	0.55392	MEADOR	0.47636		
SANDERS	0.54684	SPRAGUE	0.47396		

continued on following page

Table 9. continued

SCOTTMORTON	0.53268	SHORTLIFFE	0.46632		
DESANCTIS	0.52287	BENBASAT	0.46507		
GINZBERG	0.51209	CARLSON	0.45276		
HUBER	0.48696	ROCKART	0.45163		
GORRY	0.4629	ZMUD	0.45095		
ACKOFF	0.46077	ROBEY	0.44258		
WAGNERGR	0.45208	LUCAS	0.43767		
NEWELL	0.44604	DICKSON	0.42958		
SPRAGUE	0.43214	HENDERSON	0.42693		
ALTER	0.41896	WATSONH	0.41933		
		STOHR	0.41015		
		BENNETT	0.40566		
Variance	14.799		12.08		4.416
% Variance	22.09%		18.03%		6.59%
Factor7		Factor8		Factor9	
Strategic Management		Group Decision Making		Implementation	
MITROFF	0.70279	VANDEVEN	0.50206	SANDERS	0.63293
MASON	0.68291	DELBECQ	0.49275	COURTNEY	0.59873
ACKOFF	0.51939	DAVISGB	0.43988	HENDERSON	0.49377
JELASSI	-0.43596			KONSYNSKI	0.43351
				HUBER	0.41701
				ELAM	0.40659
				ZMUD	0.40571
Variance	3.187		3.747		4.752
% Variance	4.76%		5.59%		7.09%

Table 10. Factor structure correlations (1970-1990)

rotation method: promax; diagonal values: 3 highest off-diagonal values/2					
Factor1		Factor2		Factor3	
Foundations		GSS		Model Management	
KEEN	0.95993	GALLUPE	0.95129	WHINSTON	0.94146
SCOTTMORTON	0.95944	HILTZ	0.91792	BONCZEK	0.94119
ALTER	0.92275	GRAY	0.91571	HOLSAPPLE	0.93471
CARLSON	0.89093	BUI	0.91336	BLANNING	0.91452
SPRAGUE	0.88596	TUROFF	0.9113	ELAM	0.90118
GORRY	0.83964	KRAEMER	0.90857	HENDERSON	0.86374
GINZBERG	0.83205	KINGJL	0.89616	STOHR	0.8601
LITTLE	0.82385	APPLEGATE	0.89411	DOLK	0.82831
SIMON	0.8161	DESANCTIS	0.86535	GEOFFRION	0.78473
BENNETT	0.80797	NUNAMAKER	0.85517	DAVISR	0.75146
ANTHONY	0.80608	GEORGE	0.80845	KONSYNSKI	0.72498
KINGW	0.79081	HUBER	0.77487	SPRAGUE	0.68514
WATSONH	0.78638	JARKE	0.7159	SHORTLIFFE	0.67171
WAGNERGR	0.77806	DELBECQ	0.68703	CARLSON	0.65223
MEADOR	0.77143	VANDEVEN	0.68354	NAYLOR	0.64614
DAVISGB	0.75277	KONSYNSKI	0.66861	WATSONH	0.61578
ROCKART	0.73428	JARVENPAA	0.57399	BENNETT	0.57433
NAYLOR	0.71531	JELASSI	0.51461	COURTNEY	0.56916
COURTNEY	0.649	DICKSON	0.46365	MEADOR	0.5519
ROBEY	0.62416			SIMON	0.52894
ACKOFF	0.58481			KEEN	0.52188
ZMUD	0.57803			JARKE	0.52172
STOHR	0.56825			SCOTTMORTON	0.49578
LUCAS	0.5586			ALTER	0.47525
MINTZBERG	0.55572			JELASSI	0.45785
HENDERSON	0.54863			GINZBERG	0.41872
SANDERS	0.54786			HUBER	0.41159
CHERVANY	0.53879				
SHORTLIFFE	0.52861				
DAVISR	0.50259				
WHINSTON	0.49768				

continued on following page

Table 10. continued

HOLSAPPLE	0.4857				
MASON	0.4812				
BENBASAT	0.48099				
NEWELL	0.48089				
BONCZEK	0.47731				
MITROFF	0.43538				
HUBER	0.43043				
BLANNING	0.41798				
Variance	19.28		12.63		14.94
% of Variance	28.78%		18.85%		22.30%
Factor4		Factor5		Factor6	
User Interface		Org. Science		MCDM	
BENBASAT	0.95209	MARCH	0.8423	KEENEY	
DEXTER	0.92254	SIMON	0.83475	RAIFFA	
LUSK	0.90435	NEWELL	0.81959	GEOFFRION	
LUCAS	0.90213	MINTZBERG	0.74919	JELASSI	
DICKSON	0.88478	TVERSKY	0.74032	JARKE	
ZMUD	0.86408	ACKOFF	0.67446	STOHR	
IVES	0.85857	ANTHONY	0.55723		
CHERVANY	0.84881	KEEN	0.53412		
ROBEY	0.72918	SCOTTMORTON	0.53412		
JARVENPAA	0.67287	DAVISGB	0.50856		
MASON	0.61158	GORRY	0.49208		
MITROFF	0.60447	COURTNEY	0.49183		
KINGW	0.5693	DAVISR	0.48454		
SIMON	0.55545	MITROFF	0.4839		
DAVISGB	0.55099	SHORTLIFFE	0.48016		
KEEN	0.53746	MASON	0.46292		
COURTNEY	0.52986	HUBER	0.46097		
DESANCTIS	0.51727	SPRAGUE	0.4591		
SCOTTMORTON	0.515	MEADOR	0.44851		
MINTZBERG	0.50778	SANDERS	0.44539		
SANDERS	0.48077	CARLSON	0.43658		
GINZBERG	0.46672	BENBASAT	0.43317		

continued on following page

Table 10. continued

HUBER	0.44862	ZMUD	0.42381	
NEWELL	0.43114	LUCAS	0.42011	
GORRY	0.42748	ROBEY	0.40571	
ACKOFF	0.40943	DICKSON	0.40095	
Variance	10.96		4.42	
% of Variance	16.36%		6.60%	

Factor7		Factor8		Factor9	
Strategic Management		Implementation		Group Decision Making	
MITROFF	0.66403	SANDERS	0.68	VANDEVEN	0.44691
MASON	0.64888	COURTNEY	0.63997	DELBECQ	0.43894
ACKOFF	0.44547	HENDERSON	0.51111	NUNAMAKER	0.4124
		HUBER	0.45316		
		ZMUD	0.43705		
		ROBEY	0.41624		
		KONSYNSKI	0.41312		
Variance	2.77		5.07		4.42
% of Variance	4.13%		7.57%		6.60%

Table 11. Factor structure correlations (1970-1990)

rotation method: promax, diagonal values: raw citation					
Factor1		Factor2		Factor3	
Foundations		GSS		Model Management	
KEEN	0.95337	GALLUPE	0.97127	ELAM	0.95508
SCOTTMORTON	0.94969	HILTZ	0.9481	BLANNING	0.93597
ALTER	0.94647	GRAY	0.94778	DOLK	0.92356
CARLSON	0.90575	BUI	0.94248	HENDERSON	0.89812
SPRAGUE	0.89635	TUROFF	0.93787	BONCZEK	0.88402
GINZBERG	0.86965	KRAEMER	0.92896	HOLSAPPLE	0.8789
GORRY	0.84629	KINGJL	0.91634	WHINSTON	0.87624
WATSONH	0.83497	APPLEGATE	0.91487	KONSYNSKI	0.78573

continued on following page

Table 11. continued

MEADOR	0.82448	NUNAMAKER	0.84897	STOHR	0.78229
BENNETT	0.81656	DESANCTIS	0.84195	GEOFFRION	0.64104
WAGNERGR	0.79044	GEORGE	0.81563	SPRAGUE	0.62642
ANTHONY	0.78253	HUBER	0.74465	WATSONH	0.61297
KINGW	0.78118	DELBECQ	0.71804	DAVISR	0.60088
SIMON	0.76267	VANDEVEN	0.70653	CARLSON	0.59168
LITTLE	0.76077	JARKE	0.67825	MEADOR	0.56907
DAVISGB	0.73758	KONSYNSKI	0.63326	COURTNEY	0.56717
ROCKART	0.70395	JARVENPAA	0.5938	NAYLOR	0.54951
NAYLOR	0.69199	JELASSI	0.52977	BENNETT	0.49491
COURTNEY	0.65678	DICKSON	0.43846	SHORTLIFFE	0.47397
ROBEY	0.62592			ALTER	0.44902
SANDERS	0.59676			KEEN	0.44842
HENDERSON	0.58051			JARKE	0.43417
ZMUD	0.5805			SIMON	0.42782
STOHR	0.56494			SCOTTMORTON	0.41179
CHERVANY	0.56117			NUNAMAKER	0.40353
WHINSTON	0.52563				
HOLSAPPLE	0.52452				
LUCAS	0.51525				
BONCZEK	0.51256				
ACKOFF	0.50548				
MASON	0.48532				
BENBASAT	0.47456				
NEWELL	0.46836				
MINTZBERG	0.46653				
BLANNING	0.46257				
DAVISR	0.46239				
SHORTLIFFE	0.44487				
MITROFF	0.4382				
ELAM	0.42483				
HUBER	0.41842				
Variance	19.124		12.841		13.18

continued on following page

Table 11. continued

% Variance	28.54%		19.17%		19.67%

Factor 4	Factor 5			Factor 6	
User Interface	Organization		Science	MCDM	
DEXTER	0.94949	NEWELL	0.79055	KEENEY	0.93681
BENBASAT	0.94569	SIMON	0.75466	RAIFFA	0.88962
LUSK	0.92597	TVERSKY	0.7464	JELASSI	0.58386
DICKSON	0.87003	MARCH	0.72696	GEOFFRION	0.53896
CHERVANY	0.85572	MINTZBERG	0.60468	JARKE	0.4823
LUCAS	0.85366	COURTNEY	0.50617		
ZMUD	0.84981	ACKOFF	0.46747		
IVES	0.83815	MEADOR	0.46002		
JARVENPAA	0.72262	KONSYNSKI	0.45533		
ROBEY	0.7043	SANDERS	0.45519		
MASON	0.62112	HENDERSON	0.43379		
MITROFF	0.61351	HUBER	0.42738		
KINGW	0.53817	MITROFF	0.42058		
DESANCTIS	0.52709	KEEN	0.41001		
COURTNEY	0.52096	ELAM	0.40158		
DAVISGB	0.50329				
KEEN	0.50242				
SIMON	0.49572				
SANDERS	0.49381				
SCOTTMORTON	0.47373				
GINZBERG	0.4628				
HUBER	0.41659				
MINTZBERG	0.40827				
NEWELL	0.40173				
GORRY	0.40161				
Variance	12.433		8.482		3.644
% Variance	18.56%		12.66%		5.44%

continued on following page

Table 11. continued

Factor7		Factor8		Factor9	
Strategic Management		Artificial Intelligence		Implementation	
MITROFF	0.70311	SHORTLIFFE	0.57668	SANDERS	0.68727
MASON	0.69435	DAVISR	0.51896	ROBEY	0.58273
ACKOFF	0.62857			COURTNEY	0.52875
MINTZBERG	0.43373			SPRAGUE	0.51647
KINGW	0.40351			GINZBERG	0.51218
				VANDEVEN	0.50692
				ZMUD	0.50004
				DELBECQ	0.48091
				KEEN	0.4715
				WAGNERGR	0.47068
				CARLSON	0.46902
				DAVISGB	0.46527
				HUBER	0.44455
				SCOTTMORTON	0.41711
				WATSONH	0.41287
Variance	3.715		2.805		6.589
% Variance	5.54%		4.19%		9.83%

Section III
Alternative Approaches: Building Custom Databases

Chapter V
The Fox-Base Approach

ABSTRACT

Chapter II introduced online cocitation counts retrieval using Dialog Classic and citation index files. Certainly Dialog Classic is an attractive alternative in that the user is using the readily available bibliographic databases and retrieval software. The majority of ACA research has used ISI databases and Dialog Classic to retrieve cocitation counts. However, this approach has well-known technical limitations as discussed earlier. They include the issue of Multiple Authorship, Name-Homographs, and Synonyms. This chapter introduces an alternative approach to retrieving a co-citation count from the custom databases through the system we have designed and implemented. Custom database and retrieval systems need time and investment to develop, but they can manage most of the technical limitations discussed. The book presents two other alternative approaches that can be used to retrieve cocitation counts in lieu of using ISI citation index files and Dialog Classic. This chapter introduces the fox-base approach in developing custom databases and the cocitation matrix generation system. The first part is concerned with the design of databases. The second part describes the cocitation retrieval system. We also discuss how our system can eliminate or minimize the technical limitations of the Thomson ISI database and Dialog Classic Software system.

INTRODUCTION

In Chapter II, we have examined online data retrieval using ISI citation databases. One of the well known limitations to the indexing of social science citations index

(SSCI) is the use of first author only. With the current trend of increasing proportion of coauthored articles, it is a critical issue that cannot be disregarded. For example, in the area of management information systems, a study identified a total of 1,573 articles written by 3,411 authors. The articles are from the four major MIS journals published between 1977 and 2002. This study found that only 25 percent of the articles has a single author. Also, over the past two and a half decades, the proportion of coauthored management information systems articles has continuously increased from 40% in 1978 to more than 80% (1992-2002) (Oh, Choi, & Kim, 2005-2006).

Chapter III compared the results of using only first author and all authors. As the results indicated, there are no theoretical merits of using only first author in ACA study. The second critical limitation with SSCI is that only an author's family name and initials are indexed. To make matters worse, some records may contain only a first initial (e.g., Keen P.) and other records may include all initials (e.g., Keen PGW). To overcome these limitations, we have developed a custom bibliographic database and cocitation retrieval systems.

The cocitation matrix generation system accepts two bibliographic databases as major inputs— citing articles database and cited reference databases. Citing sources and the majority of cited references are from journals. Cited references come from books, conference proceedings, papers presented at the meetings without proceedings, doctoral dissertations and masters thesis, working papers, articles from encyclopedia, book chapters in edited books, newspaper articles, reports, personal communications, electronic sources primarily published on the Internet, unpublished working papers, etc. If hard disk storage requirements are not a major consideration, the database can include all unique fields from each source type. For example, a journal article may consist of <u>author</u> (last_name, first_name, middle_name for primary and secondary authors), <u>title</u> of article, <u>journal</u>, <u>Volume</u>, <u>issue</u>, <u>year</u>, <u>date</u>, etc. Book entry may require different fields such as <u>author</u> (last_name, first_name, middle_name for primary and secondary authors), <u>title</u> of books, <u>publisher</u>, <u>city</u>, <u>edition</u>, <u>year</u>, <u>ISBN</u>, etc. Combining all unique fields from each of different reference sources may require more than 100 fields in the database. Most of fields, however, will be "empty". Therefore, there are tradeoffs between storage requirement considerations and the maintenance of accurate field names.

DATABASE DESIGN

Considering storage requirements as the primary factor, the bibliographic database we have created consists of the following structures. It is possible to design multiple tables to create a view (interrelated tables). However, here we introduce how to build single files for citing articles and cited references without the data normalization process.

The Structure of the Bibliographic File

Since all citing articles and the majority of cited reference articles are from journals, the database structure is based on the structure suited for the journal articles. When entering bibliographical information other than journals, the "journal" field is used to enter publishers, city, state, etc. Typical bibliographical information is entered in fields 1 through 10. The remaining fields are important for ACA purposes.

Last Name

This field is to hold the last name of the first author. We realized that there is no easy way of controlling the problem of name-homograph in any bibliographic databases. For example, using common English name "Smith" or Korean name "Lee" as the cited author, there will be too many cited works of authors who are not in the list of selected authors. Each authors must be differentiated by creating a different last-name. Gordon Davis, Fred Davis, John Davis will be entered as DavisG, DavisF, and DavisJ respectively. If these authors are appeared as non-primary authors, their names in the middle name fields must be changed accordingly.

Table 1. Structure of cited reference file

Field	Field Name	Type	Width	Dec
1	Lastname	Character	18	
2	Firstname	Character	16	
3	Middlename	Character	100	
4	Title	Character	140	
5	Journal	Character	100	
6	Volume	Character	6	
7	Number	Character	5	
8	Date	Character	13	
9	Year	Character	4	
10	Pages	Character	15	
11	IDNO	Character	4	
12	Frequency	Numeric	3	
13	Citingby	Character	180	
14	Control	Character	3	

First Name

This field holds the first name of the first author.

MiddleName

The middle name field includes all the remaining authors' names. The co-citation retrieval system is designed to deal with different ways of entering the names of all authors. For example, the authors of the following record can be entered in to different ways.

Musen, M. A., Fagan, L. M., Combs, D. M., & Shortlife, E. H. (1987). Use of a Domain model to drive an Interactive Knowledge-editing tool. *International Journal of Man-Machine Studies, 12*, 63-87.

The middle name field of this record will look like:

A., Fagan, L. M., Combs, D. M., & Shortlife, E. H. or
A., L. M. Fagan, D. M. Combs, & E. H. Shortlife

IDNO

This field represents the citing article's identification numbers. Each citing article is assigned a unique number which can be numeric or alphanumeric. The example shown below includes F8 as an identification number of the citing article.

FREQUENCY

The frequency field is a three digit numeric field to record the citing frequencies of cited references. The example below also shows that this article entitled "A Survey of Decision Support System Applications (1971-April 1988)" is cited 23 times.

CITINGBY

This is the field that keeps a listing of citing articles' reference numbers. The example record indicates that the article is cited by 23 citing articles with idno from 361 through h6.

Figure 1. Example of cited reference record

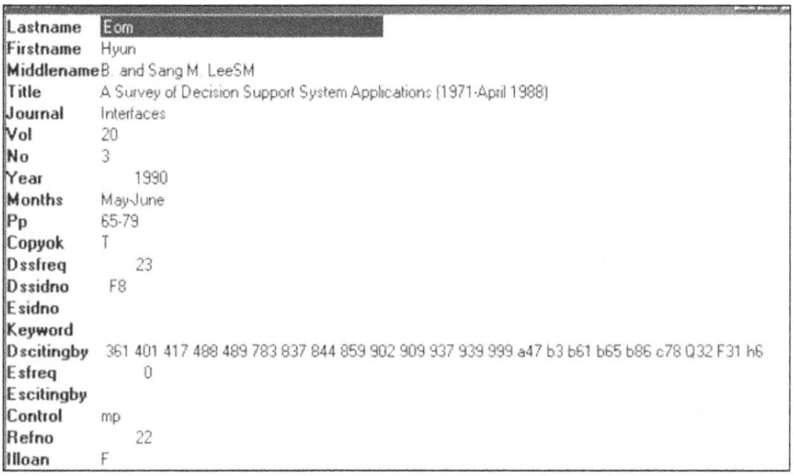

Lastname	Eom
Firstname	Hyun
Middlename	B. and Sang M. LeeSM
Title	A Survey of Decision Support System Applications (1971-April 1988)
Journal	Interfaces
Vol	20
No	3
Year	1990
Months	May-June
Pp	65-79
Copyok	T
Dssfreq	23
Dssidno	F8
Esidno	
Keyword	
Dscitingby	361 401 417 488 489 783 837 844 859 902 909 937 939 999 a47 b3 b61 b65 b86 c78 Q32 F31 h6
Esfreq	0
Escitingby	
Control	mp
Refno	22
Illoan	F

POPULATING DATABASE

Selection Criteria for Creating The Citing and cited Article Database

One advantage of the custom-databased ACA is the inclusion of articles selected by researchers. Article and journal selection can possibly affect the outcome of the ACA analysis. The ISI® *Social Sciences Citation Index* (*SSCI*®) and *Social SciSearch*® provide access to current and retrospective bibliographic information and cited references found in over 1,700 of the world's leading scholarly social sciences journals covering more than 50 disciplines. Nevertheless, it is possible that the ISI database may not include all important journals and articles in a discipline. To build a valuable bibliographic database, it is important to set selection criteria when deciding which papers to include in the database. For example, see the data and research methodology section of Chapter VI of the book.

Managing the Technical Limitations of ISI Databases

Technical limitations of ISI databases discussed earlier in Chapter III can be effectively managed by the co-citation matrix generation systems,

Multiple Authorship

The problem of ISI citation index files of retrieving records only by last name and initials of first author only can be solved by the cocitation generation system. The system retrieves the cocitation counts of all authors, regardless of the author status (first author, or non-first authors).

This is part of our coding to show that once the user enter the name of author's last name, the system checks whether the author name is part of middle name field. Our database is designed to use last name field to record the last name of the first author in the last name field. All other authors' names are entered to the middle name field.

```
@2,2 SAY " ENTER THE FIRST AUTHOR'S NAME  "
@2,35 GET LNAME
READ
LNAME=UPPER(LNAME)
@23,5 SAY "Just a moment please ..........  "
  COPY  ALL  FOR  UPPER(LNAME)=UPPER(LAST)  .OR.  "  "+TRIM(LNAME)+"
    "$UPPER(MIDDLE).OR.  "  "+TRIM(LNAME)+","$UPPER(MIDDLE)
    TO TEMP
```

Therefore, all authors, primary (first) or non-primary, will be counted when compiling cocitation frequency matrix.

Name-Homographs

SSCI indexes only the author's last name and initials. Consequently, citation records by an author of the same last name and initials may not be authored by the same person. The user should be aware the possible consequences of inexact result of query due to name-homographs (e.g., Smith, G., Smith, GA, Smith, GN, Smith, GD, Smith, GR, etc.). It is not possible for the indexers to differentiate among authors bearing the same first and middle initials.

Our system uses only the last name without the first or middle name information. It was very difficult to design a system that can use a combination of last name and first name. The question now is how to distinguish Johnson, Andrew and Johnson, John in the database if we use only the last name. The only way we could handle this situation was by adding the first initial to the last name. Records of Johnson, John are changed to JohnsonJ, whether he is the primary or non-primary author. However, in a given academic field, the database builders are familiar with the most of authors in the filed. In our database of decision support systems area, there are authors of the same last name with different first names such as Davis, G. and Davis, R. The whole databases records should be changed from Davis, G to DavisG, G, and from Davis, R. to DavisR, R.

Synonyms

To make matters more complicated, the same author's initials are recorded in many different ways in ISI index files. Some examples of synonym are Keen, P. or Keen, PGW., Lee, S. or Lee, SM. Our system uses only the last name to retrieve co-citation counts. Therefore, synonyms can be managed without difficulties.

Furthermore, there are many individuals who have the same names. Another problem is that there are no easy ways to handle name changes. The name of women authors could be changed when they marry, or some change their names for many different reasons. There are no software that can distinguish Jane Doe, whose name changed later due to her marriage. The only reasonable way this situation can be managed is by the knowledge of database builders. This is not a issue that can be handled by the design of software. In building the bibliographic databases of a specific field of study, due to the limited scope of databases, we were able to successfully manage synonyms or author's name change.

REFERENCED (CITED) ARTICLE DATABASE

As an example of ACA databases, we describe a bibliographic database built for our continuing research. A total of 1,616 *citing* articles contain 25,339 *cited* unique reference records. Each cited record has fields (dssfreq, dsscitingby) to include a total number of frequencies of citation by the citing articles and the list of identification numbers of citing articles in the decision support systems area. Each of the 1616 citing records has an average of 30 references, computed by total citation frequency/total citing records (48556/1616).

Table 2 shows that about 75% of cited articles were cited by the citing articles just once. Articles with two citations constitute approximately 13% of total cited records (25,339). Less than 2% of cited articles (502 articles) have been cited 10 or more times by the 1616 citing articles. Although there are no comparable statistics in other disciplines, the majority of cited articles (about 88% of total cited articles) were cited just once or twice by the 1616 citing articles. This fact can be a possible symptom of a fragmented research trend in the DSS area, or it could be a quality level indicator of published DSS and its reference discipline research in general.

Table 2. Citation behavior

Citing Freq.	No. of Records	Proportion	Cumulative	Total Freq.	Proportion	Cumulative
1	19077	75.29%	75.29%	19077	39.29%	39.29%
2	3095	12.21%	87.50%	6190	12.75%	52.04%
3	1147	4.53%	92.03%	3441	7.09%	59.12%
4	560	2.21%	94.24%	2240	4.61%	63.74%
5	348	1.37%	95.61%	1740	3.58%	67.32%
6	212	0.84%	96.45%	1272	2.62%	69.94%
7	178	0.70%	97.15%	1246	2.57%	72.51%
8	124	0.49%	97.64%	992	2.04%	74.55%
9	96	0.38%	98.02%	864	1.78%	76.33%
10	71	0.28%	98.30%	710	1.46%	77.79%
11	48	0.19%	98.49%	528	1.09%	78.88%
12	29	0.11%	98.60%	348	0.72%	79.59%
13	43	0.17%	98.77%	559	1.15%	80.75%
14	28	0.11%	98.88%	392	0.81%	81.55%
15	29	0.11%	99.00%	435	0.90%	82.45%
16	21	0.08%	99.08%	336	0.69%	83.14%
17	18	0.07%	99.15%	306	0.63%	83.77%
18	13	0.05%	99.20%	234	0.48%	84.25%
19	14	0.06%	99.26%	266	0.55%	84.80%
20	18	0.07%	99.33%	360	0.74%	85.54%
21	12	0.05%	99.38%	252	0.52%	86.06%
22	9	0.04%	99.41%	198	0.41%	86.47%
23	13	0.05%	99.46%	299	0.62%	87.09%
24	10	0.04%	99.50%	240	0.49%	87.58%
25	5	0.02%	99.52%	125	0.26%	87.84%
26	6	0.02%	99.55%	156	0.32%	88.16%
27	9	0.04%	99.58%	243	0.50%	88.66%
28	6	0.02%	99.61%	168	0.35%	89.00%
29	8	0.03%	99.64%	232	0.48%	89.48%
30	4	0.02%	99.65%	120	0.25%	89.73%
31	6	0.02%	99.68%	186	0.38%	90.11%

continued on following page

Table 2. continued

Citing Freq.	No. of Records	Proportion	Cumulative	Total Freq.	Proportion	Cumulative
32	6	0.02%	99.70%	192	0.40%	90.51%
33	10	0.04%	99.74%	330	0.68%	91.19%
34	2	0.01%	99.75%	68	0.14%	91.33%
35	0	0.00%	99.75%	0	0.00%	91.33%
36	2	0.01%	99.76%	72	0.15%	91.48%
37	3	0.01%	99.77%	111	0.23%	91.70%
38	3	0.01%	99.78%	114	0.23%	91.94%
39	5	0.02%	99.80%	195	0.40%	92.34%
40	2	0.01%	99.81%	80	0.16%	92.51%
41	2	0.01%	99.81%	82	0.17%	92.67%
42	1	0.00%	99.82%	42	0.09%	92.76%
43	2	0.01%	99.83%	86	0.18%	92.94%
44	1	0.00%	99.83%	44	0.09%	93.03%
45	3	0.01%	99.84%	135	0.28%	93.31%
46	2	0.01%	99.85%	92	0.19%	93.50%
47	1	0.00%	99.85%	47	0.10%	93.59%
48	2	0.01%	99.86%	96	0.20%	93.79%
49	4	0.02%	99.88%	196	0.40%	94.19%
50	2	0.01%	99.89%	100	0.21%	94.40%
51	2	0.01%	99.89%	102	0.21%	94.61%
52	1	0.00%	99.90%	52	0.11%	94.72%
53	0	0.00%	99.90%	0	0.00%	94.72%
54	0	0.00%	99.90%	0	0.00%	94.72%
55	0	0.00%	99.90%	0	0.00%	94.72%
56	2	0.01%	99.91%	112	0.23%	94.95%
57	2	0.01%	99.91%	114	0.23%	95.18%
58	0	0.00%	99.91%	0	0.00%	95.18%
59	0	0.00%	99.91%	0	0.00%	95.18%
60	1	0.00%	99.92%	60	0.12%	95.31%
61	0	0.00%	99.92%	0	0.00%	95.31%
62	1	0.00%	99.92%	62	0.13%	95.43%
63	0	0.00%	99.92%	0	0.00%	95.43%
64	1	0.00%	99.93%	64	0.13%	95.57%
65	0	0.00%	99.93%	0	0.00%	95.57%
66	1	0.00%	99.93%	66	0.14%	95.70%

continued on following page

Table 2. continued

67	0	0.00%	99.93%	0	0.00%	95.70%
68	0	0.00%	99.93%	0	0.00%	95.70%
69	1	0.00%	99.93%	69	0.14%	95.84%
Citing Freq.	**No. of Records**	**Proportion**	**Cumulative**	**Total Freq.**	**Proportion**	**Cumulative**
72	2	0.01%	99.94%	144	0.30%	96.14%
75	1	0.00%	99.94%	75	0.15%	96.29%
77	1	0.00%	99.95%	77	0.16%	96.45%
85	1	0.00%	99.95%	85	0.18%	96.63%
88	1	0.00%	99.96%	88	0.18%	96.81%
97	2	0.01%	99.96%	194	0.40%	97.21%
103	2	0.01%	99.97%	206	0.42%	97.63%
104	1	0.00%	99.98%	104	0.21%	97.85%
116	1	0.00%	99.98%	116	0.24%	98.09%
118	1	0.00%	99.98%	118	0.24%	98.33%
144	1	0.00%	99.99%	144	0.30%	98.63%
189	1	0.00%	99.99%	189	0.39%	99.02%
236	1	0.00%	100.00%	236	0.49%	99.50%
242	1	0.00%	100.00%	242	0.50%	100.00%

Cited Record 25339, Total Citation 48556, Average Citing Frequencies 30.04703, Citing Records 1616

GENERATION OF COCITED AUTHOR COUNTS

Cocitation counts can be either retrieved from commercial online bibliographical databases such as *Science Citation Index*, *Social Sciences Citation Index*, and *Arts and Humanities Citation Indexes,* or generated from the custom-built bibliographical databases. If researchers use the commercial on-line databases, cocitation frequencies will be retrieved using the query command as described in Chapter III. Several examples of those commands are given in McCain (1990).

The Cocitation Count Generation System

This book introduces an alternative approach to ACA research—the generation of cocited author counts using a custom-built bibliographic database and in-house cocitation count generation systems. FoxBase database management systems are used to enter the bibliographic records. The system is coded using Fox-Base database management systems. It computes author cocitation frequencies between any

Figure 2. Citation pattern

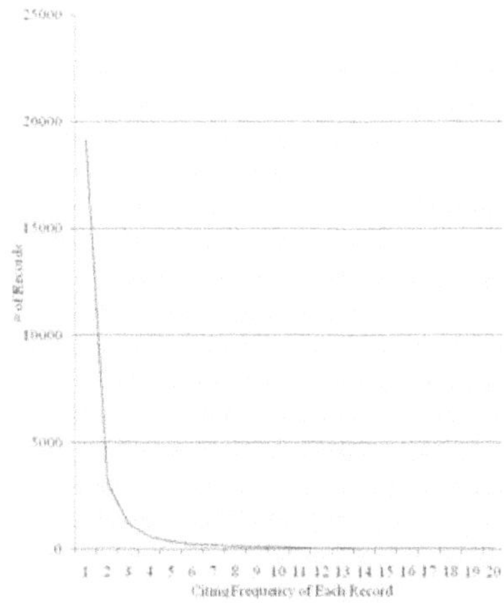

pair of all (primary and non-primary) authors under study. The author cocitation frequency generation system enables the users to overcome the problem with the Institute for Scientific Information (ISI) databases which code only the first author of a cited work. The cocitation matrix generation system we developed gives access to cited coauthors as well as first authors.

The user specifies the investigation period by entering the beginning and ending years. Currently our database consists of citing article database from 1969 to 2004. The second choice (frequency comparison) activates the systems to compute cocitation frequency counts. It has two different modes: Continuous vs manual. The continuous mode of operations will compute co-citation counts of two different author sets: Authorl and author2. If authorl and author2 databases contain 100 author last names each, the system will compute cocitation frequency counts of 5000 combinations of author pairs (100 * 100/2) and the results will be written to a file.

This screen shows 35 co-citation counts between Keen and Scott Morton during the investigation period (1970-2000). It also lists identification number (idno) of all citing articles beginning from D7 to S23. Each citing article is assigned a unique identification number. Currently, there are more than 2000 citing article records and more than 40,000 cited reference records in the decision support systems area. Initially, we begin to number each citing article from 1 to 999. After that, we begin to use prefix of various characters of upper case and lower case characters. For

Figure 3. Main menu of author cocitation frequency generation systems

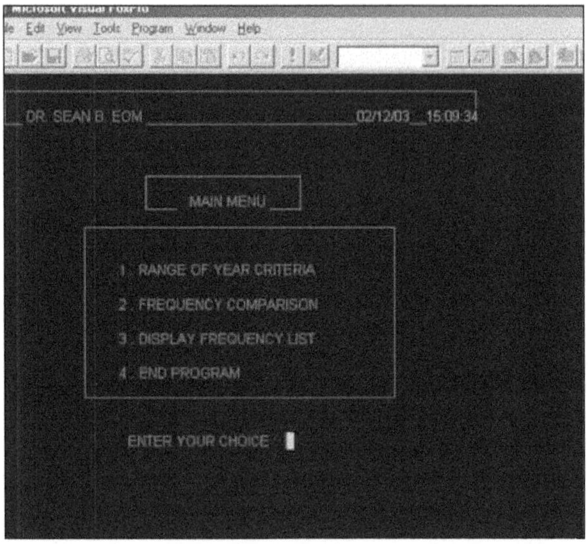

Figure 4. Retrieval of cocitation frequency between two authors

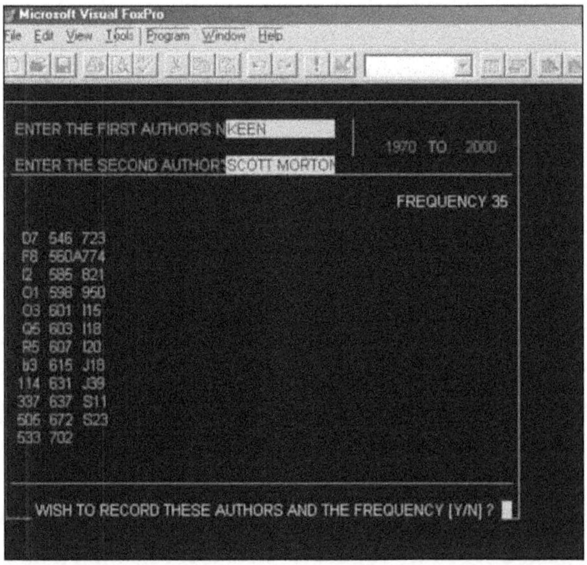

Figure 5. Displaying cocitation frequency of a file

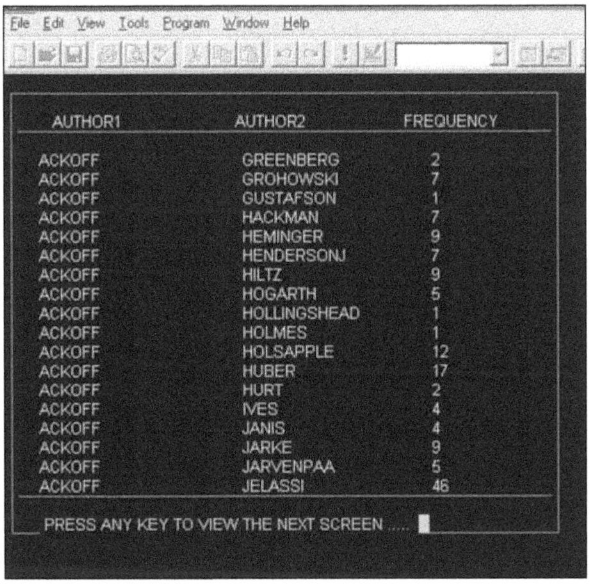

AUTHOR1	AUTHOR2	FREQUENCY
ACKOFF	GREENBERG	2
ACKOFF	GROHOWSKI	7
ACKOFF	GUSTAFSON	1
ACKOFF	HACKMAN	7
ACKOFF	HEMINGER	9
ACKOFF	HENDERSONJ	7
ACKOFF	HILTZ	9
ACKOFF	HOGARTH	5
ACKOFF	HOLLINGSHEAD	1
ACKOFF	HOLMES	1
ACKOFF	HOLSAPPLE	12
ACKOFF	HUBER	17
ACKOFF	HURT	2
ACKOFF	IVES	4
ACKOFF	JANIS	4
ACKOFF	JARKE	9
ACKOFF	JARVENPAA	5
ACKOFF	JELASSI	46

example, "D7" is one of citing article from *Decision Support Systems* journal. The prefix Q represents citing articles from *MIS Quarterly*.

This screen shot lists part of the out file created by the system.

CONCLUSION

In this chapter, we introduced an alternative approach of using Foxbase database management systems in lieu of ISI citation index files and Dialog Classic. We discussed the design of databases and cocitation counts generation systems. The system introduced here can effectively manage all of the technical limitations of ISI databases discussed earlier. It is possible to cope with the technical limitations with the development of custom databases and software primarily due to the narrow and limited scope of databases.

REFERENCES

McCain, K. W. (1990). Mapping Authors in Intellectual Space: A Technical Overview. *Journal of the American Society for Information Science, 41*(6), 433-443.

Oh, W., Choi, J. N., & Kim, K. (2005-2006). Coauthorship Dynamics and Knowledge Capital: The Patterns of Cross-Disciplinary Collaboration in Information Systems Research. *Journal of Management Information Systems, 22*(3), 265-292.

Chapter VI
Building Databases and the Cocitation Counts Generation System Using Microsoft Excel Program in Visual Basics

ABSTRACT

This chapter shows another alternative approach of building citation database and retrieval system using the spreadsheet program, Microsoft Excel. McIntyre built a custom database based on the Clothing and Textile Research Journal (CTRJ), covering from 1990 to 2006 as part of his master's thesis. The database includes all the author citations, citations sorted by article, and the top cited author's cocitation frequencies.

INTRODUCTION

The previous chapter introduced an approach of building custom databases using Foxbase database management systems. This chapter introduces an alternative approach using a spreadsheet program, Microsoft Excel. McIntire (McIntire, 2007) invented this approach as part of his Mater's thesis at the University of Columbia. His thesis is based on the International Textile and Apparel Association (ITAA) publication database. The motive for the design of database and cocitation counts system was simply that the ISI citation index files do not include the specific journal in the textile and apparel area.

DATABASE DESIGN

McIntire completed his Master's thesis in the department of textile and apparel management. He built a custom database based on the Clothing and Textile Research Journal (CTRJ), covering from 1990 to 2006. The database includes all the author citations, citations sorted by article, and the top cited author's cocitation frequencies. Using the chronological list of Clothing and Textile Research Journal (CTRJ), each article published between 1990-2006 was viewed as a PDF document. Each article was assigned a six digit number: the first four digits representing the publication year, the fifth and sixth digits representing the article number for that year beginning with one.

Example: *the article Dickerson, K.G. & Dalecki, M. Apparel Manufacturers' Perceptions of Supplier-Retailer Relationships would be assigned the article number 1990-10, since it was published in 1990 and is the tenth article published in CTRJ that year.*

A custom database is created containing the names of all cited authors and when and where they were cited. Within an Excel spreadsheet, a column is titled with the appropriate article number. The rows are labeled alphabetically with author names. Each time an author is listed in the reference section of the article, it is counted in the article column.

Figure 1. Example for article 1990-10

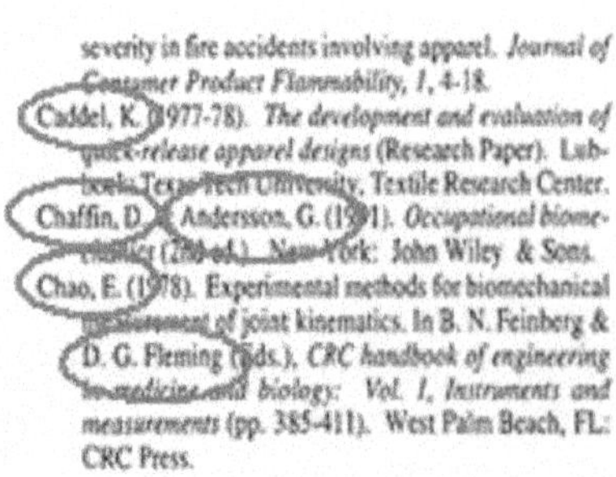
severity in fire accidents involving apparel. *Journal of Consumer Product Flammability, 1*, 4-18.
Caddel, K. (1977-78). *The development and evaluation of quick-release apparel designs* (Research Paper). Lubbock, Texas: Texas Tech University, Textile Research Center.
Chaffin, D., & Andersson, G. (1991). *Occupational biomechanics* (2nd ed.). New York: John Wiley & Sons.
Chao, E. (1978). Experimental methods for biomechanical measurement of joint kinematics. In B. N. Feinberg & D. G. Fleming (Eds.), *CRC handbook of engineering in medicine and biology: Vol. 1, Instruments and measurements* (pp. 385-411). West Palm Beach, FL: CRC Press.

	1990-10	1990-11	1990-12	1990-13
Anderson, G	1			1
Caddel, K	1			1
Chaffin, D	1	4		
Channing, R			1	
Chao, E	1		1	
D.G. Fleming	1			1

Figure 1 shows how the references of article 1990-10 are entered into the excel database. The visible portion of the references lists four authors and a coeditor. The other co-editor, B.N. Feinberg, is not one of the final author sets under study.

- It is notable that multiple authors in one article are counted regardless their position. For example, the second article has two coauthors (Chaffin and Andersson). The database equally treats the two authors by entering the two authors individually and gives an equal number of citation count of 1.
- Multiple citations are counted, i.e. four references of Chaffin in article 1990-11 would be entered as a 4 in Excel for the author in the appropriate article number column.

The citation counts compiled above are spot checked for accuracy. The final document appears as this:

The citation counts are divided into five year blocks: 1990-1995, 1996-2000, 2001-2006. An arbitrary number of top cited authors are selected, and a matrix is formed showing how often they are cited in the same documents.

The cocitation count is limited by the author cited the least. An author cited four times, and another cited twice, will only be counted as cocited twice. For example, consider the following citation count.

Once this is transformed into a cocitation count, the matrix would include the following pairs:

Figure 5 is an excerpt from the actual cocitation matrix.

THE CO-CITATION COUNTS GENERATION SYSTEM

McIntyre wrote an excel program in visual basic that computes the cocitation counts for all author pairs from a citation spreadsheet database. Figure 1 shows

Figure 2. An example of excel spreadsheet database

	Article Number	1991-01	1991-02	1991-03	1991-04	1991-05	1991-06	1991-07	1991-08
Kaiser, S.B.			1				1		6
Eicher, J.B.									2
Davis, L.L.							1		3
Roach-Higgins, M.E.									1
Drake, M.F		1		1	1		1		
Lennon, S.J									2
Darden, W.R				1					
Winakor, G.				3					
Laughlin, J.M.									
Erekosima, T.V.									2
DeLong, M.R.						2			
Shim, S.		1							
Damhorst, M.L.									2

Figure 3. Citation database that show multiple references by an author

	Article Number	1991-01	1991-02	1991-03	1991-04	1991-05	1991-06	1991-07	1991-08
Kaiser, S.B.			1				1		6
Eicher, J.B.									[illegible]
Davis, L.L.							1		3
Roach-Higgins, M.E.									1
Drake, M.F		1		1	1		1		
Lennon, S.J									2
Darden, W.R				1					
Winakor, G.				3					
Laughlin, J.M.									
Erekosima, T.V.									2
DeLong, M.R.						2			
Shim, S.		1							
Damhorst, M.L.									2

Figure 4. A sample cocitation matrix

	Drake	Winakor	Kaiser	Davis, LL
Drake		1	0	0
Winakor	1		0	0
Kaiser	0	0		4
Davis, LL	0	0	4	

Figure 5. An excerpt from the actual co-citation matrix

	Dickerson, K.G.	Dickey, L.E.	Dillman, D.A.	Douglas, S.P.	Drake, M.F
Dickerson, K.G.					
Dickey, L.E.	1				
Dillman, D.A.	2				
Douglas, S.P.		3			
Drake, M.F	2	5	3	5	
Easley, C.B.					
Eicher, J.B.					1
Engel, JF		3		3	10
Erekosima, T.V.					
Ettenson, R			2		2
Finley, E.L.					
Fiske, S.T					2
Forsythe, S.M. &			2	2	6
Francis, S.K.		2	2		6

the architecture of the co-citation generation system McIntire developed. The citation worksheet holds the citation data. His database includes all the bibliographic information from all of the articles published in only one journal, Clothing and Textile Research Journal, during the year 1990-2006. Column A lists all of the authors' name under investigation. Column B lists citation information from each article published in each year. Figure 6 shows citation counts of 5 articles published in 1991. He developed a new algorithm to transform the raw citation counts to

Figure 6. Spreadsheet-based co-citation generation systems developed by Jonah McIntire

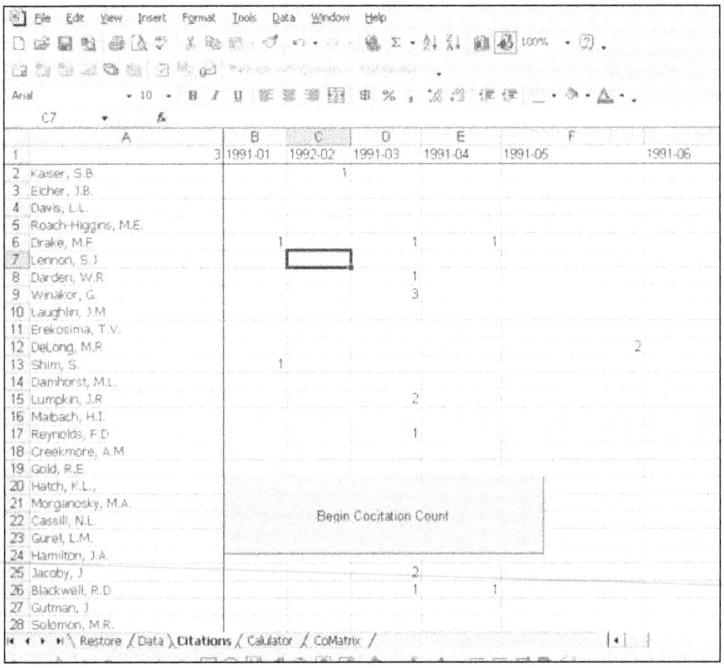

Figure 7. Excel–based spread sheet databases

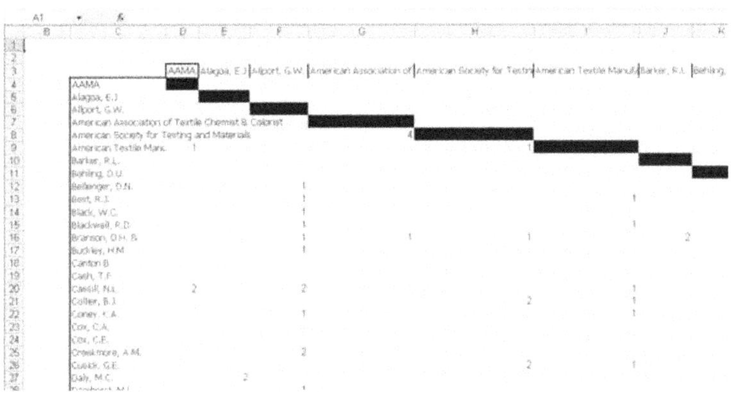

cocitation counts. Readers are referred to (McIntire, 2007) for the details of data transformation algorithm.

CONCLUSION

This chapter introduced another alternative approach of using a popular spread sheet program, Microsoft Excel program. This spreadsheet based system is also designed to handle multiple authorships by including all non-primary authors in the databases. In addition, all of the technical limitations associated with the use of the ISI citation index files can be managed effectively.

REFERENCE

McIntire, J. S. (2007). *The Clothing and Textile Research Base: An Author Cocitation Study.* Unpublished Master's Thesis, University of Missouri, Columbia, Missouri.

Section IV
Author Cocitation Analysis Procedures

Chapter VII
Overview of Author Cocitation Analysis Procedures

ABSTRACT

The previous two chapters examined the two alternative approaches of retrieving cocitation counts using custom databases and cocitation frequency counts extraction systems. The cocitaion frequency counts are the inputs to the SAS or SPSS systems for multivariate statistical analysis. The primary purpose of this chapter is to overview several important steps in author cocitation analysis. ACA consists of the six major steps beginning with the selection of author sets for further analysis, then collection of cocitation frequency counts, statistical analysis of the cocitation frequency counts, and the validation and interpretation of statistical outputs.

INTRODUCTION

ACA consists of the assembling and interpretation of bibliographical statistics taken from the cited references which are taken from the selected citing articles. (See Figure 1).

1. Selection of Author
2. Collection of Cocitation Frequencies
3. Preparation of SAS Input Files

4. Multivariate Statistical Analysis of Correlation Matrix
5. Output Preparation
6. Validation and Interpretation

SELECTION OF AUTHORS/COCITATION THRESHOLDS

The goal of ACA is to produce empirical maps of prominent authors in an academic discipline. The first step is to select the eminent scholars in the academic discipline. Basically, two approaches are available: starting a predetermined list of authors in a given field (the subjective approach), or compiling a set of authors from scratch (the objective approach). An important purpose of ACA is an overall examination of the intellectual structure of an academic discipline. Therefore, it is critical to establish a diversified list of authors. McCain (1990, p.433) states:

In the aggregate, this author set defines the scholarly landscape being mapped. If the authors are not chosen to capture the full range of variability in subject specializations, methodologies, political orientations, etc., these aspects of structure cannot be determined.

SUBJECTIVE (TOP-DOWN) APPROACH

The first approach may start with a pre-determined list of authors or with the selection of a list of authors to be searched as cited references in the ISI databases as discussed by McCain (1990). The predetermined list of authors can be compiled by:

- Personal knowledge
- Consultation with researchers in the area to be studied
- Conducting surveys
- Using directories
- Organizational membership
- Conference attendance rosters
- Lists of awards, etc.

Compiling a predetermined list of authors inevitably involves subjective judgments. This approach can be efficient because no lengthy time is spent finalizing a list of authors for further analysis. A critical weakness of this approach is that it may often fail to identify emerging scholars in a given area of an academic discipline. The majority of previous research in this area has used the subjective approach of using a predetermined list of eminent scholars (M. J. Culnan, 1986; McCain, 1986). The reason for doing so was not because this approach is superior but because most of the previous research used commercial online bibliographic databases to retrieve the cocitation frequency matrix. It is virtually impossible to scan the entire citation databases to screen and select an author set without using the personal judgments of researchers.

Figure 1. Author cocitation analysis steps

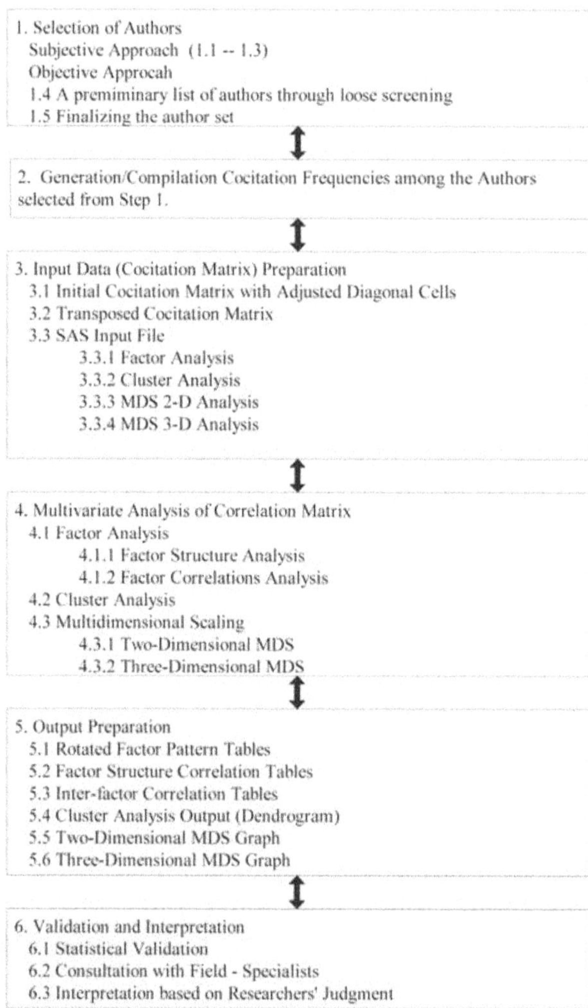

OBJECTIVE (BOTTOM-UP) APPROACH

The objective approach does not start with a predefined list of authors at all. No subjective judgments are necessary. It will be very difficult to apply this objective approach to the commercial databases, such as social science citation index, due to the size of the databases and high costs. However, it is relatively manageable to screen the whole custom-built databases to select a list of authors for further analysis. ACA, using custom databases, allows researchers to unobtrusively select authors for the study. Emerging scholars are more likely to be included in the selection process, unlike the subjective approach discussed earlier.

The other important advantage of using custom-built databases includes the identification of reference disciplines. It will be virtually impossible to include all the predefined lists of authors in the reference disciplines prior to conducting actual analysis due to the search time and financial constraints. As explained below, the objective approach we are introducing is suitable when using the custom-built databases, since researchers do not have to start with a list of authors in the reference disciplines.

The Loose Screening to Produce a Preliminary List

Unlike ACA using commercial on-line databases such as ISI citation index files, the selection process starts with no predefined list of authors. Instead, each author in the in-house bibliographic databases is initially included for further screening. The first stage of loose screening is based on the citation frequency of each record. Table 2 of Chapter V reveals an interesting citation pattern.

- About 75% of all cited works are cited just once.
- Only about 7% of all cited works are cited three or more times.
- About 5% of all cited works are cited five or more times.

Based on this citation pattern of the database, deleting all records whose citation frequencies are ten or less (98.31%), we have now less than 2% of all cited reference records for further analysis. Therefore, this stage yields a list of a significantly reduced number of authors. The actual preliminary list of authors can be compiled via identifying unique names in smaller number of records. These records may contain different numbers of unique authors, whether primary or non-primary, after removing redundant names from the first author fields of each record and adding names from the non-first author fields (the middle name fields in our database).

Finalizing the List of Author via Detailed Analysis

Citation counts of the individual authors in the preliminary list can be used to finalize the list of authors. This step further filters the preliminary list compiled from the previous step into a set of authors based on author *cocitation* counts (frequencies). Although it rarely happens, theoretically it is possible for authors with higher citation counts to have very low cocitation counts with other authors. In this case, these authors may appear in the final author set. However, they will carry lower factor loadings that can be interpreted as insignificant authors to the formation of the intellectual structure of the academic discipline under study. To objectively examine the intellectual structure of a discipline, personal judgment must be avoided in selecting authors by objectively counting the frequency of each name from the data bases.

Using in-house databases, author selection criteria must be established in regard to the number of citations received. As the citation behavior figure shows, the higher number of threshold means the smaller number of authors for further analysis. The optimal number of authors is primarily dependent on the number of cited database records. In addition, there may be some differences in the citation behavior across the different academic disciplines. There are no quantitative tools that can be blindly applied in deciding the number of authors. A study of author cocitation of a journal in consumer research over the 15 year period used 4 citations as selection criteria to compile the list of authors (Hoffman & Holbrook, 1993). In other studies of ACA that mapped the intellectual structures of decision support systems, the adjusted diagonal cell values of 25 or more were used (Eom, 1995, 1996). Once the threshold values are used to select the author list, it is important to apply the same criteria consistently to the subsequent studies to be followed to trace the changing structures of a discipline.

Ad Hoc Criteria for Further Screening of Authors

Due to the possible instability of small cocitation counts, author cocitation analysis researchers introduced several ad hoc criteria for further screening a large pool of candidate authors to finalize a list of authors. The criteria include a *mean cocitation rate* above a certain lower limit per author in each time period (e.g., nine for 10 years of Social Scisearch data), cocitation with at least one-third of the entire author set, or restricting the final author set to the 20% receiving the highest number of citations and cocitations in initial retrieval trials. For further details on several different approaches to compiling a predetermined list of authors, see (McCain, 1990). However, author cocitation analysis researchers suggested that all these criteria be applied to the commercial on-line databases such as SCISEARCH and SOCIAL SCISEARCH. The nature of the database we created makes it meaningless to use those criteria suggested by ACA researchers.

Our databases are significantly different from those commercial databases in terms of record size. Besides, the cocitation matrix generation system we developed gives access to cited coauthors as well as first authors. Due to these differences, we could not follow the suggested criteria by McCain (1990), such as "a mean cocitation rate of 'x' or more cocitations in each time period." Rather, we had to invent a new criterion through the method of trial and error. We experimented with the sensitivity of changing the cocitation threshold on the final outcomes (number of meaningful factors to accurately represent research subspecialties). With our databases, we conclude that the number of cocitations of an author with himself/herself can be a better criterion to determine the final author set due to its simplicity. Applying the mean cocitation criteria to worksheet files of any spreadsheet program (the output from the cocitation counts generation system) involves too many computations. For example, whenever we delete/add an author to the final author set, we need to compute the mean cocitation rate of each author again. Using the cocitation rate of 25 with the author himself/herself in the investigation period, the final set of authors was chosen. In our previous papers (Eom, 1995; Eom & Farris, 1996), the cocitation rate of 25 with the author himself/herself was applied to finalize a list of authors for further analysis.

Regardless of the nature of bibliographic databases (commercial vs. custom-built), determining the threshold cocitation rate is not the result of a structured process; rather, it is an unstructured process requiring the investigator's personal judgments. An exact quantitative basis for deciding the threshold cocitation rate has not been developed. Lowering the threshold in general increases the number of authors to be included in a study, which in turn may or may not change the number of meaningful factors in the study. Also, it is important to point out that cocitation thresholds themselves, as sole connection criteria, are suspect in a highly multi-disciplinary area. One should look at the overall connectedness and the focused cocitation counts as well.

Raw Cocited Author Counts

In ACA, cocitation frequency (cocited author counts) is the prime input data. The cocitation count generation system produces a raw author cocitation frequency table. If the number of authors under study is n, all of the possible number of cocitation counts to be produced by the system is computed by $n*(n+1)/2$. For example, Table 2 lists 210 cocitation counts with 20 authors. The 45 cocitation counts include 20 diagonal cell values. On the other hand, the possible number of cocitation counts

Figure 2. Cocitation frequency matrix preparation steps in ACA

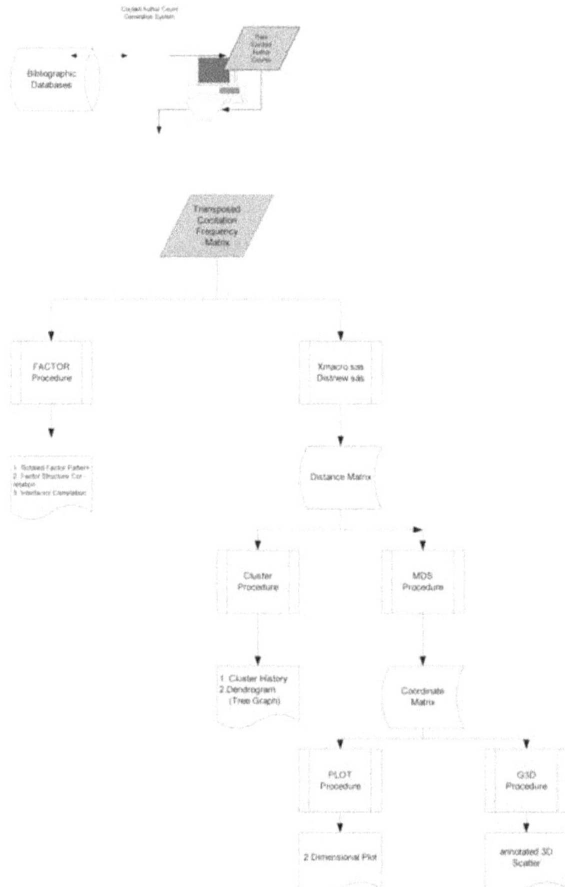

to be retrieved by the commercial system is computed by n*(n-1)/2. There is a difference between the two because the system we developed produces diagonal cell values while the diagonal values are not retrieved by the commercial system.

Creation of Raw Cocitation Matrix

The next step in preparing the SAS input is to create a raw cocitation matrix. Table 3 shows an example of diagonal cell value adjusted cocitation matrix in Microsoft Excel format. The cocitation generator output can be imported into Microsoft Excel

Table 1. Part of the cocitation count generation system output

AUTHOR1	AUTHOR2	COCITATION FREQUENCY
ALTER	ALTER	121
ALTER	BONCZEK	25
ALTER	CARLSON	66
ALTER	HUBER	23
ALTER	KEEN	92
ALTER	SCOTTMORTON	83
ALTER	SIMON	38
ALTER	SPRAGUE	67
ALTER	WHINSTON	27
BONCZEK	BONCZEK	103
BONCZEK	CARLSON	53
BONCZEK	HUBER	23

program and need tedious manual "cut and paste" operations to create raw cocitation matrix in the shape of a triangle.

The example of the cocitation matrix data used in this chapter was taken from the actual data used in (Eom, 1995; Eom & Farris, 1996). The data was gathered from a total 692 citing articles in the decision support systems area over the period of 1971-1990. The bibliographic database used here consists of 15,030 cited reference records. Our data consist of only 20 authors help the beginners understand the ACA process more thoroughly and clearly.

Conversion of Raw Cocitation Matrix

It is not suitable to use the raw citation matrix in a triangle shape as an input to the SAS system for ACA study. The next step is adjusting diagonal cell values in the raw cocitation matrix created.

Diagonal Cell Values

Raw cocitation frequencies of row and column authors fill the off-diagonal cells. The off-diagonal cell value is the total number of the cocitation count between these two authors. The term "author" in author cocitation analysis is neither an individual nor individuals. It refers to a body of writings by a person.

Diagonal values in the cocitation frequency counts matrix are a fundamental issue in ACA study. Diagonal values are the co-citation frequency counts between the author himself/herself excluding self-citation. Retrieving exact values of diagonal values in the co-citation matrix requires a manual and time consuming procedure. For that reason, ACA researchers suggested many different approaches to create, instead of retrieving the real values, the diagonal cells in the cocitation matrix. They include the mean cocitation count, missing values, zeroes, highest off-diagonal counts, adjusted off-diagonal values, and the number of times the author is cocited with himself/herself. The majority of ACA researchers seems to prefer using either the adjusted value approach by adding three highest off-diagonal values and divided

Table 2. Raw cocitation matrix

Author	No	1	2	3	4	5	6	7	8	9	10	11	12	13
ALTER	1	109												
BENBASAT	2	16	71											
BONCZEK	3	25	13	93										
CARLSON	4	66	23	53	163									
DESANCTIS	5	8	33	18	18	82								
DICKSON	6	15	43	13	17	43	84							
GINZBERG	7	34	13	15	32	8	13	58						
GORRY	8	30	13	11	34	7	9	15	67					
HENDERSON	9	22	16	34	33	16	13	14	8	67				
HOLSAPPLE	10	25	13	84	49	16	13	14	12	33	87			
HUBER	11	23	22	23	34	49	35	21	12	22	23	93		
KEEN	12	92	44	46	112	33	45	53	46	45	46	47	235	
KONSYNSKI	13	16	12	34	24	33	19	8	4	37	32	35	29	77
LUCAS	14	14	28	11	18	18	29	14	11	7	11	16	39	6
MASON, R.	15	15	24	8	22	16	23	13	16	16	8	20	40	16
MITROFF	16	12	22	8	18	17	23	10	15	17	8	17	36	16
SCOTT-MORTON	17	83	38	42	101	27	39	42	62	34	41	39	174	23
SIMON	18	38	31	33	54	28	34	17	39	26	33	37	82	28
SPRAGUE	19	67	31	59	133	27	26	38	32	44	56	49	126	35
WHINSTON	20	27	13	93	54	19	14	15	13	34	87	24	50	34

continued on following page

Table 2. continued

Author	No	14	15	16	17	18	19	20
ALTER	1							
BENBASAT	2							
BONCZEK	3							
CARLSON	4							
DESANCTIS	5							
DICKSON	6							
GINZBERG	7							
GORRY	8							
HENDERSON	9							
HOLSAPPLE	10							
HUBER	11							
KEEN	12							
KONSYNSKI	13							
LUCAS	14	60						
MASON, R.	15	18	66					
MITROFF	16	16	56	63				
SCOTT-MORTON	17	27	37	33	206			
SIMON	18	22	24	27	82	143		
SPRAGUE	19	20	25	22	105	58	182	
WHINSTON	20	11	8	8	45	37	61	101

by two or the missing value approach. Chapter IV empirically examined the impact of these different approaches on the ACA outcomes.

A possible solution for the diagonal value problem is suggested by Ahlgren, Jarneving, and Rousseau (2003, p.551.) and White (Howard D. White, 2003). The diagonal values should be the number of articles in the bibliographic database that cite at least two (different) works authored or coauthored by the author. Ahlgren, Jarneving, and Rousseau suggest excluding self-citations. This number is very difficult to get when using ISI databases. The custom database and cocitation matrix generation system we have developed allowed us to retrieve the numbers so that we can yield a co-citation frequency matrix, which Ahlgren, Jarneving, and Rousseau describe as "a mathematically complete matrix."

Six Alternative Approaches

The best way to count the diagonal value of author$_i$ is manually examining the references of each citing article to see whether each citing article contains at least two contributions of author$_{i,}$ excluding self-citations. Needless to say, this is an extremely time-consuming procedure because all citing articles' references should be examined one by one. Furthermore, it is virtually impossible to identify all citing articles during the investigation period, if we use ISI databases. In the case of using custom databases, it is a doable but extremely time-consuming task. The cocitation matrix generation system we developed can easily identify the total number of citing articles that contain at least one contribution by author$_i$. For example, our system processes the citing database of N (e.g.,1000) citing articles to identify n (e.g., 300) citing articles with at least one contribution by author$_i$. Next, each of n articles must be processed manually to see whether the cocitation between the author and himself or herself occurs. It is difficult to develop a system that is intelligent enough to distinguish each citation as self-citations or not. Furthermore, it may be considered a simple task for the cocitation generation system to identify the two contributions of author$_i$ in a citing article, but it is actually a complex task. Because of this reason, ACA researchers invented the following 6 possible approaches to fill the diagonal cells in cocitation counts matrix.

1. Missing values (McCain, 1990, 1991; Ramos-Rodríguez & Ruíz-Navarro, 2004; H.D. White & McCain, 1998).
2. The mean cocitation count for each author (McIntire, 2007).
3. Zero.
4. Highest off-diagonal cocitation counts.
5. Three highest off-diagonal values divided by two (M. J. Culnan, 1986; M.J. Culnan, O'Reilly, & Chatman, 1990; Eom & Farris, 1996; H. D. White & Griffith, 1981).
6. Raw Cocitation.

Based on the results of this study, if the pure cocitation counts are not used, the next best alternatives are as follows. They are the missing value approach, mean cocitation value approach, and the highest off-diagonal value approach in the order of the highest total variance explained.

Table 3 illustrates that alternative 5 is chosen to represent the diagonal values, three highest off-diagonal values are added and divided by two.

Table 3. Diagonal cell value adjusted cocitation matrix (1970-1990)

Author	No	1	2	3	4	5	6	7	8	9	10	11
ALTER	1	121										
BENBASAT	2	16	62.5									
BONCZEK	3	25	13	118								
CARLSON	4	66	23	53	173							
DESANCTIS	5	8	33	18	18	62.5						
DICKSON	6	15	43	13	17	43	65.5					
GINZBERG	7	34	13	15	32	8	13	66.5				
GORRY	8	30	13	11	34	7	9	15	73.5			
HENDERSON	9	22	16	34	33	16	13	14	8	63		
HOLSAPPLE	10	25	13	84	49	16	13	14	12	33	110	
HUBER	11	23	22	23	34	49	35	21	12	22	23	72.5
KEEN	12	92	44	46	112	33	45	53	46	45	46	47
KONSYNSKI	13	16	12	34	24	33	19	8	4	37	32	35
LUCAS	14	14	28	11	18	18	29	14	11	7	11	16
MASON, R.	15	15	24	8	22	16	23	13	16	16	8	20
MITROFF	16	12	22	8	18	17	23	10	15	17	8	17
SCOTT-MORTON	17	83	38	42	101	27	39	42	62	34	41	39
SIMON	18	38	31	33	54	28	34	17	39	26	33	37
SPRAGUE	19	67	31	59	133	27	26	38	32	44	56	49
WHINSTON	20	27	13	93	54	19	14	15	13	34	87	24
the largest		92	44	93	133	49	45	53	62	45	87	49
second largest		83	43	84	112	43	43	42	46	44	84	49
third largest		67	38	59	101	33	43	38	39	37	49	47
sum		242	125	236	346	125	131	133	147	126	220	145
adjusted diagonal		121	62.5	118	173	62.5	65.5	66.5	73.5	63	110	72.5

continued on following page

Table 3. continued

Author	No	12	13	14	15	16	17	18	19	20
ALTER	1									
BENBASAT	2									
BONCZEK	3									
CARLSON	4									
DESANCTIS	5									
DICKSON	6									
GINZBERG	7									
GORRY	8									
HENDERSON	9									
HOLSAPPLE	10									
HUBER	11									
KEEN	12	206								
KONSYNSKI	13	29	53.5							
LUCAS	14	39	6	48						
MASON, R.	15	40	16	18	66.5					
MITROFF	16	36	16	16	56	62.5				
SCOTT-MORTON	17	174	23	27	37	33	190			
SIMON	18	82	28	22	24	27	82	111		
SPRAGUE	19	126	35	20	25	22	105	58	182	
WHINSTON	20	50	34	11	8	8	45	37	61	121
the largest		174	37	39	56	56	174	82	133	93
second largest		126	35	29	40	36	105	82	126	87
third largest		112	35	28	37	33	101	58	105	61
sum		412	107	96	133	125	380	222	364	241
adjusted diagonal		206	53.5	48	66.5	62.5	190	111	182	121

Transposing Adjusted Diagonal Cell Value Cocitation Matrix

Table 4 is the transposed cocitation matrix converted from Table 3. The SAS system does not take the triangle shape data format.

GETTING THE DATA INTO A SAS DATA SET

Before analyzing cocitation frequency data using multivariate statistical tools, the data file must be imported from the original sources. The data can be retrieved either from commercial online bibliographical databases or custom programs us-

Table 4. Transposed cocitation frequency counts matrix (1970-1990)

Author	No	1	2	3	4	5	6	7	8	9	10	11
ALTER	1	121	16	25	66	8	15	34	30	22	25	23
BENBASAT	2	16	62.5	13	23	33	43	13	13	16	13	22
BONCZEK	3	25	13	118	53	18	13	15	11	34	84	23
CARLSON	4	66	23	53	173	18	17	32	34	33	49	34
DESANCTIS	5	8	33	18	18	62.5	43	8	7	16	16	49
DICKSON	6	15	43	13	17	43	65.5	13	9	13	13	35
GINZBERG	7	34	13	15	32	8	13	66.5	15	14	14	21
GORRY	8	30	13	11	34	7	9	15	73.5	8	12	12
HENDERSON	9	22	16	34	33	16	13	14	8	63	33	22
HOLSAPPLE	10	25	13	84	49	16	13	14	12	33	110	23
HUBER	11	23	22	23	34	49	35	21	12	22	23	72.5
KEEN	12	92	44	46	112	33	45	53	46	45	46	47
KONSYNSKI	13	16	12	34	24	33	19	8	4	37	32	35
LUCAS	14	14	28	11	18	18	29	14	11	7	11	16
MASON, R.	15	15	24	8	22	16	23	13	16	16	8	20
MITROFF	16	12	22	8	18	17	23	10	15	17	8	17
SCOTT-MORTON	17	83	38	42	101	27	39	42	62	34	41	39
SIMON	18	38	31	33	54	28	34	17	39	26	33	37
SPRAGUE	19	67	31	59	133	27	26	38	32	44	56	49
WHINSTON	20	27	13	93	54	19	14	15	13	34	87	24

continued on following page

Table 4. continued

Author	No	12	13	14	15	16	17	18	19	20
ALTER	1	92	16	14	15	12	83	38	67	27
BENBASAT	2	44	12	28	24	22	38	31	31	13
BONCZEK	3	46	34	11	8	8	42	33	59	93
CARLSON	4	112	24	18	22	18	101	54	133	54
DESANCTIS	5	33	33	18	16	17	27	28	27	19
DICKSON	6	45	19	29	23	23	39	34	26	14
GINZBERG	7	53	8	14	13	10	42	17	38	15
GORRY	8	46	4	11	16	15	62	39	32	13
HENDERSON	9	45	37	7	16	17	34	26	44	34
HOLSAPPLE	10	46	32	11	8	8	41	33	56	87
HUBER	11	47	35	16	20	17	39	37	49	24
KEEN	12	206	29	39	40	36	174	82	126	50
KONSYNSKI	13	29	53.5	6	16	16	23	28	35	34
LUCAS	14	39	6	48	18	16	27	22	20	11
MASON, R.	15	40	16	18	66.5	56	37	24	25	8
MITROFF	16	36	16	16	56	62.5	33	27	22	8
SCOTT-MORTON	17	174	23	27	37	33	190	82	105	45
SIMON	18	82	28	22	24	27	82	111	58	37
SPRAGUE	19	126	35	20	25	22	105	58	182	61
WHINSTON	20	50	34	11	8	8	45	37	61	121

ing database or spreadsheets. The next step is to get the data into a SAS data set. Basically the data can be imported from spreadsheet files such as Microsoft Excel files or the user creates a data set.

1. Importing the data from Spreadsheet files.
2. Creating a SAS data set manually.

Importing the Data from Spreadsheet Files

Assuming that the data is created by using Microsoft Excel format (e.g., Table 3), the SAS version 9.1.3 has import data capability. Under the file dropdown menu, Import data option is available.

Figure 3. SAS import data option

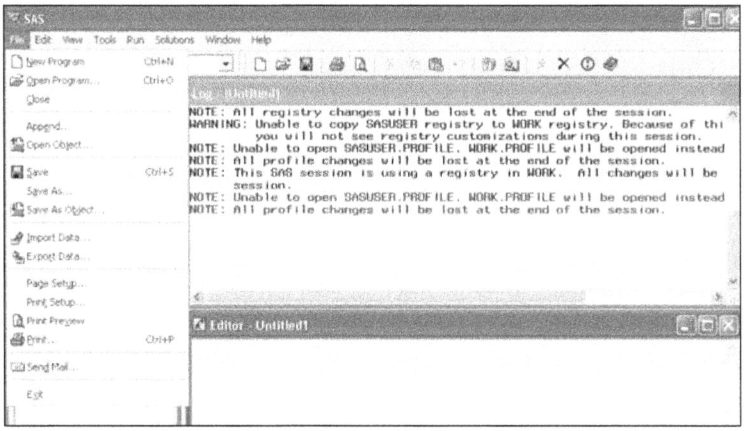

Figure 4. SAS file import wizard

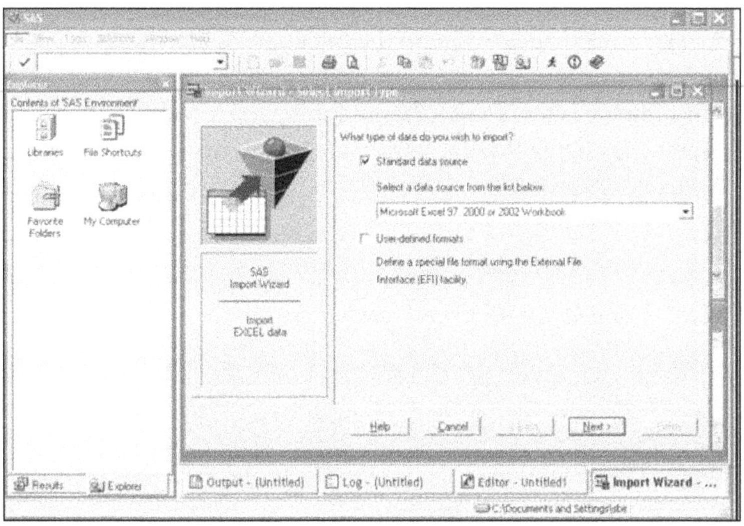

Clicking "import data" options activates the file import wizard.

The file import wizard can import data from standard data source from the following:

All types of Microsoft Excel files (xls) from Excel 97, 2000, or 2002 workbook or all previous versions.

All types of Microsoft ACCESS files (mdb) from ACESS 97, 2000, or 2002.

Comma separated values (*.csv) files

Tab delimited files (*.txt)

Delimited files (*.*)
dBase files (*.dbf)
JMP files (*.jmp)
LOTUS 123 Spreadsheet files (*.wk1)
LOTUS 123 Spreadsheet files (*.wk3)
LOTUS 123 Spreadsheet files (*.wk4)
Choose "MicroSoft Excel 97, 2000, 2002 workbook and Click "Next".
Then, the "connect to MS excel" window pops out.

If you know the name of the file, you may type it in or you may browse your computer. The workbook refers to the name of excel file that contains your data. Co-citation frequency files are stored in bookdata.xls file.

Click the "Browse" button if you are not sure of the correct name and location of the cocitation data file.

Click the "open" button. The next screen shows that the cocitation workbook file is selected.

click OK

Selecttheworksheetfileifyourfilecontainsmultipleworksheets.Theworksheetfileis SASDATA. The wizard adds $ sign to the file name. Click next.

When a SAS session starts, the SAS system gives the user choices to choose from. The first choice is the WORK library, followed by other user defined libraries. Usually, the files in the WORK library are temporary files. They are deleted at the end of your SAS session.

Figure 5. SAS "Connect to MS Excel" dialog box

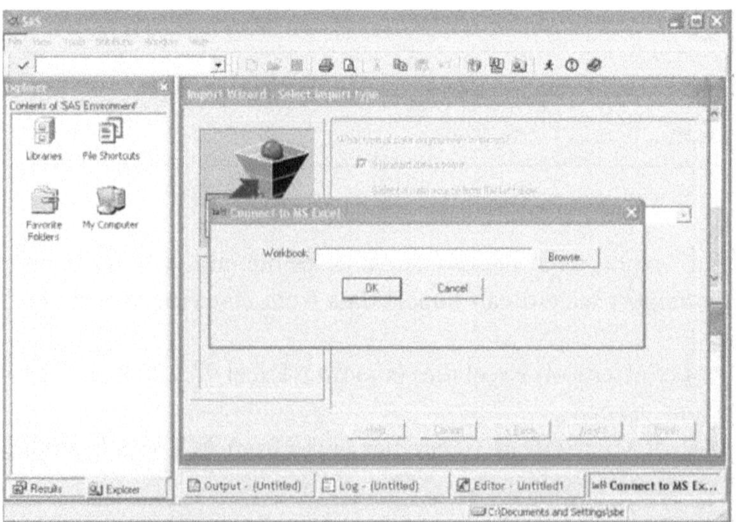

Figure 6. Selecting a file to import to SAS

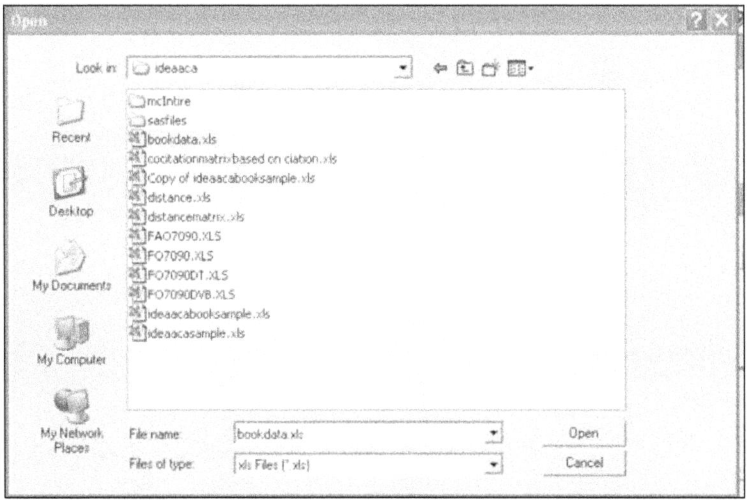

Figure 7. Selecting the workbook

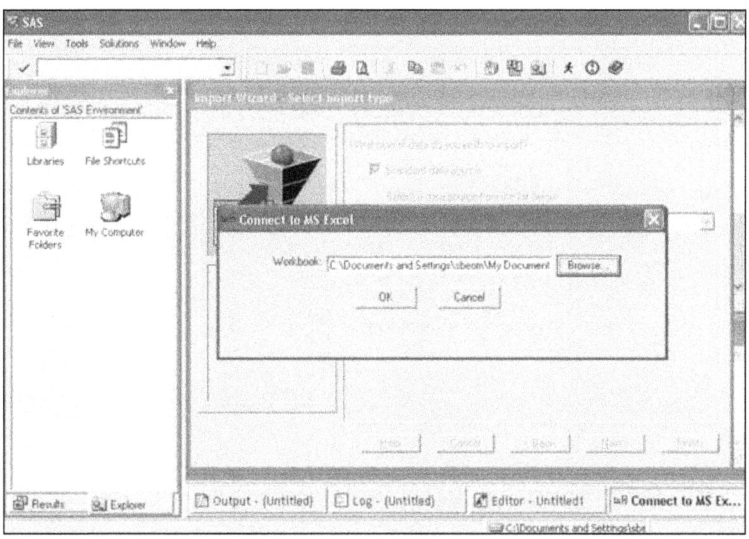

Figure 8. Import wizard after selecting the worksheet file

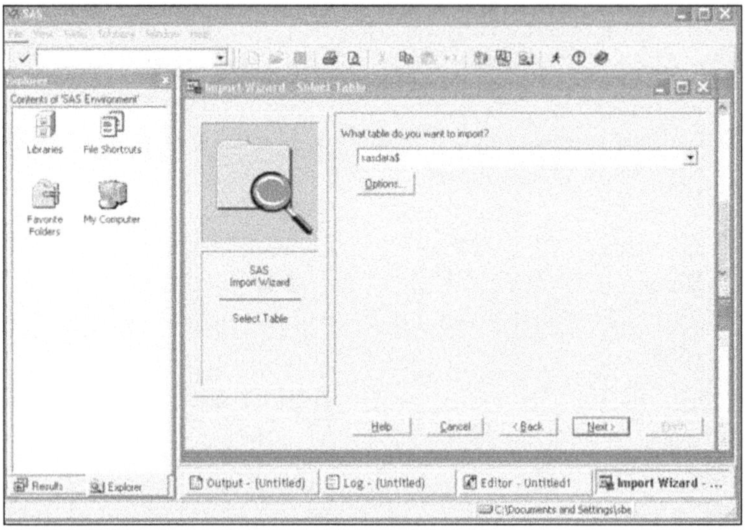

Figure 9. Choosing the SAS destination (library name)

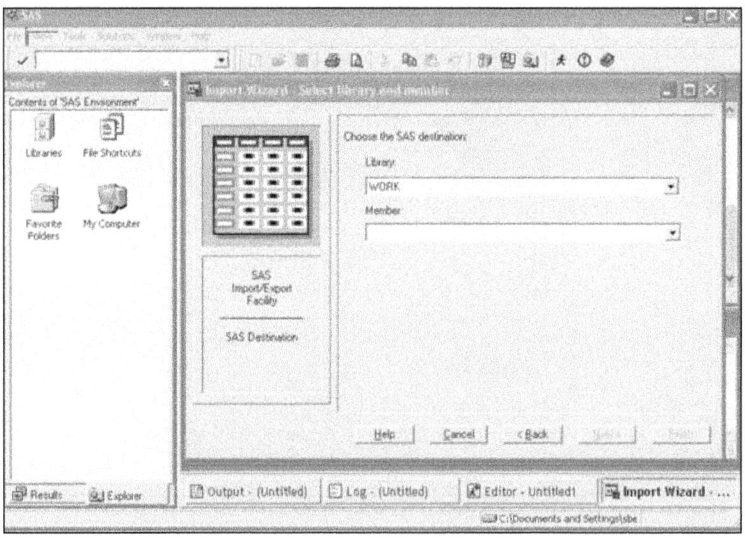

Figure 10. Specifying a file name (member) in the chosen library

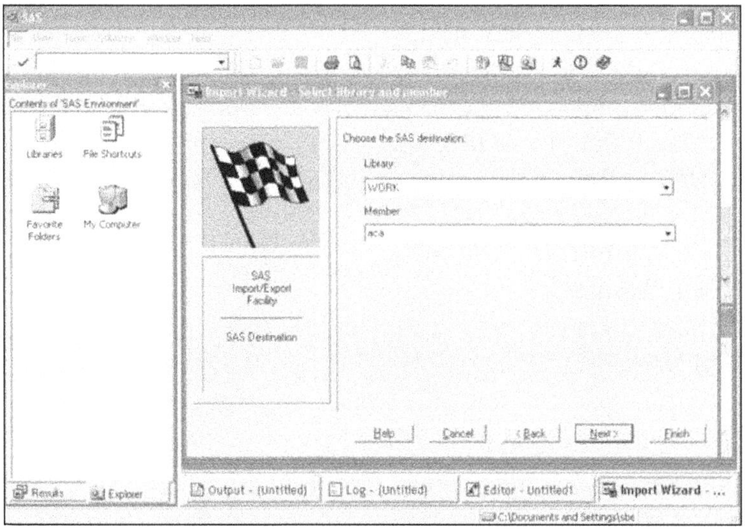

Figure 11. Creating SAS Statements for data import

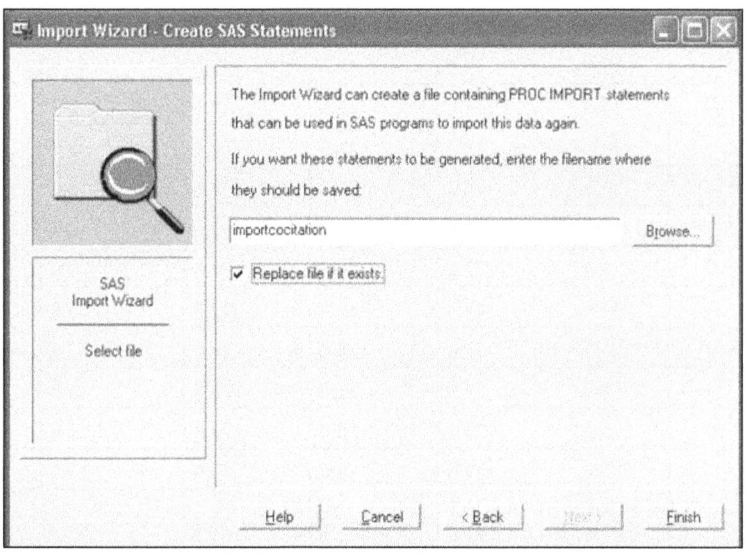

Type in the name of the file (e.g., aca) in the member area. A SAS data library is a collection of SAS files. Each file is a member of the library. You are asked to enter the name of the file Click, next.

The steps taken by import wizard are recorded and produced a file (importcocitation) and saved at C:\Documents and Settings\sbeom\ the following:

```
PROC IMPORT OUT= WORK.Aca
        DATAFILE= "C:\Documents and Settings\sbeom\My Documents\idea
            aca\bookdata.xls"
        DBMS=EXCEL REPLACE;
    SHEET="sasdata$";
    GETNAMES=YES;
    MIXED=YES;
```

Table 5. SAS dataset in Microsoft excel format

Author	ALTER	BENBASAT	BONCZEK	CARLSON	DESANCTIS	DICKSON
ALTER	121.0	16.0	25.0	66.0	8.0	15.0
BENBASAT	16.0	62.5	13.0	23.0	33.0	43.0
BONCZEK	25.0	13.0	118.0	53.0	18.0	13.0
CARLSON	66.0	23.0	53.0	173.0	18.0	17.0
DESANCTIS	8.0	33.0	18.0	18.0	62.5	43.0
DICKSON	15.0	43.0	13.0	17.0	43.0	65.5
GINZBERG	34.0	13.0	15.0	32.0	8.0	13.0
GORRY	30.0	13.0	11.0	34.0	7.0	9.0
HENDERSON	22.0	16.0	34.0	33.0	16.0	13.0
HOLSAPPLE	25.0	13.0	84.0	49.0	16.0	13.0
HUBER	23.0	22.0	23.0	34.0	49.0	35.0
KEEN	92.0	44.0	46.0	112.0	33.0	45.0
KONSYNSKI	16.0	12.0	34.0	24.0	33.0	19.0
LUCAS	14.0	28.0	11.0	18.0	18.0	29.0
MASON_R	15.0	24.0	8.0	22.0	16.0	23.0
MITROFF	12.0	22.0	8.0	18.0	17.0	23.0
SCOTT_ MORTON	83.0	38.0	42.0	101.0	27.0	39.0
SIMON	38.0	31.0	33.0	54.0	28.0	34.0
SPRAGUE	67.0	31.0	59.0	133.0	27.0	26.0
WHINSTON	27.0	13.0	93.0	54.0	19.0	14.0

```
    SCANTEXT=YES;
    USEDATE=YES;
    SCANTIME=YES;
RUN;
```

The data import statement has the following syntax.

```
PROC IMPORT
DATAFILE="filename" | TABLE="tablename"
OUT=<libref.>SAS-data-set <(SAS-data-set-options)>
<DBMS=identifier><REPLACE>;
RUN;
```

We assume that your aca data is on your PC with Excel Format.

Required Arguments

DATAFILE="*filename*"

specifies the complete path and filename or a fileref for the input PC file, spreadsheet, or delimited external file. If you specify a fileref or if the complete path and filename does not include special characters (such as the backslash in a path), lowercase characters, or spaces, you can omit the quotation marks. A fileref is a SAS name that is associated with the physical location of the output file. To assign a fileref, use the FILENAME statement. For more information about PC file formats, see ***SAS/ACCESS for PC Files: Reference****.*

```
PROC IMPORT
DATAFILE="C:\mydocuments\ideaaca\ideaacasample.xls"
OUT=<work.acadata.>SAS-data-set <(SAS-data-set-options)>
<DBMS=EXCEL REPLACE;
Proc print data=acadata;

RUN;
```

Once the ACA dataset is prepared, there are three important steps to be followed. They are multivariate analysis of ACA data, output preparation, validation and interpretation. Chapters IX, X, and XI discuss three multivariate analysis tools (factor analysis, cluster analysis, and multi-dimensional scaling) respectively. The last two chapters (Chapters XII and XIII) of this book include ACA applications in the management information systems area to demonstrate how to interpret the output of statistical procedures. The following section briefly overview SAS Multivariate analysis Procedures and Outputs.

OVERVIEW OF SAS MULTIVARIATE ANALYSIS PROCEDURES AND OUTPUTS

As Figure 2 shows, ACA data analysis requires the 7 different procedures of the SAS system. There is one necessary input (cocitation frequency matrix), seven procedures (Factor, Xmacro, distnew, cluster, MDS, PLOT, and G3D), and several outputs from 3 multivariate statistical procedures. All three techniques used in the ACA aim at grouping/classifying all variables into several subgroups with common underlying hidden structures, characteristics, and/or attributes. The hidden structures/characteristics/attributes are given different terms: factors in factor analysis, clusters in cluster analysis, and dimensions in multidimensional scaling. Although all three techniques seek to summarize and simplify a large number of variables, there are also some important differences in their outputs. Factor procedures produce three outputs— rotated factor pattern, factor structure correlation, and interfactor correlation. Cluster procedures generate cluster history and dendrogram (tree graph). Multidimensional scaling procedures result in iteration history, convergence status, coordinate matrix of each author (first column), and configuration of each author on multidimensional spaces (dimension 1-column 2, dimension 2-column 3, dimension 3-column 4, etc).

The Factor Procedure

The transposed, diagonal-cell-adjusted cocitation matrix of the authors (Table 4) can be analyzed by the factor analysis program in most popular statistical software such as the SAS system or Statistical Package for the Social Sciences (SPSS) to ascertain the underlying intellectual structure of an academic discipline. Table 5 is converted from Table 4 as an input file to the SAS system. The objective of this analysis is to group (condense) a large number of selected variables (authors) into a smaller set of composite dimensions (factors) representing research subspecialties and contributing disciplines of an academic discipline. In doing so, each variable (author) is viewed as a dependent variable that is a function of a set of latent (underlying) factors.

The factor procedure produces three types of outputs as listed in Figure 2. The factors in a matrix of factor loadings correspond to research subspecialties or schools of thought. The matrix of factor loadings (factor pattern) consists of multiple columns of factors and communality and multiple numbers of author variable rows, eigenvalue, and variance explained by each factor. Factor loadings represent the correlation between author variables and the factors. The eigenvalue (the latent root) of a factor is the column sum of the squared factor loadings of all variables on that factor. Variance explained by each factor is identical to the eigenvalue of each factor divided by the number of authors (variables), which is often expressed as percentage. Final communality estimates of each author (observed variable) are the sum of the square of factor loadings for that variable.

The Cluster Procedure

Cluster analysis is a data reduction technique for grouping various entities (individuals, variables, objects) into clusters so that the entities in the same cluster are similar in respect to some predetermined selection criteria (Everitt, 1980; Joseph F. Hair, Jr., Anderson, Tatham, & Black, 1992). Therefore, it is necessary to convert the raw cocitation frequency matrix into a measure of similarity or distance. To do so, SAS institute has developed the DISTANCE macro for computing various measures of distance, dissimilarity, or similarity between the observations of a SAS data set.

The cluster procedure (PROC CLUSTER), which produces cluster history, illustrates the process of hierarchical clustering all of the authors into a new aggregate cluster. Factors and clusters correspond to research subspecialties or schools of thoughts. However, the cluster procedure shows the process of how each author is combined into a new aggregate cluster until all authors are grouped into the final cluster. The tree graph shows a linkage among various clusters, and influence could be inferred based on the joining sequence in the tree graph. Each author belongs to only one cluster whereas an author usually loads on more than one factor.

Multidimensional Scaling

Multidimensional scaling (MDS) is a class of multivariate statistical techniques/procedures to produce two or three dimensional pictures of data (geometric configuration of points) using proximities among any kind of objects as input. Proximity data consist of one or more square symmetric or asymmetric matrices of similarities or dissimilarities between objects or stimuli (Kruskal & Wish, 1978, pp. 7-11). The MDS outputs consist of a spatial representation of data which shows underlying relationships on a two or three dimensional map. The MDS map helps clarify relationships visually using the ratio of distances on a map to corresponding data values such as a map of a country showing cities. The magnitude of number indicates how similar or dissimilar two objects are.

The purposes of MDS are to help researchers identify the "hidden structures" in the data and visualize the relationships among/within the hidden structures to give clearer explanations of these relationships to others (Joseph F. Hair, Jr., Anderson, & Tatham, 1987; Kruskal & Wish, 1978). Three SAS procedures (MDS, PLOT, and G3D) are necessary to convert the author cocitation frequency matrix to two or three dimensional pictures of data.

The distance matrix produced earlier by using xmacro.sas and distnew.sas programs should be converted to a coordinate matrix. A distance matrix is the input to the multidimensional scaling procedure, PROC MDS, and the G3D procedure of the SAS system (release 8.0). The PLOT and G3D procedures process the coordinate matrix to visualize the similarity and dissimilarity within each group and among the various subspecialties of an academic discipline. The annotate facility in the SAS system produces figures with the name of the author on each data point.

The MDS maps show the big picture of inter-cluster relationships – central and peripheral research specialties. The placement of a cluster on the center of the map indicates its centrality to an academic discipline. Similarly, the placement of an author on the center of a cluster on the map indicates its centrality of that author within that school of thought in that the author is linked with a substantial portion of the author set, with relatively high correlations. Placement near the periphery represents a more focused linkage.

Because of some difficulties in interpreting the nature of dimensionality in MDS, domain visualization has recently become one of the new exciting research areas to reveal the essence of a knowledge domain. It uses several techniques including pathfinder networks, visualization and animation of high-dimensional abstract spaces (Chen, Paul, & O'Keefe, 2001). These domain visualization tools used with traditional multivariate statistical tools such as factor analysis will significantly enhance ACA researchers' ability to trace the development of academic disciplines, as demonstrated by Chen and others (2001).

REFERENCES

Chen, C., Paul, R. J., & O'Keefe, B. (2001). Fitting the Jigsaw of Citation: Information Visualization in Domain Analysis. *Journal of the American Society For Information Science and Technology, 52*(4), 315-330.

Culnan, M. J. (1986). The Intellectual Development of Management Information Systems, 1972-1982: A Co-Citation Analysis. *Management Science, 32*(2), 156-172.

Culnan, M. J., O'Reilly, C. A., & Chatman, J. A. (1990). Intellectual Structure of Research in Organizational Behavior, 1972-1984: A Co-Citation Analysis. *Journal of the American Society for Information Science, 41*, 453-458.

Eom, S. B. (1995). Decision Support Systems Research: Reference Disciplines and a Cumulative Tradition. *Omega: The International Journal of Management Science, 23*(5), 511-523.

Eom, S. B. (1996). Mapping the Intellectual Structure of Research in Decision Support Systems through Author Cocitation Analysis (1971-1993). *Decision Support Systems, 16*(4), 315-338.

Eom, S. B., & Farris, R. (1996). The Contributions of Organizational Science to the Development of Decision Support Systems Research Subspecialties. *Journal of the American Society for Information Science, 47*(12), 941-952.

Everitt, B. S. (1980). *Cluster Analysis.* London: Heinemann Educational Books Ltd.

Hair Jr., J. F., Anderson, R. E., Tatham, R., & Black, W. C. (1992). *Multivariate Data Analysis with Readings* (3rd ed.). New York: Macmillan Publishing.

Hair, Jr., J. F., Anderson, R. E., & Tatham, R. L. (1987). *Multivariate Data Analysis with Readings* (2nd ed.). New York: Macmillan Publishing Company.

Hoffman, D. L., & Holbrook, M. B. (1993). The Intellectual Structure of Consumer Research: A Bibliometric Study of Author Cocitations in the First 15 Years of the Journal of Consumer Research. *Journal of Consumer Research, 19*, 505-517.

Kruskal, J. B., & Wish, M. (1978). *Multidimensional Scaling.* Beverly Hills, CA: Sage Publications.

McCain, K. W. (1986). Cocited Author Mapping as a Valid Representation of Intellectual Structure. *Journal of the American Society for Information Science, 37*(3), 111-122.

McCain, K. W. (1990). Mapping Authors in Intellectual Space: A Technical Overview. *Journal of the American Society for Information Science, 41*(6), 433-443.

McCain, K. W. (1991). Mapping Economics through the Journal Literature: An Experiment in Journal Cocitaion Analysis. *Journal of the American Society for Information Science, 42*(4), 290-296.

McIntire, J. S. (2007). *The Clothing and Textile Research Base: An Author Cocitation Study.* Unpublished Master's Thesis, University of Missouri, Columbia, Missouri.

Ramos-Rodríguez, A.-R., & Ruíz-Navarro, J. (2004). Changes in the Intellectual Structure of Strategic Management Research: A Bibliometric Study of the Strategic Management Journal, 1980-2000. *Strategic Management Journal, 25*, 981-1004.

White, H. D. (2003). Author Co-Citation Analysis and Pearson's R. *Journal of the American Society for Information Science and Technology, 54*(13), 1250-1259.

White, H. D., & Griffith, B. C. (1981). Author Cocitation: A Literature Measure of Intellectual Structure. *Journal of the American Society for Information Science, 32*(3), 163-171.

White, H. D., & McCain, K. (1998). Visualizing a Discipline: An Author Cocitation Analysis of Information Science, 1972-1995. *Journal of the American Society for Information Science, 49*, 327-335.

Chapter VIII
Principal Component Analysis Using the Factor Procedure

ABSTRACT

This chapter describes the factor procedure. The first section of the chapter begins with the definition of factor analysis. This is the statistical techniques whose common objective is to represent a set of variables in terms of a smaller number of hypothetical variables (factor). ACA uses principal component analysis to group authors into several catagories with similar lines of research. We also present many different approaches of preparing datasets including manual data inputs, in-file statement, and permanent datasets. We discuss each of the key SAS statements including DATA, INPUT, CARDS, PROC, and RUN. In addition, we examine several options statements to specify the followings: method for extracting factors; number of factors, rotation method, and displaying output options.

INTRODUCTION

This chapter describes principal component analysis using the factor procedure in the SAS system. The first section of the chapter explains why ACA uses principal component analysis of the factor procedures in the SAS system. In addition, the chapter discusses the following topics.

- Overview of Input, Procedures, and Outputs of ACA SAS Analysis
- Definition of Factor Analysis
- Getting the Data into a SAS Data Set
- Preparing an ACA Proc Factor Program
- Processing the Inputs
- Interpreting the Factor Procedure Outputs

Numerous multivariate analysis tools exist. The appropriate research methods are determined by three important questions regarding the characteristics of the variables under study (Cooper & Emory, 1995, p. 521).

1. Are there dependent variables in the problems?
2. Is there more than one dependent variable?
3. Are the variables metric or nonmetric?

The variables in author cocitation analysis are the authors. The author is defined as "a body of writings by a person" or "a body of contributions by a person." The term "a person" refers to a single author or one of multiple authors. Authors are not dependent on each other, although one author may exercise some influence on other authors. The second important question is whether the variables are metric (quantitative) or nonmetric (qualitative). Metric variables are measured by ratio and interval measurements; nonmetric variables refer to data that are nominal or ordinal. Nominal data indicate classes and categories that are not measurable by the quantitative units such as kg, meter, miles, pounds, frequency, etc. Examples of nominal data include an individual's religion, nationality, gender, marital status, union membership status, etc. Since our data has no dependent variables and is metric data, three multivariate analysis tools are applied in this research: factor analysis, cluster analysis and multidimensional scaling.

All three techniques used in the ACA aim at grouping/classifying all variables into several subgroups with common underlying hidden structures, characteristics and/or attributes. The hidden structures/characteristics/attributes are given different terms: factors in factor analysis, clusters in cluster analysis, and dimensions in multidimensional scaling. Although all three techniques seek to summarize/simplify a large number of variables, there are some distinctive differences among these techniques. The basic concepts and some differences among the three techniques are discussed in Chapters VIII, IX, and X.

OVERVIEW OF INPUT, PROCEDURES, AND OUTPUTS

As Figure 1 shows, ACA data analysis requires the following 7 different procedures of the SAS system.

- Factor Procedure
- Distance Procedure
- Cluster Procedure
- MDS Procedure
- Plot Procedure
- G3D Procedure

Each procedure requires inputs to produce outputs. It is extremely important for ACA researchers to prepare input files for these procedures in a most efficient manner. There is one necessary input (cocitation frequency matrix), seven procedures (Factor, Xmacro, distnew, cluster, MDS, PLOT, and G3D), and several outputs from 3 multivariate statistical procedures. Factor procedures produce three outputs—rotated factor pattern, factor structure correlation, and interfactor correlation. Cluster procedures generate cluster history and dendrogram (tree graph). Multidimensional scaling procedures result in iteration history, convergence status, coordinate matrix of each author (first column), and configuration of each author on multidimensional spaces (dimension 1-column 2, dimension 2-column 3, dimension 3-column 4, etc).

DEFINITION OF FACTOR ANALYSIS

Factor analysis is defined (Kim & Mueller, 1978) as "a variety of statistical techniques whose common objective is to represent a set of variables in terms of a smaller number of hypothetical variables." For those who are not familiar with this technique, see (Child, 1970; Hair, Anderson, & Tatham, 1987; Kim & Mueller, 1978) for the basics of factor analysis.

The transposed cocitation matrix of authors can be analyzed by the factor analysis program in most popular statistical software such as SAS (statistical analysis systems) or SPSS to ascertain the underlying intellectual structure of an academic discipline. The objective of this analysis is to group (condense) a large number of selected variables (authors) into a smaller set of composite dimensions (factors) representing research subspecialties/subdisciplines and contributing disciplines of a discipline. In doing so, each variable (author) is viewed as a dependent variable that is a function of a set of latent (underlying) factors. Factor loadings represent

Figure 1. The factor procedure in ACA

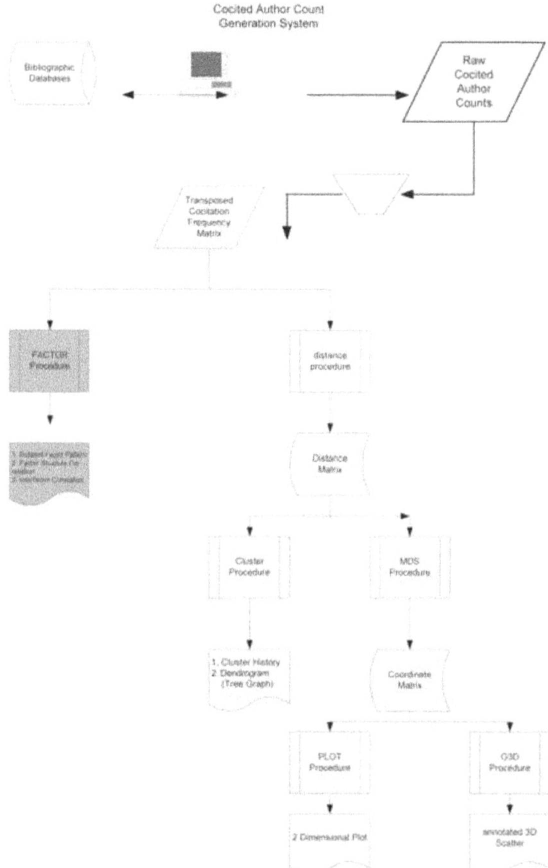

the correlations between the original variables and the underlying factors. An eigenvalue (latent root or extracted variance) represents the amount of common variance accounted by a factor. The percentage of total variation is an index used to determine how the variables in each factor are related to each other. A high percentage variance means a high degree of interrelatedness among the variables with significant loadings.

THE FACTOR PROCEDURE

The FACTOR procedure in the SAS system performs a variety of data reduction procedures including factor and principal component analyses. These two techniques are very similar, but not identical. Readers are referred to (Hatcher, 1994) to understand the similarities and the differences between the two. Both factor analysis and principal component analysis can be done by the factor procedure in the SAS system.

Similarity between Factor Analysis and Principal Component Analysis

- Both are data reduction techniques that can be used to identify groups of observed variables.
- Both are performed with the factor procedure in the SAS system.
- Both follow almost identical SAS procedures.
- Both use the same data such as a correlation matrix, a covariance matrix, etc.
- Often, both produce very similar results.

Conceptual Differences between Factor Analysis and Principal Component Analysis

- Factor analysis assumes that factors appeared in the factor analysis exerts influence on each variable that constitutes latent factor that cannot be measured and among the factors there exist assumed causal influence, whereas principal component analysis assumes that there is no such assumption.

PROC FACTOR STATEMENT TO CONDUCT PRINCIPAL COMPONENT ANALYSIS

Basically there are two approaches in preparing SAS data files. The first approach is importing from existing data files such as Microsoft excel. The second approach is manual data inputs.

In Chapter VII, I introduced how to import ACA data stored in Microsoft Excel format. SAS can import data in many different format in various formats such as Excel, Access, Lotus, etc. SAS statements are different with each different type of data (imported data and inputted data).

Processing an Imported SAS Data Set

Understanding how the data is inputted to the SAS program is crucial in becoming a good SAS user. Throughout this book, the primary reference to SAS programming is SAS Online Documentation Version 9.1.3. The user can access this online documentation at http://support.sas.com/onlinedoc/913/docMainpage.jsp. Even if I fail to give proper credits at all times, all credits go to *SAS Onlinedoc®* 9.1.3 and other SAS publications (SAS Institute Inc., 2007). Figure 2 shows the Proc factor procedure to process the imported data from excel file. Table 1 is the contents of the dataset imported from an Excel data file and stored as temporary data file.

Figure 2 contains 2 SAS procedures. The first procedure produces dataset, ACA, for factor analysis. The dataset is shown below (Table 1). The second procedure is to do factor analysis using the dataset, ACA. The SAS import procedure creates the dataset that does not need to specify the INPUT and CARD statements.

Processing a Directly Entered or Copied SAS Data Set

The beginner who does not know how to import the data in excel form into the data editor may often copy the dataset from Excel data files and paste it on the SAS data editor. If so, the user may have to add several SAS statements. The SAS program consists of the following statements.

- Data statements
- Input statements
- Cards statements
- Proc statements
- Run statements

To enter data directly or copy and paste into the SAS system, SAS statements use the DATA, INPUT, and CARDS as shown in Table 2.

One important statement that saves ACA researchers time when interpreting the output of multivariate statistical techniques is the inclusion of author's names. When interpreting the outcomes of factor analysis as shown in a later chapter, it is important to include names of authors in the INPUT statement instead of observation numbers or variable numbers. Including authors' names in the CARDS statement will not print the authors' name on the SAS outputs.

Depending on the statistical packages used, input formats will be different. Transposed cocitation matrix needs further modification to be an input to SAS factor procedures as described below.

Several SAS statements must be added to the transposed cocitation matrix file as shown in Table 2.

Figure 2. The proc factor procedure with an imported data set

```
PROC IMPORT OUT= WORK.ACA
         DATAFILE= "C:\Documents and Settings\sbeom\My Documents\idea
         aca\spss\bookdata.xls"
         DBMS=EXCEL REPLACE;
       SHEET="sasdata$";
       GETNAMES=YES;
       MIXED=NO;
       SCANTEXT=YES;
       USEDATE=YES;
       SCANTIME=YES;
RUN;
proc print data=aca;
run;
proc factor data=aca method=principal mineigen=1 rotation=promax;
run;
```

Table 1. The dataset (ACA)

The SAS System 12:54 Mo

Obs	Author	ALTER	BENBASAT	BONCZEK	CARLSON	DESANCTIS	DICKSON	GINZBERG	GORRY	HENDERSON	HOLSAPPLE	HUBER	KEEN	KONSYNSKI	LUCAS	MASON_R
1	ALTER	121	16.0	25	66	8.0	15.0	34.0	30.0	22	25	23.0	92	16.0	14	15.0
2	BENBASAT	16	62.5	13	23	33.0	43.0	13.0	13.0	16	13	22.0	44	12.0	28	24.0
3	BONCZEK	25	13.0	118	53	18.0	13.0	15.0	11.0	34	84	23.0	46	34.0	11	8.0
4	CARLSON	66	23.0	53	173	18.0	17.0	32.0	34.0	33	49	34.0	112	24.0	18	22.0
5	DESANCTIS	8	33.0	18	18	62.5	43.0	8.0	7.0	16	16	49.0	33	33.0	18	16.0
6	DICKSON	15	43.0	13	17	43.0	65.5	13.0	9.0	13	13	35.0	45	19.0	29	23.0
7	GINZBERG	34	13.0	15	32	8.0	13.0	66.5	15.0	14	14	21.0	53	8.0	14	13.0
8	GORRY	30	13.0	11	34	7.0	9.0	15.0	73.5	8	12	12.0	46	4.0	11	16.0
9	HENDERSON	22	16.0	34	33	16.0	13.0	14.0	8.0	63	33	22.0	45	37.0	7	16.0
10	HOLSAPPLE	25	13.0	84	49	16.0	13.0	14.0	12.0	33	110	23.0	46	32.0	11	8.0
11	HUBER	23	22.0	23	34	49.0	35.0	21.0	12.0	22	23	72.5	47	35.0	16	20.0
12	KEEN	92	44.0	46	112	33.0	45.0	53.0	46.0	45	46	47.0	206	29.0	39	40.0
13	KONSYNSKI	16	12.0	34	24	33.0	19.0	8.0	4.0	37	32	35.0	29	53.5	6	16.0
14	LUCAS	14	28.0	11	18	18.0	29.0	14.0	11.0	7	11	16.0	39	6.0	48	18.0
15	MASON_R	15	24.0	8	22	16.0	23.0	13.0	16.0	16	8	20.0	40	16.0	18	66.5
16	MITROFF	12	22.0	8	18	17.0	23.0	10.0	15.0	17	8	17.0	36	16.0	16	56.0
17	SCOTT_MORTON	83	38.0	42	101	27.0	39.0	42.0	62.0	34	41	39.0	174	23.0	27	37.0
18	SIMON	38	31.0	33	54	28.0	34.0	17.0	39.0	26	33	37.0	82	28.0	22	24.0
19	SPRAGUE	67	31.0	59	133	27.0	26.0	38.0	32.0	44	56	49.0	126	35.0	20	25.0
20	WHINSTON	27	13.0	93	54	19.0	14.0	15.0	13.0	34	87	24.0	50	34.0	11	8.0

The DATA Statement

The first line of the data file begins with the DATA statement. General form is DATA data-set-name, ending with semi-colon (;). The data set in Table 3 is named directinput. Of course, other names are acceptable such as ACAauthor, D1, Dataone, etc. The data set name begins with a character, and no special characters or blank spaces are allowed.

Table 2. Processing a SAS dataset copied from Excel file

```
data directinput;
input @11 ALTER    BENBASAT    BONCZEK CARLSON DESANCTIS    DICKSON GINZBERG
GORRY   HENDERSON   HOLSAPPLE   HUBER   KEEN    KONSYNSKI   LUCAS   MASON_R
MITROFF SCOTT_MORTON    SIMON   SPRAGUE WHINSTON;
cards;
ALTER           121.0   16.0    25.0    66.0    8.0     15.0    34.0    30.0    22.0    25.0
BENBASAT        16.0    62.5    13.0    23.0    33.0    47.0    13.0    13.0    16.0    13.0
BONCZEK         25.0    13.0    118.0   53.0    18.0    13.0    15.0    11.0    34.0    84.0
CARLSON         66.0    23.0    53.0    173.0   18.0    17.0    32.0    34.0    33.0    49.0
DESANCTIS       8.0     33.0    18.0    18.0    62.5    43.0    8.0 7.0 16.0    16.0    49.0
DICKSON         15.0    43.0    13.0    17.0    43.0    65.5    13.0    9.0     13.0    13.0
GINZBERG        34.0    13.0    15.0    32.0    8.0     13.0    66.5    15.0    14.0    14.0
GORRY           30.0    11.0    11.0    34.0    7.0     9.0     15.0    73.5    8.0     12.0
HENDERSON       22.0    16.0    34.0    33.0    16.0    13.0    14.0    8.0     63.0    33.0
HOLSAPPLE       25.0    12.0    84.0    49.0    16.0    13.0    14.0    12.0    33.0    110.0
HUBER           23.0    22.0    23.0    34.0    49.0    35.0    21.0    12.0    22.0    23.0
KEEN            92.0    44.0    46.0    112.0   33.0    45.0    53.0    46.0    45.0    46.0
KONSYNSKI       16.0    12.0    34.0    24.0    33.0    19.0    8.0     4.0     37.0    32.0
LUCAS           14.0    29.0    11.0    18.0    18.0    29.0    14.0    11.0    7.0     11.0
MASON_R         15.0    24.0    6.0     22.0    16.0    23.0    13.0    16.0    16.0    6.0
MITROFF         12.0    22.0    8.0     18.0    17.0    23.0    10.0    15.0    17.0    8.0
SCOTT_MORTON    83.0    38.0    42.0    101.0   27.0    39.0    42.0    62.0    34.0    41.0
SIMON           38.0    31.0    33.0    54.0    28.0    34.0    17.0    39.0    26.0    33.0
SPRAGUE         67.0    31.0    59.0    133.0   27.0    26.0    38.0    32.0    44.0    56.0
WHINSTON        27.0    13.0    93.0    54.0    19.0    14.0    15.0    13.0    34.0    87.0
;
proc factor method=principal mineigen=1 rotate=promax;
run;
```

The INPUT Statement

Line 2 begins with the INPUT statement. It provides a name for the variables to be included in the data set. The INPUT statement has the following general syntax form.

> INPUT Variable $ beginning column-number –
> ending column-number #line-number
> @column-number (variable-name)
> @column-number (variable-name)
> @column-number (variable-name)
> @column-number (variable-name)

- Variable-name is to be displayed as part of outputs. Authors names in the cards statement will not be part of the SAS output. The variable names defined in the INPUT statement (variable-name) are to be printed on the SAS outputs. Therefore, the SAS outputs include author variables as x1, x2, … x9.
- **$** indicates to store a variable value as a character value rather than a numeric value.
- The INPUT statement includes two pointer control statements: line pointer and column pointer (#line-number ***n*** moves the pointer to line *n*. @column ***n*** moves the pointer to column *n*.

Table 3. Creating a temporary data set with line pointer control statements

```
DATA Table4;
INPUT @6 X1-X15
   #2 @6 X16-X30 #3 @6 X31-X45
   #4 @6 X46-X60 #5 @6 X61-X67;
CARDS;
1    43  10   9   8  12   3   6  10   5  16  10   8   8   2   8
      9   1  16   6   7   6   2   6   7  13   4  10   6  11  13
      4   3   2   1  33   1   6   8  13   5   7   7   2   8  19
      6  14  18   3  11   9   7   2   4   4  27   3  26  13   4
      7   5   8   3   6  11  10

more data for author variable 2 through 66 here

67   10  17   4   3  26   3   4   8   4  20  16  13   8   2   6
     17  19  25   3   6   4   0   2  15  12   2  15   5   8  20
     10   2  10   1  35   0   2  10   7   4   3  18  17   2  19
      4  13  17   2   8   3   1  14   5  10  25   2  21  26   4
      5   5   7   6   7   8  44
;
PROC PRINT;
RUN;
```

Table 3 includes only the column pointer control statement because the data of each variable can be listed on one line. Table 4 shows both the line control and column control statements because each variable has 67 data that can be listed on 5 lines. The first data-line statement (#1) is optional and omitted in the table.

The INPUT statement ends with a semicolon. The difference between Tables 3 and 4 is whether the names of authors are printed as part of the SAS outputs. Table 4 produces outputs using numerical variable names, while Table 3 prints the author's names as part of the SAS outputs. Including authors' name in the INPUT statement as in Table 4 is recommended since the outputs with authors' names helps ACA researchers improve the output interpretations in assigning the names of factors.

The CARDS Statement

The CARDS Statement signals SAS to read the raw data following the CARDS statement. The raw data begins after the CARDS statement with a semicolon. The DATA statement signals the SAS system to begin a step to create a SAS data set. The DATA, INPUT, and CARDS statements create a dataset that begins with

WORK. Any dataset whose name begins with WORK is a temporary SAS data set, which is automatically deleted at the end of the session. Tables 2 and 3 show two different ways of creating temporary data sets.

The last part of the SAS program is the PROC (Procedure) step. This step includes programming statements requesting specific statistical analyses of data. The methods for factor extraction available in the factor procedure include principal component analysis, principal factor analysis, iterated principal factor analysis, unweighted least-squares factor analysis, maximum likelihood (canonical) factor analysis, alpha factor analysis, image component analysis, and Harris component analysis. Of these, ACA use principal component analysis to classify authors into several groups with similar lines of research themes/subspecialties.

The general statement to perform a principal component analysis is as follows:

PROC FACTOR < options > ;

There are 9 categories of options available in the PROC FACTOR Statement based on Tasks. They are to specify (1)Datasets, (2)Extract factors and communalities, (3) Data Analysis method, (4)Number of factors, (5)Numerical properties, (6)Rotation method, (7)Displayed outputs, (8)the exclusion of the correlation matrix from the OUTSTAT=data set, and (9) Miscellaneous. The PROC FACTOR statement can be used with a number of options as shown above. There are several important options in analyzing cocitation data.

Data Options (Data=)

DATA=data-set-name

Factor Extraction Options (Method=)

METHOD=PRINCIPAL

This option specifies the method for extracting factors. The default is METHOD=PRINCIPAL or PRIN. The two most frequently used factor extraction methods are principal component analysis (or simply component analysis) and common factor analysis. The principal component analysis is used to summarize most of the original information in a minimum number of factors. ACA research uses the principal component analysis to identify the intellectual structure of an academic discipline.

Specifying Number of Factors Options (MINEIGEN=, NFACTORS=)

MINEIGEN=:

Eigenvalue is the column sum of squares for a factor. It represents the amounts of variance accounted by a factor. Minimum eigenvalue criterion is one of several criteria used for the number of factors to be extracted such as latent root (eigenvalue) criterion, a priori criterion, percentage of variance criterion, and scree test criterion. Readers are referred to (Hair, Anderson, & Tatham, 1987) for the details of these criteria. The latent root criterion is the most commonly used technique and simple to apply. According to this criterion, only the factors that have eigenvalues greater than one are considered significant. Otherwise, the analyst may interpret the factors with less than 1 eigenvalues to be insignificant and exclude them for further interpretation. To do so, the analyst may specify MINEIGEN=1.

NFACT=:

There is no exact quantitative basis for deciding the number of factors to extract as the final solution. In addition to these two criteria (the scree test and the minimum eigenvalue criteria), another important criterion is the meaningfulness of the factor. Very often, the factors considered to be acceptable under the eigenvalue and/or scree test criteria fails to be included in the final solution because it is difficult to assign a useful name to the factor that represents the common characteristics of all authors under that factor. Thus, it is very important to include the option of specifying the number of factors to extract.

This option specifies the number of factors to be included in the output. N is the number of factors to extract.

NFACT=n

Specifying Rotation Method Option

The rotation of factors is a procedure used to achieve a more meaningful factor solution. In most cases, rotation of the factors improves the interpretation of the factor loadings obtained from the same data by reducing some of the ambiguity of an initial un-rotated factor pattern. The two most frequently used factor rotation options are orthogonal and oblique. When rotating the factors using the orthogonal method, the axes are maintained at 90 degrees (right angles). Therefore, the factors are mathematically independent and they are assumed to be uncorrelated with each other. On the other hand, the oblique rotation method assumes that the factors are correlated with each other. Consequently, the oblique solution provides us with the inter-factor correlation information among the factors.

The PROMAX rotation specification (ROTATE = PROMAX) provides both orthogonal and oblique rotations with only one invocation of PROC FACTOR. Out of the two major rotation options, most ACA studies use an oblique rotation method. Compared to an orthogonal rotation method, the oblique factor rotation is "more desirable because it is theoretically and empirically more realistic"(Hair, Anderson, & Tatham, 1987). It allows a more natural rotation without the imposition of orthogonal factors. If the orthogonal rotation method is desirable, the "ROTATE = VARIMAX" option can be specified.

Displaying Output Options (Scree)

The SCREE option produces a plot which displays the size of the eigenvalues associated with each factor. The scree tail test is an approach that determines the optimum number of factors that can be extracted before the amount of unique variance begins to dominate the common variance structure (Cattell, 1966). The scree test involves the plotting of the latent roots (eigenvalues) against the number of factors in their order of extraction. To decide the maximum number of factors extracted, we need to connect factor 1 (at the upper left hand side) and factor 2 (at the middle left hand side) using a straight line, then connect factor 2 and factor 3. This process continues until all factors are connected.

EXAMPLES OF ACA PROC FACTOR STATEMENTS

The SAS Display Manager

Three windows that make up the SAS display manager system (Release 9) will be visible. The windows consist of the Results Window at the left hand side, the Log Windows at the top, right-hand side, and the Editor Window at the bottom, right-hand side. SAS input files are placed in the editor window for processing.

The first step is entering the SAS input files in the Editor window area. To do so, click the rectangle shaped icon at the left hand side of "Editor" title area. A drop-down menu appears showing a submenu of Menu, Move, Size, etc. Click the Menu, File, and Open. The Open window will appear on the screen. One can now choose the subdirectory where his or her SAS input file is stored. Figure 4.3 shows the SAS Display Manager System Screen with the Editor Window filled with the input file, fac9INPUTname.sas.

The output to be discussed in this chapter is based on the SAS input file of 9 authors. All the SAS input files discussed in this chapter includes only 9 authors for the sake of simplicity. For complete cocitation matrix data, refer to (Eom, 1995). This cocitation matrix is compiled from a total of 692 citing articles in the decision support systems (DSS) area over the past 20 years (1969-1990). For a detailed discussion on these datasets, see Eom (Eom, 1995; Eom & Farris, 1996). To execute the SAS input files entered, click the "Run" menu, followed by "Submit" menu. After the program has run, the SAS display manager shows the Output screen with minimized "results", "data", and "log" screen buttons.

There is a great chance of making errors when learning to use the SAS system. The LOG window displays important information including error messages, number of observations, number of variables, the amounts of processing time used by the SAS system, number of factor to be retained by the NFACTOR criterion.

INTERPRETING THE FACTOR PROCEDURE OUTPUTS

As shown in Figure 4, the SAS factor procedure produced 19 different outputs under three major headings: initial factor method: principal components; prerotation method: varimax; and rotation method: promax.

Initial Factor Method: Principal Components

The transposed cocitation matrix of 9 authors is analyzed by the principal components analysis with the latent root criterion (eigenvalue 1 criterion) applied to obtain the initial solution of 4 factors.

The first segments of the output are as follows. ACA analysis uses the principal components analysis to reduce the number of variables into a smaller number of observed variables that tend to hang together. The SAS FACTOR procedure performs both principal components analysis and factor analysis. Although these two techniques are similar, they are not identical due to the assumptions of an underlying causal structure. While principal components analysis assumes no underlying causal structure between the observed variables and latent variables, factor analysis is based on the assumption that "the covariation in the observed variables is due to the presence of one or more latent variables (factors) that exert causal influence on these observed variables."(Hatcher, 1994, p.9).

Figure 3. The SAS display manager system screen

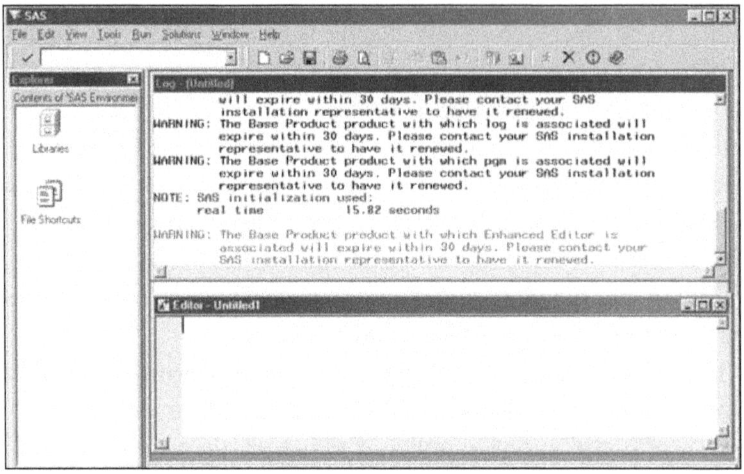

Figure 4. The SAS display manager system screen with factor analysis data and corresponding outputs

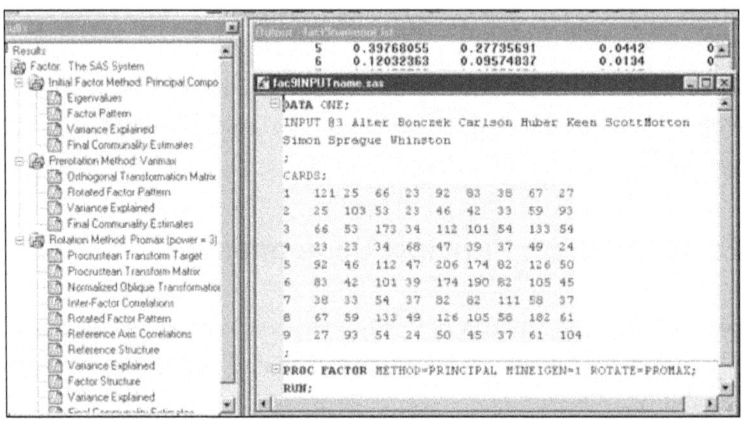

The difference between principal component analysis and factor analysis is that the former analyzes all types of variability for each observed variable, while the latter analyzes only shared variability, ignoring unique and error variability. Therefore, factor analysis analyzes covariance and principal components analysis analyzes variance. Although SAS outputs do not use the term "components", the resulting linear combinations produced by principal components are often referred to as "components". Table 5 below includes "factor pattern", which is equivalent to component matrix produced by SPSS in Chapter XI.

Outputs under the heading of "initial factor method: Principal components" list four elements of information that can be summarized as in Table 4.12. Eigenvalue (the latent root) of a factor is the column sum of squares for that factor. For example, eigenvalue of factor 1 is the sum of squared factor loadings of all variables on that factor and is computed as follows.

$0.73621^2 + (-0.58395)^2 + 0.66579^2 + 0.3274^2 + 0.92069^2 + 0.89789^2 + 0.59921^2 + 0.68328^2 + (-0.55898)^2 = 4.225729$

Factor pattern is a matrix of factor loadings (factor pattern) consisting of three columns of factors 1, 2, and 3 and nine author variable rows. Often "factor pattern" is referred to as "component matrix". Factor loadings represent the correlation between original author variables and the factors. Variance explained by each factor is identical to the eigenvalue of each factor, often expressed as percentage. The last row of Table 5 shows the variance of each factor in terms of percentage, which is the variance of each factor divided by the number of authors (variables). For example, the variance explained by factor 1 (46.95%) is computed as follows.

$(4.225729/9)*100 = 46.9525\%$

Final communality estimates of each author (observed variable) are the sum of the square of factor loadings for that variable. For example, communality for Alter is computed as the sum of the square of each of three factor loadings for that author.

$(0.736212^2 + 0.072712^2 + (-0.53374)^2 = .542005+.005287+.284878=.83217.$

Optimal Number of Factors

Before proceeding to the next section of output (Prerotation Method: Varimax), it is very important to understand the different approaches to find the optimal number of factors in factor analysis. There are three commonly used methods.

Eigenvalue One Criterion

In Table 5, the first column represents the number of principal components extracted from 9 authors (variables). Eigenvalue associated with each variable is listed under the second column with heading "Eigenvalue". Notice that the first component's eigenvalue (4.22) accounts for 46.95% of the total variance. Each of succeeding component accounts for progressively smaller amounts of variance.

Table 5. Initial factor method: principal components

The SAS System 11:46 Friday, January 25, 2002
The FACTOR Procedure Initial Factor Method: Principal Components
Prior Communality Estimates: ONE
Eigenvalues of the Correlation Matrix: Total = 9 Average = 1

	Eigenvalue	Difference	Proportion	Cumulative
1	4.22575393	2.01474495	0.4695	0.4695
2	2.21100899	1.13125294	0.2457	0.7152
3	1.07975604	0.15566117	0.1200	0.8352
4	0.92409487	0.52641432	0.1027	0.9378
5	0.39768055	0.27735691	0.0442	0.9820
6	0.12032363	0.09574837	0.0134	0.9954
7	0.02457526	0.00776854	0.0027	0.9981
8	0.01680672	0.01680672	0.0019	1.0000
9	0.00000000		0.0000	1.0000

3 factors will be retained by the MINEIGEN criterion.

Factor Pattern

	Factorl	Factor2	Factor3
Alter	0.73621	0.07271	-0.53374
Bonczek	-0.58395	0.77515	0.08406
Carlson	0.66579	0.55143	0.24243
Huber	0.32740	-0.54845	0.73192
Keen	0.92069	0.23074	-0.05239
Scott Morton	0.89789	0.19807	-0.12872
Simon	0.59921	-0.15880	-0.05081
Sprague	0.68328	0.51844	0.40691
Whinston	-0.55898	0.78327	0.07652

Variance Explained by Each Factor

Factorl	Factor2	Factor3
4.2257539	2.2110090	1.0797560

Final Communality Estimates: Total = 7.516519

Alter	Bonczek	Carlson	Huber	Keen
0.83216903	0.94893043	0.80613171	0.94369172	0.90365786
Scott Morton	Simon	Sprague	Whinston	
0.86201759	0.38685474	0.90123407	0.93183182	

From the first part of output (Table 5), ACA researchers derive a very important piece of information — "3 factors will be retained by the MINEIGEN criterion." When preparing input file, ACA researchers begin with no idea with regard to the number of factors to be derived. Using the eigenvalue = 1 criterion, three factor solution is a starting point for finding the meaningful number of factors. The meaningful solution may consist of more or less factors than the three-factor solution. To find the meaningful final solution, ACA researchers need to carefully interpret all of the solutions to ascertain the meaning of each factor. At this stage, there is no need to examine all factors. Rather, attention should be given to the last one or two factors to assess the meaning of the factor. Therefore, ACA analysts may specify NFACT options to include 2, 3, and 4 factor solutions. In ACA of using custom databases, the meaning factor solution is usually smaller than the number of factors identified by the Eigenvalue criterion.

```
PROC FACTOR METHOD=PRINCIPAL NFACT=2 ROTATE=PROMAX;
PROC FACTOR METHOD=PRINCIPAL NFACT=3 ROTATE=PROMAX;
PROC FACTOR METHOD=PRINCIPAL NFACT=4 ROTATE=PROMAX;
```

Specifying different number of factors may result in different factor loadings. All the factor loadings under each factor of 1, 2, and 3 may be different from 3 factor solution using NFACT=3 and 4 factor solution using NFACT=4 option.

Table 6. A matrix of factor loadings with eigenvalue and communality

	Factor1	Factor2	Factor3	Communality
Alter	0.73621	0.07271	-0.53374	0.8321703
Bonczek	-0.58395	0.77515	0.08406	0.94892121
Carlson	0.66579	0.55143	0.24243	0.80612367
Huber	0.3274	-0.54845	0.73192	0.94369505
Keen	**0.92069**	0.23074	-0.05239	0.90365574
Scott Morton	**0.89789**	0.19807	-0.12872	0.86200702
Simon	0.59921	-0.1588	-0.05081	0.38685172
Sprague	0.68328	0.51844	0.40691	0.90122734
Whinston	-0.55898	0.78327	0.07652	0.93182584
Eigenvalue	4.225729	2.210999	1.07975	7.51647788
Variance %	46.95%	24.56%	11.99%	83.51%

Scree Test

The scree test is another approach for deciding the optimal number of factors. Figure 5 shows that the lines connecting the eigenvalues of factors 1 through 9. Notice that as the number of factors increases, the slope of the lines connecting two successive factors becomes more gentle and almost a horizontal line. This example is taken from another dataset in our previous research. By applying the scree test, a seven-factor solution can be the final solution, even though the minimum eigenvalue 1 criterion suggests that nine factors would be acceptable. Needless to say, identifying the intellectual structure of an academic discipline is not a well-structured process.

The Subjective Approach

The third approach is the subjective judgment-based approach. Factor analysis in ACA should be a supporting tool that must be used with expert judgments in regard to the interpretability of each factor. Specifically, ACA analyst should have a reasonable level of understanding as to what each author in a discipline under investigation has done in the development of that field. The data that can be used in this process include frequently cited articles in the cited bibliographic databases. ACA analysts should make a judgment on each factor regarding the meaning of the factor. This evaluation of the interpretability of each factor begins from the factors with smaller eigenvalues.

Prerotation Method: Varimax

The second part of the output under the "Prerotation Method: Varimax" heading in Table 7 lists orthogonal transformation matrix, rotated factor pattern, variance explained, and final communality estimates. The second output from the factor procedure is a rotated factor matrix. The rotation of the factor matrix makes the factor solution easier to interpret by reducing the ambiguities in interpreting the factor matrix. Readers are referred to Chapter VI of (Hair, Anderson, & Tatham, 1987) to understand the concept of rotation. Comparing Tables 5 (unrotated factor matrix) and 7 (rotated factor matrix), the rotated factor matrix includes many variables (authors) whose factor loadings are grater than .8.

The factor loadings matrix can be rotated by using an orthogonal or oblique method. A varimax rotation is the most commonly used orthogonal rotation in ACA study, and it maximizes the variance of a column of the factor pattern matrix, not a row of the matrix. The term "orthogonal" involves right (90 degree) angles, while an "oblique" rotation involves acute or obtuse angles. A rotated factor pattern that uses a varimax rotation is assumed to have no inter-factor correlations. Notice that

Figure 5. Scree plot of eigenvalues

```
                    The SAS System  14:42 Friday, January 4, 2002
                    The FACTOR Procedure
              Initial Factor Method: Principal Components

Scree Plot of Eigenvalues
    ,
 25 ˆ
    ,
    ,
    ,  1
    ,
    ,
    ,
 20 ˆ
    ,
    ,
    ,
    ,
E   ,
i   ,
g 15 ˆ
e   ,
n   ,
v   ,
a   ,  2
l   ,
u   ,
e 10 ˆ
s   ,
    ,  3
    ,
    ,
    ,
    ,
  5 ˆ
    ,
    ,   4
    ,    5
    ,     6
    ,      79
    ,       0124567
  0 ˆ              9012456790124567901245679012456790124567
    ,
    Šfffˆffffffffˆffffffffˆffffffffˆffffffffˆffffffffˆffffffffˆffffff
fˆffff
    0    10    20    30    40    50    60    70

                    Number
```

Table 7. Rotated factor pattern (varimax)

The SAS System 11:46 Friday, January 25, 2002
The FACTOR Procedure Prerotation Method: Varimax

Orthogonal Transformation Matrix

	1	2	3
1	0.80457	-0.59156	-0.05215
2	0.55076	0.77614	-0.30704
3	0.22210	0.21831	0.95027

Rotated Factor Pattern

	Factorl	Factor2	Factor3
Alter	0.51383	-0.49560	-0.56791
Bonczek	-0.02424	**0.96542**	-0.12766
Carlson	**0.89323**	0.08705	0.02635
Huber	0.12391	-0.45956	**0.84684**
Keen	**0.85621**	-0.37700	-0.16865
Scott Morton	**0.80292**	-0.40553	-0.22996
Simon	0.38336	-0.48882	-0.03078
Sprague	**0.92566**	0.08701	0.19186
Whinston	-0.00135	**0.95531**	-0.13862

Variance Explained by Each Factor

Factorl	Factor2	Factor3
3.4594143	2.8621496	1.1949551

Final Communality Estimates: Total = 7.516519

Alter	Bonczek	Carlson	Huber	Keen
0.83216903	0.94893043	0.80613171	0.94369172	0.90365786

Scott Morton	Simon	Sprague	Whinston
0.86201759	0.38685474	0.90123407	0.93183182

rotating factor loading matrix changes (1) the factor loadings of each author and (2) the variance explained by each factor, but it does not change (1) the final communality estimate and (2) the communality of each variable (author). Rotation of the factor loadings matrix does not change the total variance of the data set (83.51 percent), while redistributing the proportion of variance that can be explained by each factor.

ACA researchers seem to agree that research subspecialties in an academic discipline represented by factors are correlated. Therefore, the last two tables are usually included as a basis of inferring the intellectual structure of a discipline.

Rotated Factor Pattern

The next step is deciding which variables to include in the result tables as a basis of inferring the intellectual structure of a discipline. There are three important tables that interest ACA researchers:

1. Rotated factor pattern tables with variance explained to be taken from the "Prerotation Method: Varimax" section of the output.
2. Factor structure (Correlations) with variance explained from the "Rotation Method: Promax (power -3)" section
3. Inter-factor correlations from the "Rotation method: Promax (power -3) section".

When reporting the result, ACA researchers recommend including only variables with factor loadings whose absolute value exceeds ± .40. Any variable whose absolute value is greater than ±.70 should be used to determine the meanings of each factor.

Rotation Method: Promax

Interfactor Correlations

Since the majority of ACA researchers seem to agree that subspecialties in an academic discipline are interrelated, it is not necessary to include the rotated factor pattern tables. However, it may help ACA researchers by providing supplementary information.

The PROMAX rotation specification provides both orthogonal and oblique rotations with only one invocation of the PROC FACTOR. Compared to an orthogonal rotation method, the oblique factor rotation is "more desirable because it is theoretically and empirically more realistic" (Hair, Anderson, & Tatham, 1987 p. 245). It allows a more natural rotation without the imposition of orthogonal factors. Moreover, it generates additional information about the correlations between the factors (Table 9). The oblique rotation method provides important information that cannot be found with an orthogonal rotation method. For oblique rotation-based factor solutions, proper interpretation of a set of factors requires examining the factor pattern (the weight matrix to calculate variable standard scores from factor standard scores), the factor structure (the correlations of the variables with the factors), and the reference structures (the correlations between the variables and the factors when the variance attributable to all other factors has been removed). This is the interfactor correlations (Table 8). It provides ACA researchers with very useful information about the influence each factor has on the development of other factors. To assess the impact of one research subspecialty on the other subdisciplines, ACA researchers often construct major factor intercorrelation network. The intercorrelation network shows that organization science, a reference discipline, has made important contributions to the development of decision support systems research subspecialties, especially in the areas of foundations, model management, and user-interfaces. For further details on this subject, refer to (Eom & Farris, 1996).

Factor Structure Correlations

See Table 9.

Table 8. Interfactor correlations (1970-1990)

Output - (Untitled)

Inter-Factor Correlations

	Factor1	Factor2	Factor3
Factor1	1.00000	-0.29358	-0.21026
Factor2	-0.29358	1.00000	0.10096
Factor3	-0.21026	0.10096	1.00000

Table 9. Factor structure correlations

Output - (Untitled)

Factor Structure (Correlations)

	Factor1	Factor2	Factor3
Alter	0.61546	-0.58566	-0.65800
Bonczek	-0.15499	0.95336	-0.07645
Carlson	0.86775	-0.04786	-0.09608
Huber	0.13411	-0.44518	0.79776
Keen	0.91071	-0.50712	-0.30594
ScottMorton	0.86616	-0.52927	-0.36040
Simon	0.45105	-0.54179	-0.10811
Sprague	0.88912	-0.04741	0.06299
Whinston	-0.13023	0.93956	-0.09101

Variance Explained by Each Factor Ignoring Other Factors

Factor1	Factor2	Factor3
3.7643555	3.1682109	1.3318792

Final Communality Estimates: Total = 7.516519

Alter	Bonczek	Carlson	Huber	Keen
0.83216903	0.94893043	0.80613171	0.94369172	0.90365786

REFERENCES

Cattell, R. B. (1966). The Scree Test for the Number of Factors. *Multivariate Behavioral Research, 1*, 245-276.

Child, D. (1970). *The Essentials of Factor Analysis.* London: Holt, Rinehart and Winston, Inc.

Cooper, D. R., & Emory, C. W. (1995). *Business Research Methods* (Fifth ed.). Chicago, IL: Irwin.

Eom, S. B. (1995). Decision Support Systems Research: Reference Disciplines and a Cumulative Tradition. *Omega: The International Journal of Management Science, 23*(5), 511-523.

Eom, S. B., & Farris, R. (1996). The Contributions of Organizational Science to the Development of Decision Support Systems Research Subspecialties. *Journal of the American Society for Information Science, 47*(12), 941-952.

Hair Jr., J. F., Anderson, R. E., & Tatham, R. L. (1987). *Multivariate Data Analysis with Readings* (2nd ed.). New York: Macmillan Publishing Company.

Hatcher, L. (1994). *A Step-by-Step Approach to Using the SAS System for Factor Analysis and Structural Equation Modeling* (3rd ed. Printing ed.). Carry, NC: SAS Institute Inc.

Kim, J., & Mueller, C. W. (1978). *Factor Analysis: Statistical Methods and Practical Issues.* Beverly Hills, CA: Sage Publications, Inc.

SAS Institute Inc. (2007). *SAS Onlinedoc® 9.1.3.* . Cary, NC: SAS Institute Inc. .

Chapter IX
The Distance and Cluster Procedure

ABSTRACT

This chapter describes the distance and cluster procedure of the SAS system. SAS version 9 introduced the proc distance procedure. All previous versions of SAS used two programs (xmacro.sas and distnew.sas) to process a transposed cocitation matrix (input) to produce a distance matrix (output). Cluster analysis is a data reduction technique for grouping various entities (individuals, variables, objects) into clusters so that the entities in the same cluster have more similarity to each other with respect to some predetermined selection criteria. The first section of this chapter explains the creation of a distance matrix, which is the input to the cluster procedure. The second part of this chapter focuses on the PROC CLUSTER statement which sets out the CLUSTER procedure steps. This chapter also includes the discussions of interpreting results of cluster analysis.

INTRODUCTION

SAS version 9 introduced the proc distance procedure. All previous versions of SAS used two programs (xmacro.sas and distnew.sas) to process a transposed cocitation matrix (input) to produce a distance matrix (output). The input to the cluster and multi-dimensional scaling analysis is a proximity matrix. The cocitation frequency counts matrix must be converted into a distance or similarity matrix. SAS version 9 created a new procedure, the distance procedure, to compute various measures of distance, dissimilarity, or similarity between the authors under investigation. The distance matrix is the input to the CLUSTER and MDS procedures.

There are many different ways of measuring inter-object similarity, including distance measures (proximity/difference between each pair of objects) and the correlation coefficient between a pair of objects. The higher cocitation frequencies between a pair of authors represent a higher level of cognitive linkages or similarities between them. In ACA, the cocitation frequency count matrix, correlation coefficient matrix, and distance matrix represent three different outputs in the same transformation process (see Table 1). Understanding input and output relations in the process helps us select the correct options in the distance and MDS procedures.

Table 1 highlights the input/output relationships in many PROC statements. The cocitation frequency counts matrix is the original input to all other procedures in ACA. The PROC CORR statement processes the cocitation frequency counts matrix to produce the correlations matrix. The third column heading of output/input indicates that the correlations matrix is the output of PROC CORR and it is also the input to the PROC FACTOR. The bold faced outcomes (factor pattern, clusters, two and three dimensional MDS maps) are the final outputs which are not going to be used as the input to other procedures. Therefore, distance matrix and MDS configuration coordinates are the outputs of previous stages as well as the inputs to the following stages.

The following section begins with the discussion of the four levels of measurements. The variables of the cocitation frequency matrix are authors, represented by labels (Alter, Keen, Scott Morton, etc.) or numerical algebraic expressions (x_1-x_{100}). Variables have certain characteristics that define the type of statistical analysis to be performed. These characteristics are referred to as the level of measurement of the variables. Understanding the four levels of measurement are critical to conduct ACA accurately. Some PROC statements such as PROC distance and PROC MDS ask ACA analysts to specify the measurement level of the data. The remaining sections of this chapter discuss the following topics.

- Creating permanent distance matrix
- The distance procedures
- The cluster procedures
- Interpreting results of cluster analysis

THE FOUR LEVELS OF MEASUREMENT

Variables have certain characteristics that define the type of statistical analysis to be performed. These characteristics are referred to as the level of measurement of

Table 1. Summary of input/output relationships in various PROC statements

	Original input	Output/input	Input/output	output
	Cocitation frequency matrix			
Proc Factor		**Factor pattern/ structure correlations**		
Proc Distance		Distance matrix		
Proc Cluster			**Clusters**	
Proc MDS			MDS Configuration Coordinates	
Proc plot				**Two dimensional MDS maps**
Proc G3D				**Three dimensional MDS maps**

the variables and proposed by (Stevens, 1946). Understanding the four levels of measurement are critical to conducting ACA accurately.

A *Categorical/Nominal* Variable

A categorical/nominal variable can be sorted according to the characteristics of one or more of its attributes. Names are assigned to variables as labels.

- The values of a nominal variable measure qualitative differences, not quantitative differences.
- Nominal variables categorize observations into discrete categories (gender: male vs. female; union membership: union member vs. non union member; college major: finance vs. accounting, etc).
- Nominal variables can be assigned a code in the form of numbers. People can be categorized based on their ethnic background. This categories can be represented by numbers (African =1, Asian=2, Hispanic =3, Caucasian=4) The numbers are simply labels. Therefore, any calculations such as computing an average would be meaningless.

An *Ordinal/Rank* Variable

An ordinal variable expresses degree, quality, satisfaction, rank order, or position in a series.

- Ordinal variables are used to categorize and rank order observations.
- The values of ordinal variables can be used to compare whether they are equal or not (inequality).
- The values of ordinal variables can be used to compare whether they are greater or less.
- Addition or subtraction of rank variables does not make sense and is meaningless. For example, Students can be ranked in the order of completing an exam. They can be ranked first, second, third, ... and last. But, it is not true to say that first + second=third.
- An ordinal/rank variable has a rating scale attached, and it can be ranked.
- Students may be asked to evaluate the instructor's effectiveness of teaching to express the level of their agreement on a scale of 1 (strongly disagree), 2 (disagree), 3 (neither disagree nor agree), 4 (agree), and 5 (strongly agree). An ordinal data have a rating scale attached and it can be ranked.

An *Interval* Variable

An interval variable has a unit of measure and is used to measure the distance between any two adjacent units of measurement (or 'intervals'). An interval variable is similar to an ordinary variable. The only difference is that the intervals between the values of the interval variable are equal. An interval variable is used when the difference between two values is meaningful. Good examples of interval variables are years in a calendar and degrees in temperature in Celsius or Fahrenheit. Only interval and ration variables have units of measurement. In these examples, year 0 and temperature 0 are arbitrary numbers. It does not mean the absence of time or temperature. Therefore, the interval variable measurement can only be added and/or subtracted to measure only differences between the two measurement points. Therefore, the ratio between temperature 0 and 5 either in Celsius or Fahrenheit is NOT zero (0/5). Consequently, division or multiplication of interval variables is not possible and meaningless.

- The values of interval variables can be categorized and rank ordered.
- The values of interval variables can be meaningfully added or subtracted.
- The value of interval variables cannot be meaningfully multiplied or divided.
- The value, 0, of interval variables does not represent the absence of the variable.

Table 2. The level of measurement and computations allowed

	Nominal	Ordinal	Interval	Ratio
Examples	Male/Female	First, second, …, last of a marathon	Year , temperature in Celsius or Fahrenheit	Height, weight, age
Frequency distribution	Yes	Yes	Yes	Yes
Median and percentile	No	Yes	Yes	Yes
Addition or subtraction	No	No	Yes	Yes
Ratio obtained by multiplication and/ or division	No	No	No	Yes

A *Ratio* Variable

A ratio variable has all the properties of an interval variable as described above. The only difference is the value of zero (0). The value of zero means that none of that variable is in an observation. When the variable equals 0.0, there is none of that variable. Ratio variables include physical quantities such as height, weight, age, annual income, etc.

THE DISTANCE PROCEDURE

A welcome addition to SAS version 9 is the new DISTANCE procedure, which computes various measures of distance, dissimilarity, or similarity between the observations (rows) of a SAS data set. In prior version 8, two %INCLUDE statements to bring entire external files (xmacro.sas and distnew.sas). In version 9, the new DISTANCE procedure is introduced to compute distance between observations of a data set. These proximity measures are to be used as input to the CLUSTER and MDS procedures (SAS Institute Inc., 2007).

The following statements are available in the DISTANCE procedure.

PROC DISTANCE < options > ;
BY variables ;
COPY variables ;
FREQ variable ;
ID variable ;

VAR level(variables < / opt-list >) ;
WEIGHT variable **;**

A SAS procedure consists of two types of statements: required and optional statements.

Required Statements

Both the PROC DISTANCE statement and the VAR statement are required.

The following distance procedure includes only essential minimum statement.

```
Proc distance data=aca out=dist method=dcorr SHAPE=SQUARE;
Var ratio(Alter—Whinston);
id author;
run;
```

Figure 1 shows dataset WORK.ACA imported from an Excel file. The variables in WORK.ACA is ratio variable. Practically, WORK.ACA can be treated as either a ratio or an interval variable. The outputs will be identical with Var ratio or Var interval. Figures 2 and 3 show distance matrices without and with ID statements. The ID statement generate names for the distance variables.

VAR Statement

VAR | VARIABLES *level (variables < / opt-list >)*
< level (variables < / opt-list >)
level (variables < / opt-list >)
.
.
.
level (variables < / opt-list >) >

where the syntax for the *opt-list* is:

< ABSENT = value >
< MISSING = miss-method or value>
< ORDER = order-option >
< STD = std-method >
< WEIGHTS = weight-list >

Figure 1. Data set WORK.ACA imported from an Excel file

Results
- Print: The SAS System
  - Data Set WORK.ACA
- Factor: The SAS System
  - Initial Factor Method: Pr
  - Prerotation Method: Var
  - Rotation Method: Prom
- Print: The SAS System
  - Data Set WORK.DISTOR
- Cluster: The SAS System
- Tree: The SAS System
- Print: The SAS System
- Cluster: The SAS System
- Tree: The SAS System

The SAS System

Obs	Author	ALTER	BENBASAT	BONCZEK	CARLSON	DESANCTIS	DICKSON	GINZBERG	GORRY	HENDERSON	HOLSAPPLE	HUBER
1	ALTER	121	16.0	25	66	8.0	15.0	34.0	30.0	22	25	23.0
2	BENBASAT	16	62.5	13	23	33.0	43.0	13.0	13.0	16	13	22.0
3	BONCZEK	25	13.0	118	53	18.0	13.0	15.0	11.0	34	84	23.0
4	CARLSON	66	23.0	53	173	18.0	17.0	32.0	34.0	33	49	34.0
5	DESANCTIS	8	33.0	18	18	62.5	43.0	8.0	7.0	16	16	49.0
6	DICKSON	15	43.0	13	17	43.0	65.5	13.0	9.0	13	13	35.0
7	GINZBERG	34	13.0	15	32	8.0	13.0	66.5	15.0	14	14	21.0
8	GORRY	30	13.0	11	34	7.0	9.0	15.0	73.5	8	12	12.0
9	HENDERSON	22	16.0	34	33	16.0	13.0	14.0	8.0	63	33	22.0
10	HOLSAPPLE	25	13.0	84	49	16.0	13.0	14.0	12.0	33	110	23.0
11	HUBER	23	22.0	23	34	49.0	35.0	21.0	12.0	22	23	72.5
12	KEEN	92	44.0	46	112	33.0	45.0	53.0	46.0	45	46	47.0
13	KONSYNSKI	16	12.0	34	24	33.0	19.0	8.0	4.0	37	32	35.0
14	LUCAS	14	28.0	11	18	18.0	29.0	14.0	11.0	7	11	16.0
15	MASON_R	15	24.0	8	22	16.0	23.0	13.0	16.0	16	8	20.0
16	MITROFF	12	22.0	8	18	17.0	23.0	10.0	15.0	17	8	17.0
17	SCOTT_MORTON	83	38.0	42	101	27.0	39.0	42.0	62.0	34	41	39.0
18	SIMON	38	31.0	33	54	28.0	34.0	17.0	39.0	26	33	37.0
19	SPRAGUE	67	31.0	59	133	27.0	26.0	38.0	32.0	44	56	49.0
20	WHINSTON	27	13.0	93	54	19.0	14.0	15.0	13.0	34	87	24.0

Figure 2. Distance matrix produced without ID statement

The SAS System

Obs	Dist1	Dist2	Dist3	Dist4	Dist5	Dist6	Dist7
1	0.00000	0.96195	0.91970	0.53008	1.11453	1.02127	0.57378
2	0.96195	0.00000	1.14941	0.95528	0.69551	0.40476	0.96830
3	0.91970	1.14941	0.00000	0.76511	1.07288	1.16246	0.98787
4	0.53008	0.95528	0.76511	0.00000	1.06152	1.03914	0.67495
5	1.11453	0.69551	1.07288	1.06152	0.00000	0.49583	1.11008
6	1.02127	0.40476	1.16246	1.03914	0.49583	0.00000	1.01272
7	0.57378	0.96830	0.98787	0.67495	1.11008	1.01272	0.00000
8	0.64562	0.94294	1.02940	0.70749	1.12545	1.01804	0.77408
9	0.82913	1.06621	0.66956	0.71739	1.01893	1.09921	0.91383
10	0.91366	1.14767	0.22986	0.76786	1.07883	1.16113	0.98726
11	0.90354	0.85367	0.98116	0.83295	0.48318	0.71157	0.90079
12	0.42370	0.80098	0.90435	0.45931	0.99783	0.87052	0.55536
13	1.02124	1.10188	0.68990	0.91066	0.76167	1.02001	1.09649
14	0.92267	0.55339	1.13808	0.94479	0.87810	0.59732	0.89397
15	0.98179	0.81667	1.19624	0.98891	0.99976	0.85288	0.98391
16	1.01929	0.82623	1.19565	1.02331	0.98691	0.84820	1.02861
17	0.44281	0.81562	0.91428	0.49430	1.01484	0.88352	0.59797

Figure 3. Distance matrix produced with ID statement

The SAS System

Obs	Author	ALTER	BENBASAT	BONCZEK	CARLSON	DESANCTIS	DICKSON
1	ALTER	0.00000	0.96195	0.91970	0.53008	1.11453	1.02127
2	BENBASAT	0.96195	0.00000	1.14941	0.95528	0.69551	0.40476
3	BONCZEK	0.91970	1.14941	0.00000	0.76511	1.07288	1.16246
4	CARLSON	0.53008	0.95528	0.76511	0.00000	1.06152	1.03914
5	DESANCTIS	1.11453	0.69551	1.07288	1.06152	0.00000	0.49583
6	DICKSON	1.02127	0.40476	1.16246	1.03914	0.49583	0.00000
7	GINZBERG	0.57378	0.96830	0.98787	0.67495	1.11008	1.01272
8	GORRY	0.64562	0.94294	1.02940	0.70749	1.12545	1.01804
9	HENDERSON	0.82913	1.06621	0.66956	0.71739	1.01893	1.09921
10	HOLSAPPLE	0.91366	1.14767	0.22986	0.76786	1.07883	1.16113
11	HUBER	0.90354	0.85367	0.98116	0.83295	0.48318	0.71157
12	KEEN	0.42370	0.80098	0.90435	0.45931	0.99783	0.87052
13	KONSYNSKI	1.02124	1.10188	0.68990	0.91066	0.76167	1.02001
14	LUCAS	0.92267	0.55339	1.13808	0.94479	0.87810	0.59732
15	MASON_R	0.98179	0.81667	1.19624	0.98891	0.99976	0.85288
16	MITROFF	1.01929	0.82623	1.19565	1.02331	0.98691	0.84820

Table 3. Summary of PROC DISTANCE statement options

Task/Statement	Options	Description
standardizing variables	ADD=	specifies the constant to add to each value after standardizing and multiplying by the value specified in the MULT= option
	FUZZ=	specifies the relative fuzz factor for writing the output
	INITIAL=	specifies the method for computing initial estimates for the A-estimates
	MULT=	specifies the constant to multiply each value by after standardizing
	NORM	normalizes the scale estimator to be consistent for the standard deviation of a normal distribution
	SNORM	normalizes the scale estimator to have an expectation of approximately 1 for a standard normal distribution
	VARDEF=	specifies the variances divisor
generating distance matrix	ABSENT=	specifies the value to be used as an absence value for all the asymmetric nominal variables
	METHOD=	specifies the method for computing proximity measures
	PREFIX=	specifies a prefix for naming the distance variables in the OUT= data set
	RANKSCORE=	specifies the method of assigning scores to ordinal variables
	SHAPE=	specifies the shape of the proximity matrix to be stored in the OUT= data set
	UNDEF=	specifies the numeric constant used to replace undefined distances
	VARDEF=	specifies the variances divisor
missing values	NOMISS	replaces missing data by the location measure (does not standardize the data); generates missing distance for observations with missing values
	REPLACE	replaces missing data by zero in the standardized data
	REPONLY	replaces missing data by the location measure (does not standardize the data)
specifying data set details	DATA=	specifies the input data set
	OUT=	specifies the output data set
	OUTSDZ=	specifies the output data set for standardized scores

Source: Source: SAS Onlinedoc 9.1.3 at http://support.sas.com/onlinedoc/913/docMainpage.jsp

level is required. It declares the levels of measurement for those variables specified within the parentheses. Available values for *level* are ANOMINAL, NOMINAL, ORDINAL, INTERVAL, and RATIO.

Variables are ratios, and only numeric variables are allowed. Ratio variables should always contain positive measurements.

Optional Statements

Although Table 3 lists all PROC DISTANCE Statement Options, the following discuss some of the important statements.

Specify the Method for Computing Distance Measures (Method= options)

The "method = " specifies the method for computing proximity measures If the user chose incorrect method option values, then all of the cluster, MDS procedures outputs will be incorrect. This option value is to be chosen in the procedures, depending on the type of variables ACA procedures are using. The co-citation frequencies are ratio variables. Unless specified otherwise, each method for computing distance or similarity measures allows only numeric variables. The method is one of the following 15 methods. ACA uses Method = DCORR. DCORR is correlation transformed to Euclidean distance as sqrt(1-CORR).

Method = DCORR

EUCLID	Euclidean distance
SIZE	Size distance
SHAPE	Shape distance
COV	Covariance
CORR	Correlation
DCORR	Correlation transformed to Euclidean distance as sqrt (1-CORR)
SQCORR	Squared correlation
DSQCORR	One minus squared correlation
L(p)	Minkowski L(p) distance, where p is a positive numeric value
SIMRATIO	Similarity ratio (if variables are binary, this is the Jaccard coefficient)
DISRATIO	One minus similarity ratio
JACCARD	Jaccard similarity coefficient computed from

numeric variables where 0=absence, 1=presence,

and intermediate values are allowed

DJACCARD	Jaccard dissimilarity coefficient
MATCH	Simple matching coefficient; allows mixed variables
DMATCH	Simple matching coefficient transformed to Euclidean distance; allows mixed variables

For use in PROC CLUSTER, the D... transformations of similarity measures should be used; for example, METHOD=DCORR rather than METHOD=CORR.

DCORR is correlation transformed to Euclidean distance as sqrt(1-CORR).

$$d_9(x, y) = \sqrt{1 - s_8(x, y)}$$

Euclidian distance is the most commonly used measure of similarity between two objects in two dimensions. Suppose there are two points, point A (X_1, Y_1) and point $B(X_2, Y_2)$ on a two dimensional space (X-axis and Y-axis). The distance is measured by the length of the hypotenuse of a right angle, as calculated by square root of $[(X_2 - X_1)^2 + [(Y_2 - Y_1)^2]$.

ID Statement

ID *variable*

The ID statement specifies a single variable to be copied to the OUT= data set and used to generate names for the distance variables. The ID variable must be a character.

THE CLUSTER PROCEDURE

Cluster analysis is a data reduction technique for grouping various entities (individuals, variables, objects) into clusters so that the entities in the same cluster have more similarity to each other with respect to some predetermined selection criteria (Everitt, 1980; Hair, Anderson, Tatham, & Black, 1992). Three main stages must be followed in applying cluster analysis to any problem. They are partitioning, interpretation, and profiling stages. The partitioning stage is concerned with the separation of a whole data set into several groups (clusters). In this stage, three key questions must be carefully considered (Hair, Anderson, Tatham, & Black, 1992).

1. Similarity measures: How should the inter-object similarity be measured?
2. Clustering algorithms: What procedure (clustering algorithms) should be used to place similar object into clusters?
3. Selecting the number of clusters: How many clusters should be selected?

Similarity Measure

Most of the issues in similarity measure are discussed in the proc distance procedure discussion. The "method = " specifies the method for computing proximity measures If the user chose an incorrect method option value, then all of the cluster and MDS procedures outputs will be incorrect. This option values to be chosen in the procedures is dependant on the type of variables ACA procedures are using. The co-citation frequencies are ratio variables. Unless specified otherwise, each method for computing distance or similarity measures allows only numeric variables. The method is one of the following 15 methods. ACA uses Method = DCORR. DCORR is correlation transformed to Euclidean distance as sqrt(1-CORR).

Clustering Algorithms

After the measure of similarity is determined between author variables, what set of rules and procedures should be used to group all authors under study? These rules and procedures are also called clustering algorithms. There are two types of clustering algorithms: hierarchical and non-hierarchical. ACA uses hierarchical procedures to join all authors into one whole cluster. In the beginning, each author (observation) is a cluster. Each author in n author dataset matrix starts out its own cluster (n clusters). Based on the similarity measures discussed earlier, two authors who cocited together most frequently will be grouped as the first cluster. Now the total number of clusters becomes n-1. This process of clustering continues to form until the last one cluster from n clusters. This procedure is called agglomerative hierarchical clustering. The other contrasting procedure is divisive hierarchical clustering, which starts out from whole cluster of entire observations and proceeds to divide it to make each observation a cluster in itself. The PROC CLUSTER procedure uses 11 clustering algorithms to compute the distance measures between two clusters. The algorithms are AVERAGE, CENTROID, COMPLETE, DENSITY, EML, FLEXIBLE, MCQUITTY, MEDIAN, SINGLE, TWOSTAGE, and WARD. ACA researchers found that Ward's method and complete linkage are the two most popular options that produce a better fitting result.

Figure 4. Cluster procedure in ACA

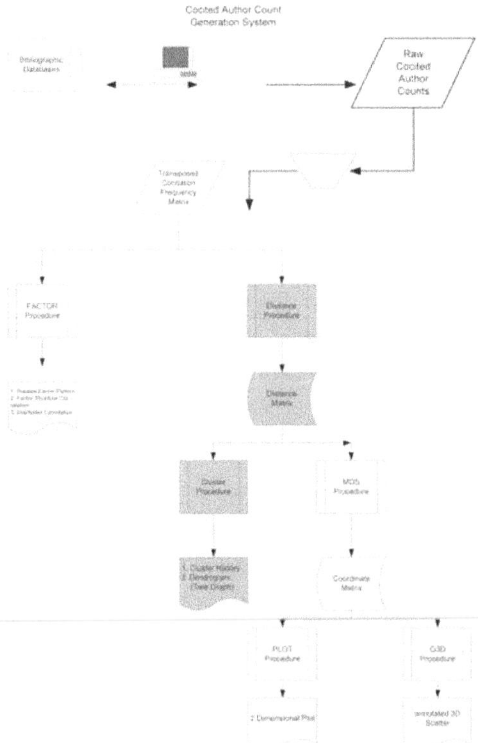

Figure 4 presents the whole picture of ACA processes and highlights the cluster procedure in the ACA processes. The basic syntax consists of PROC CLUSTER statement with options. This procedure builds a tree-like structure (dendrogram/tree graph), which can be a very useful tool for understanding the evolution of thought in an academic discipline. The dendrogram shows a chronological order of the emergence of each subspecialty as well as their interdependency on one another. While factor analysis output produces a snap-shot of invisible/visible colleagues interacting together, cluster analysis output (dendrogram) shows a historical pattern of interaction/cooperation among the colleagues.

It is necessary to convert the raw cocitation frequency matrix into a measure of similarity or distance. To do so, SAS version 8 used the DISTANCE macro for computing various measures of distance, dissimilarity, or similarity between the observations of a SAS data set. SAS version 9 introduced the PROC DISTANCE.

Table 4. Options in the PROC CLUSTER statement

Tasks	Options
Specify input and output data sets	
specify input data set	DATA=
Create output data set	OUTTREE=
Specify clustering methods	
specify clustering method	METHOD=
beta for flexible beta method	BETA=
minimum number of members for modal clusters	MODE=
penalty coefficient for maximum-likelihood	PENALTY=
Wong's hybrid clustering method	HYBRID
Control data processing prior to clustering	
suppress computation of eigenvalues	NOEIGEN
suppress normalizing of distances	NONORM
suppress squaring of distances	NOSQUARE
standardize variables	STANDARD
omit points with low probability densities	TRIM=
Control density estimation	
Dimensionality for estimates	DIM=
number of neighbors for *k*th-nearest-neighbor	K=
Radius of sphere of support for uniform-kernel	R=
Suppress checking for ties	NOTIE
Control display of the cluster history	
display cubic clustering criterion	CCC
suppress display of ID values	NOID
specify number of generations to display	PRINT=
display pseudo F and t^2 statistics	PSEUDO
display root-mean-square standard deviation	RMSSTD
display R^2 and semipartial R^2	RSQUARE
Control other aspects of output	
suppress display of all output	NOPRINT
display simple summary statistics	SIMPLE

Options in the PROC CLUSTER Statements

Table 4 summarizes the options in the PROC CLUSTER statement (Source: SAS Online Doc, Version 9). Of these options available, the PROC Cluster Statement in Table 3 uses only DATA=, METHOD=, Pseudo options, and ID variables. Table 3 shows how to perform the cluster procedure using the permanent data set (mysave. dist). Table 4 shows a detailed procedure which begins the DATA, INPUT, CARDS, DISTANCE, CLUSTER, and TREE procedures.

Specify Input and Output Data Sets

DATA= SAS-data-set

Input to the PROC CLUSTER procedure is the distance matrix.

Table 5. The Proc cluster procedure statement with entered data file

```
DATA newbook;
INPut Author $1-13 @14 Alter Bonczek Carlson Huber Keen ScottMorton
Simon Sprague Whinston
;
CARDS;
Alter        121    25    66    23    92    83    38    67    27
Bonczek      25     103   53    23    46    42    33    59    93
Carlson      66     53    173   34    112   101   54    133   54
Huber        23     23    34    68    47    39    37    49    24
Keen         92     46    112   47    206   174   82    126   50
ScottMorton 83      42    101   39    174   190   82    105   45
Simon        38     33    54    37    82    82    111   58    37
Sprague      67     59    133   49    126   105   58    182   61
Whinston     27     93    54    24    50    45    37    61    104
;
Proc print data=newbook;
run;
Proc factor data=newbook method=principal mineigen=1 rotate=promax;
Proc distance data=newbook out=dist method=dcorr SHAPE=SQUARE;
Var ratio(Alter--Whinston);
id author;
run;
Proc cluster method=ward data=dist pseudo;
ID author;
proc tree space=2;
run;
```

Table 6. The Proc cluster procedure with imported data

```
PROC IMPORT OUT= WORK.ACA
            DATAFILE= "C:\Documents and Settings\sbeom\My Documents\idea
aca\spss\bookdata.xls"
            DBMS=EXCEL REPLACE;
     SHEET="sasdata$";
     GETNAMES=YES;
     MIXED=NO;
     SCANTEXT=YES;
     USEDATE=YES;
     SCANTIME=YES;
RUN;
proc print data=aca;
run;
proc factor data=aca method=principal mineigen=1 rotate = promax;
run;
Proc distance data=aca out=dist method=dcorr SHAPE=SQUARE;
Var ratio(Alter--Whinston);
id author;
run;
proc cluster method=ward data=dist pseudo;
ID author;
proc tree space=2;
run;
```

Specify Clustering Methods: Complete Linkage or Ward Method

METHOD= {AVERAGE, CENTROID, COMPLETE, DENSITY, EML, FLEXIBLE, MCQUITTY, MEDIAN, SINGLE, TWOSTAGE, WARD}

The PROC CLUSTER procedure uses 11 clustering algorithms to compute the distance measures between two clusters. The algorithms are AVERAGE, CENTROID, COMPLETE, DENSITY, EML, FLEXIBLE, MCQUITTY, MEDIAN, SINGLE, TWOSTAGE, and WARD.

The METHOD= name specifies the clustering method used by the procedure. Most of the commonly used clustering algorithms are classified as either hierarchical or non-hierarchical. The hierarchical procedures construct a tree-like structure. In constructing a tree-like structure, there are two possible approaches—agglomerative or divisive. Most statistical packages, including the SAS system, use agglomerative methods. Although the SAS version 9.1.3 manual includes 11 different options, here are the five most commonly used agglomerative methods. They are the single link-

Table 7. The Proc cluster procedure statement with the version 8 SAS system

```
DATA ACA9VAR;
INPUT Author $ 1-12 @13 x1-x9;
CARDS;
ALTER          121   25    66    23    92    83    38    67    27
BONCZEK        25    103   53    23    46    42    33    59    93
CARLSON        66    53    173   34    112   101   54    133   54
HUBER          23    23    34    68    47    39    37    49    24
KEEN           92    46    112   47    206   174   82    126   50
SCOTTMORTON    83    42    101   39    174   190   82    105   45
SIMON          38    33    54    37    82    105   58    182   61
WHINSTON       27    93    54    24    50    45    37    61    104
;
%INCLUDE 'c:\program files\SAS Institute\SAS\V8\stat\sample\xmacro.sas';
%INCLUDE 'c:\program files\SAS Institute\SAS\V8\stat\sample\distnew.sas';
%DISTANCE ( data=aca9var,
      id=author,
      options=print,
      method=dcorr,
      var=x1-x9,
      out=dist);
RUN;

PROC CLUSTER METHOD=WARD DATA=dist PSEUDO;
ID author;
PROC TREE SPACE=2;
RUN;
```

age, complete linkage, average linkage, centroid, and Ward's methods. Of these five clustering methods, ACA researchers recommend to use either complete linkage or Ward method. They found that these two methods consistently provide similar interpretable results(McCain, 1990).

SINGLE | SIN

The single linkage procedure (a.k.a., the nearest-neighbor approach) finds the shortest distance between two objects (each author) from the distance matrix. The distance matrix below (Figure 5) shows the shortest distance is .13646 between Bonczek and Whinston, followed by .16624 between Keen and Scott Morton. Figure 6 shows that Bonczek and Whinston formed CL8 and Keen and Scott Morton are joined to make a new cluster, CL7.

Figure 6 illustrates that all distance measures between authors are sorted in an ascending order from shortest to longest. The first three clusters (CL8, CL7, and CL6) are formed according to the shortest distance from the distance matrix. The next cluster, CL5, joins Alter and Keen because the distance of .5631 in the distance matrix indicates the distance between Alter and Keen. Notice that Keen is already part (member) of cluster 7 (Keen and Scott Morton). Therefore, this cluster is formed by merging Alter and Cluster 7. The distance between any two clusters is the shortest distance between from any point in one cluster to any point in the second cluster. Row 7 has distance of .58664 between Alter and Scott Morton. But Column C (Cluster Formed) contains "xxxx", indicating no clusters formed using the distance of .58664. This is because the distance is the distance between two members in Cluster 4.

Figure 5. Distance matrix

Obs	Author	Alter	Bonczek	Carlson	Huber	Keen	Scott Morton	Simon	Sprague	Whinston
1	Alter	0								
2	Bonczek	1.21223	0							
3	Carlson	0.74515	1.01124	0						
4	Huber	1.07661	1.23667	0.98788	0					
5	Keen	0.5631	1.14216	0.61729	0.89764	0				
6	ScottMortc	0.58664	1.14668	0.66094	0.92865	0.16624	0			
7	Simon	0.91242	1.17918	0.89841	0.91234	0.66143	0.62572	0		
8	Sprague	0.76448	1.00294	0.35056	0.88388	0.58349	0.65439	0.89201	0	
9	Whinston	1.20648	0.13646	1.00781	1.23657	1.12845	1.13344	1.16098	0.99648	0

Figure 6. Clustering process

	Name Box	B	C	D
1	Distance	Cluster Joined	CL Former	Cluster Members
2	0.13646	Bonczek-Whinston	CL8	Bonczek,Whinston
3	0.16624	Keen-Scott Morton	CL7	Keen, Scott Morton
4	0.35056	Carlson-Sprague	CL6	Carlson, Sprague
5	0.5631	Alter-Keen[CL7]	CL5	Alter, Keen, Scott Morton
6	0.58349	Keen-Sprague [cl5][cl6]	CL4	Alter, Keen, Scott Morton, Carlson, Sprague
7	0.58664	Alter- Scott Morton	xxxx	
8	0.61729	Keen-Carlson	xxxx	
9	0.62572	Scott Morton[CL4]-Simon	CL3	Alter, Keen, Scott Morton, Carlson, Sprague, Simon
10	0.65439	Scott Moorton[CL4]- Sprague	xxxx	
11	0.66094	Scott Morton- Carlson	xxxx	
12	0.66143	Keen-Simon	xxxx	
13	0.74515	Alter-Carlson	xxxx	
14	0.76448	Sprague-Alter	xxxx	
15	0.88388	Sprague(CL3)-Huber	CL2	Alter, Keen, Scott Morton, Carlson, Sprague, Simon, Huber
16	0.89201	Simon-Carlson	xxxx	
17	0.89764	Huber-Keen	xxxx	
18	0.89841	Simon-Carlson	xxxx	
19	0.91234	Huber-Simon	xxxx	
20	0.91242	Simon-Alter	xxxx	
21	0.92865	Huber-Scott Morton	xxxx	
22	0.98788	Huber-Carlson	xxxx	
23	0.99648	Sprague (CL2)-Whinston(CL8)	CL1	Alter, Keen, Scott Morton, Carlson, Sprague, Simon, Huber, Bonczek, Whinston

Figure 7. Cluster history with complete linkage method

Cluster History

NCL	-------Clusters	Joined--------	FREQ	Norm Max Dist	Tie
8	Bonczek	Whinston	2	0.1569	
7	Keen	ScottMorton	2	0.1912	
6	Carlson	Sprague	2	0.4031	
5	Alter	CL7	3	0.6746	
4	CL5	CL6	5	0.8791	
3	Huber	Simon	2	1.0491	
2	CL4	CL3	7	1.238	
1	CL2	CL8	9	1.422	

COMPLETE | COM

The complete linkage method is also known as furthest neighbor, maximum method, diameter method, or rank order typal analysis, because it is based on maximum distance. To form a cluster by joining two single observations (CL8, CL7, and CL6 in Figure 7), single or complete linkage method produces the identical result. The distance between the two single points is identical regardless the method used. The distance between an observation and a cluster with more than 2 observations is computed as follow.

Using the same distance matrix, Figure 7 shows that Alter and cluster 7 is merged to form cluster 5. To understand this process, it is necessary to computer distances from clusters 6, 7, and 8 to every single observation (Alter, Huber, and Simon) which is not part of any clusters and other clusters (6, 7, and 8).

Figures 8, 9, and 10 list distances from each cluster already formed to all other observations or clusters. Each figure contains bold-faced distance indicating maximum distance. The next cluster to be formed will be selected by choosing the smallest distance among those bold-faced distances in the three figures. Figure 9 contains the smallest figure, the distance between Scott Morton (Cluster 7 member) and Alter **(.58664 ****).** Consequently, the cluster history shows that the next cluster is formed by joining Cluster 7 and Alter. The rest of the cluster will be formed by following the same procedures repeatedly.

AVERAGE | AVE

The average linkage method uses the average distance instead of the minimum or maximum value. The average distance of a cluster that has multiple members is computed by group average, unweighted pair-group method using arithmetic aver-

Figure 8. Distance from CL6

From	To	Distance
Carlson	Alter	.74515
Sprague	Alter	**.76448**
Carlson	Huber	**.98788**
Sprague	Alter	.89764
Carlson	Simon	**.89841**
Sprague	Simon	.89201
Carlson	Keen	.61729
Carlson	Scott Morton	**.66094**
Sprague	Keen	.58349
Sprague	Scott Morton	.65439
Carlson	Bonczek	**1.01124**
Carlson	Whinston	1.00781
Sprague	Bonczek	1.00294
Sprague	Whinston	.99648

Figure 9. Distance from CL7

From	To	Distance
Keen	Alter	.5631
Scott Morton	Alter	**.58664** ****
Keen	Carlson	.61729
Keen	Sprague	.58349
Scott Morton	Carlson	**.66094**
Scott Morton	Sprague	.65439
Keen	Huber	.89764
Scott Morton	Huber	**.92865**
Keen	Simon	**.66143**
Scott Morton	Simon	.62572
Keen	Bonczek	1.14216
Keen	Whinston	1.12845
Scott Morton	Bonczek	**1.14688**
Scott Morton	Whinston	1.13344

Figure 10. Distance from CL8

From	To	Distance
Bonczek	Simon	1.17918
Bonczek	Huber	**1.23667**
Whinston	Simon	1.16098
Whinston	Huber	1.23657
Bonczek	Carlson	**1.01124**
Bonczek	Sprague	1.00294
Whinston	Carlson	1.00781
Whinston	Sprague	.99648
Bonczek	Keen	1.14216
Bonczek	Scott Morton	**1.14688**
Whinston	Keen	1.12845
Whinston	Scott Morton	1.13344
Bonczek	Alter	1.14216
Whinston	Alter	**1.14668**

ages, UPGMA). Distance data are squared unless you specify the NOSQUARE option.

Clusters 8, 7, and 6 are identical regardless of the clustering method selected because single observation has the same minimum, maximum, and average. Figures 7.9, 7.10, and 7.11 shows average distances and the shortest distance is between CL7 and Alter.

CENTROID | CEN

In the centroid method, the distance between two clusters is computed using the Euclidian distance between the centroid (means) of the two clusters. Distance data are squared unless you specify the NOSQUARE option.

WARD | WAR

In Ward's minimum-variance method, the distance between two clusters is the *ANOVA* sum of squares between the two clusters added up over all the variables.

This method minimizes the sum of squares of any pair of clusters to be formed at each step. This is done by computing the sum of squared Euclidean distances from each variable in a cluster to the mean of all variables. Distance data are squared unless you specify the NOSQUARE option. To reduce distortion by outliers, the TRIM= option is recommended. See the NONORM option.

The sums of squares are easier to interpret when they are divided by the total sum of squares to give proportions of variance (squared semipartial correlations).(SAS

Figure 11. Cluster history with average linkage method

Cluster History

NCL	-------Clusters Joined--------		FREQ	PSF	PST2	Norm RMS Dist	Tie
8	Bonczek	Whinston	2	51.3	.	0.149	
7	Keen	ScottMorton	2	48.0	.	0.1816	
6	Carlson	Sprague	2	23.2	.	0.3829	
5	Alter	CL7	3	10.2	15.6	0.628	
4	CL5	CL6	5	6.1	4.3	0.7363	
3	CL4	Simon	6	5.8	2.4	0.8826	
2	CL3	Huber	7	6.4	2.7	1.0378	
1	CL2	CL8	9	.	6.4	1.2363	

Figure 12. Average distance from CL6

From	To	Distance	Average distance
Carlson	Alter	.74515	
Sprague	Alter	**.76448**	0.754815
Carlson	Huber	**.98788**	
Sprague	Alter	.89764	0.94276
Carlson	Simon	**.89841**	0.89521
Sprague	Simon	.89201	
Carlson	Keen	.61729	
Carlson	Scott Morton	**.66094**	
Sprague	Keen	.58349	
Sprague	Scott Morton	.65439	0.629028
Carlson	Bonczek	**1.01124**	
Carlson	Whinston	1.00781	
Sprague	Bonczek	1.00294	
Sprague	Whinston	.99648	1.004618

Figure 13. Average distance from CL7

From	To	Distance	Average Distance
Keen	Alter	.5631	
Scott Morton	Alter	**.58664 ******	**.57487**
Keen	Carlson	.61729	
Keen	Sprague	.58349	
Scott Morton	Carlson	**.66094**	
Scott Morton	Sprague	.65439	.629028
Keen	Huber	.89764	
Scott Morton	Huber	**.92865**	.913145
Keen	Simon	**.66143**	
Scott Morton	Simon	.62572	.643575
Keen	Bonczek	1.14216	
Keen	Whinston	1.12845	
Scott Morton	Bonczek	**1.14688**	
Scott Morton	Whinston	1.13344	1.137683

Figure 14. Average distance from CL8

From	To	Distance	Average Distance
Bonczek	Simon	1.17918	
Bonczek	Huber	**1.23667**	
Whinston	Simon	1.16098	
Whinston	Huber	1.23657	1.20335
Bonczek	Carlson	**1.01124**	
Bonczek	Sprague	1.00294	
Whinston	Carlson	1.00781	
Whinston	Sprague	.99648	1.004618
Bonczek	Keen	1.14216	
Bonczek	Scott Morton	**1.14688**	
Whinston	Keen	1.12845	
Whinston	Scott Morton	1.13344	1.137683
Bonczek	Alter	1.14216	
Whinston	Alter	**1.14668**	1.14442

Institute Inc., 1988). All these procedures maximize the differences between clusters relative to the variations within clusters. The ratio of the between-cluster variation to the average within-cluster variation is then comparable to the F-ratio in analysis of variance (Hair, Anderson, Tatham, & Black, 1992).

As said earlier, since clusters 6, 7, and 8 combine only two single observations, regardless of the clustering methods, all methods produce identical clusters 8, 7, and 6. The next cluster is to be formed to minimize within cluster-error sum of squares. To do so, we need to compute within the cluster-error sum of squares. To evaluate a given clustering, we compute the mean of the data in each cluster, then add up the sum of squared differences between each point and its cluster mean.

Control Display of the Cluster History

PSEUDO is an option to print pseudo F and t^2 statistics. This option is effective only when the data are coordinates or METHOD=AVERAGE, CENTEROID, OR WARD.

Print the Name of All Authors in the Outputs

The ID option allows printing the text values of the variables. The variables in ACA are all authors. The ID statement is used to print the name of all authors in the outputs. This option is an important tool that can facilitate the interpretation of outputs easier and faster. With the ID statement, the cluster history and tree graph will use the names of authors instead of OB*n*, where *n* is the observation number. The ID statement specifies variables used to identify observations. The ID variables must be variables in the DATA=

ID *variable ;*

Figure 15. Cluster history using the Ward method

Cluster History

NCL	-------Clusters Joined--------		FREQ	SPRSQ	RSQ	PSF	PST2	Tie
8	Bonczek	Whinston	2	0.0028	.997	51.3	.	
7	Keen	ScottMorton	2	0.0041	.993	48.0	.	
6	Carlson	Sprague	2	0.0183	.975	23.2	.	
5	Alter	CL7	3	0.0644	.910	10.2	15.6	
4	CL5	Simon	4	0.1068	.804	6.8	3.1	
3	CL4	CL6	6	0.1446	.659	5.8	3.0	
2	CL3	Huber	7	0.1825	.477	6.4	2.7	
1	CL2	CL8	9	0.4766	.000	.	6.4	

List Additional Options

Additional options are PRINT, NOMISS, REPLACE, and REPONLY.

Options= PRINT

Specify the Variables in the Input Data Sets Using VAR Statement

VAR *variables ;*

The VAR statement lists the numeric variables to be used in the cluster analysis. If you omit the VAR statement, all numeric variables not listed in other statements are used. The VAR statement is used to specify the variables in the input data set that are to be analyzed by the procedure.

Specify the Output Data Set to be Produced

The output data set produced by the OUT= option of the distance procedure. The distance matrix (dist) contains a triangle shape of distance matrix converted from correlations coefficients.

Out = Output data-set name

These proximity measures are stored as a lower triangular matrix in an output data set that can then be used as input to the CLUSTER, multidimensional scaling (MDS), or MODECLUS procedures. The input data set may contain numeric variables, character variables, or both, depending on which proximity measure is used. ACA uses the measurements of inter-objective similarities using correlation transformed to Euclidean distance using METHOD=DCORR option as an input to the PROC CLUSTER procedure.

INTERPRETING RESULTS OF CLUSTER ANALYSIS

Outputs of cluster analysis are cluster history and dendrogram/tree graph.

Cluster History

The next output shows cluster history (Table 9).

Table 9. Cluster history

NCL	------Clusters	Joined-------	FREQ	SPRSQ	RSQ	PSF	PST2	Tie
8	BONCZEK	WHINSTON	2	0.0028	.997	51.3	.	
7	KEEN	SCOTTMORTON	2	0.0041	.993	48.0	.	
6	CARLSON	SPRAGUE	2	0.0183	.975	23.2	.	
5	ALTER	CL7	3	0.0644	.910	10.2	15.6	
4	CL5	SIMON	4	0.1068	.804	6.8	3.1	
3	CL4	CL6	6	0.1446	.659	5.8	3.0	
2	CL3	HUBER	7	0.1825	.477	6.4	2.7	
1	CL2	CL8	9	0.4766	.000	.	6.4	

The PROC CLUSTER output displays the following. Column 1 (NCL — the Number of Clusters): With N number of variable datasets, the maximum number of clusters is N-1. Table 9 shows the maximum number of cluster 8. Columns 2 and 3 (the names of the Clusters Joined): The observations are identified by the formatted value of the ID variable, if any; otherwise, the observations are identified by OB*n*, where *n* is the observation number. Clusters of two or more observations are identified as CL*n*, where *n* is the number of clusters existing after the cluster in question is formed. Column 4 (FREQ) represents the number of observations in the new cluster). The first row's FRQ is 2. Cluster 8 is formed by joining 2 observations (Bonczek and Whinston). The last row, cluster 1, has all 9 authors in it. Column 5 shows SPRSQ, which stands for Semipartial R-Squared. SPRSQ indicates the decrease in the proportion of variance accounted for resulting from joining the two clusters. This equals the between-cluster sum of squares divided by the corrected total sum of squares.

Column 6 (RSQ is the squared multiple correlation, R-Squared). R^2 is the proportion of variance accounted for by the clusters. For example, all 8 clusters account for 99.7% of variance in the dataset. Column 7 lists PSF (Pseudo F). The pseudo F statistic measures the separation among all the clusters at the current level. Column 8 (PST2 is Pseudo t^2). The pseudo t^2 statistic measures the separation between the two clusters most recently joined.

Selecting the Number of Clusters

The proc cluster procedure produces a cluster history beginning with the number of cluster 1 to N-1 with the N number of variable dataset. Table 5.7 exhibits the number of clusters ranging 8 through 1, since the data set has 9 authors. One of the important decisions for ACA analysts is the selection of the optimal number of clusters. There are no structured procedures in determining the number of clusters. The optimal number of clusters can be chosen by using one or more of the following: (1) the cubic clustering criterion (CCC), (2) the pseudo F statistic, (3) the pseudo t^2 statistic. The cubic clustering criterion (CCC) can only be applied if the data are coordinates. Since the ACA uses distance data,) the pseudo F statistic and the pseudo t^2 statistic are the most frequently used.

Plot of the Pseudo F Statistic Against Number of Clusters

Plots of the pseudo F statistic against the number of clusters are useful to select the number of clusters. Figure 15 shows the plot. The peaks of pseudo F values in the plot can be used to decide the number of clusters. Depending on the number of variables, ACA analysts may ignore the pseudo F values of the last 2 or 3 clusters (e.g., clusters 7 and 8). Figure 15 has peaks at 2 and 8 clusters. Ignoring the last cluster, this criterion suggests that only two clusters to be used. In ACA, the important criteria that cannot be ignored is the interpretability of clusters. The first cluster include all authors in the area of foundations. They all belong to the founding fathers of decision support systems area who provided basic concepts and frameworks. The second cluster seems to include the authors in model management area. Due to small number of variables used to produce the plot, this example may not be good for beginners in ACA. Figure 16 is based on the data set used in SAS/STAT® User's Guide (Release 6.03 Edition). In this example, there are three peaks at clusters 3, 8, and 12, ignoring the last two clusters (14 and 15).

Figure 15. Plot of pseudo f statistic against number of clusters

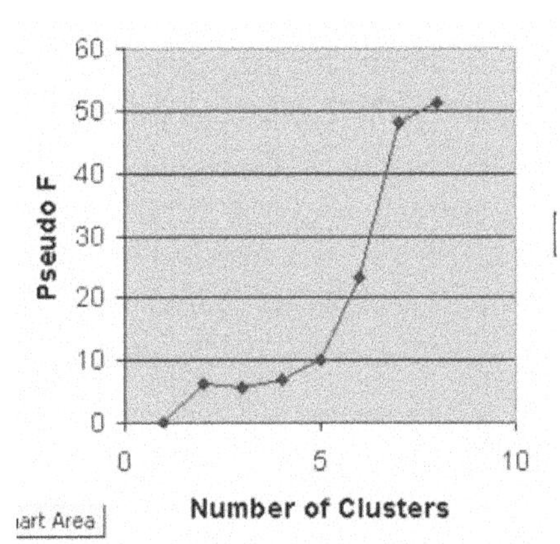

Figure 16. Plot of pseudo f statistic against number of clusters using birth and death rates in 74 countries

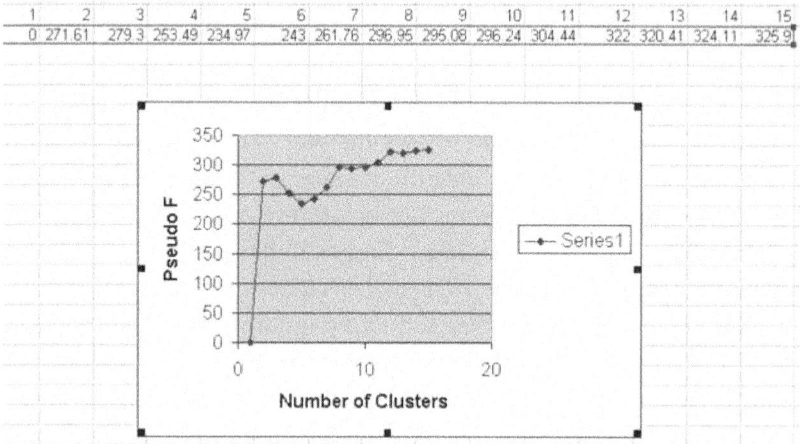

Plot of the Pseudo t^2 Statistic Against Number of Clusters

Plots of the pseudo t^2 statistic against the number of clusters are also useful to select the number of clusters (Figures 17 and 18). The valleys of pseudo F values in the plot can be used to decide the number of clusters. Depending on the number of variables, ACA analysts may ignore the values of the last 2 or 3 clusters. Figure 17 has a valleys at 2 cluster. In Figure 18, the pseudo t^2 statistic indicates 3, 8, and 12 clusters.

Considering the pseudo t^2 statistic and the pseudo F statistic simultaneously narrows down the number of clusters for final selection. In the case of the 9 variable ACA example we have used, the 2 cluster solutions are suggested for further evaluation. The cluster analysis of the birth and death rates data in 74 countries suggests 3, 8, and 12 cluster solution for further evaluation. The interpretability of the cluster solutions is the decisive criteria in selecting the number of cluster. ACA analysts should be able to name each cluster based on the contents of each author under each cluster.

Proc Tree Graph Output

The final output is the tree graph (dendrogram). The dendrogram illustrates hierarchical clustering of eight authors of decision support systems researchers. Figure 19 shows both the cluster structure and the joining sequence to show how each of the authors in the study is combined into a new aggregate cluster until all 9 authors are grouped into the final one cluster, cluster one (CL1).

Figure 5, which displays the distance among each author, shows that the shortest distance between two authors is .13646 between Bonczek and Whinston, followed by .16624 between Keen and Scott Morton, and .35056 between Sprague and Carlson. Based on the proximity of these author sets, clusters 8, 7, and 6 are formed successively. In the next subsequent step, cluster 3 is combined with cluster 5 to form a new cluster (CL2). The dendrogram shows a linkage among various

Figure 17. Plot of pseudo t2 statistic against number of clusters

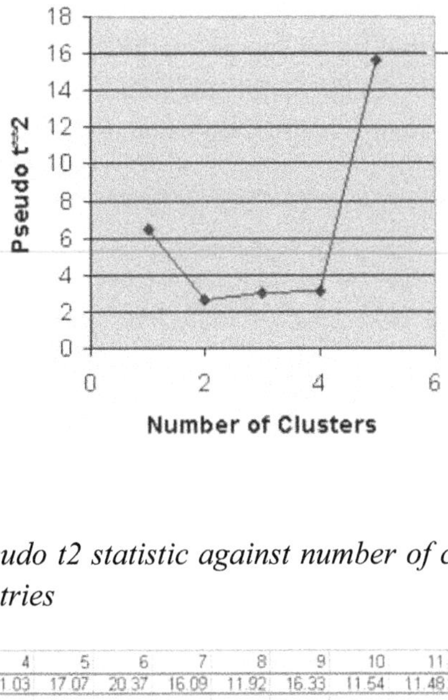

Figure 18. Plot of pseudo t2 statistic against number of clusters using birth and death rates in 74 countries

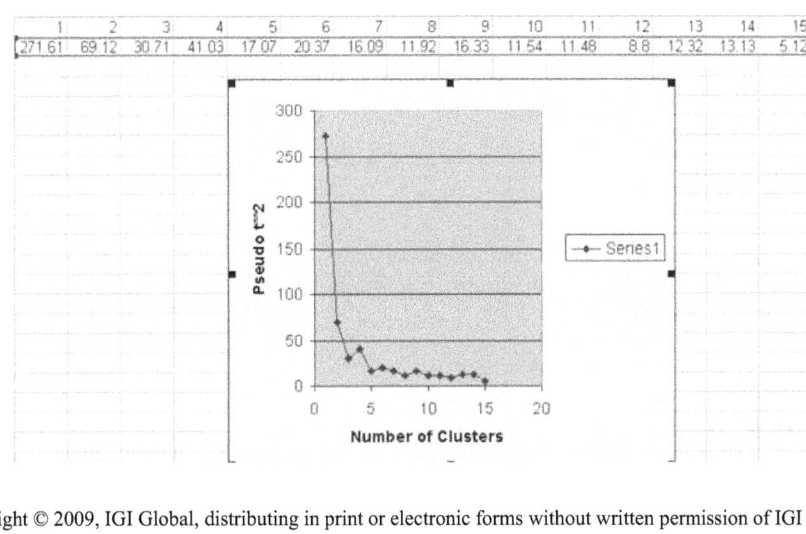

clusters, and the influence could be inferred based on the close examination of the works of authors in the clusters. In the final agglomerative clustering procedure, Cluster 8 (Bonczek and Whinston) and cluster 2 (cluster of all other authors except two authors in Cluster 8) are the last clusters to be joined to form the final cluster (CL1), which indicates the heterogeneity of the authors of cluster 2.

Comparison of the two solutions from the factor analysis and cluster analysis provides some valuable information on the similarities and differences of the two solutions. This helps us reach a better interpretation of the results of multivariate analysis. In this example of small variables, factor solution and cluster solution coincide very well. Cluster 3 (6 authors) corresponds to factor 1. Factor 2 coincides with cluster 8. Factor 3 is a factor with only one author (Huber). In ACA, a single factor solution is not generally acceptable. Statistically, it may be an acceptable solution, but in the study of author cocitation, we are interested in the linkage among authors. Therefore, if the single author factor appears in the final solution in any ACA study, ACA researchers do not accept the factor solution. In that case, the tree graph from cluster analysis can provide a useful guide when assigning a single author to any

Figure 19. Dendrogram (tree graph) depicting cluster structure and joining sequences

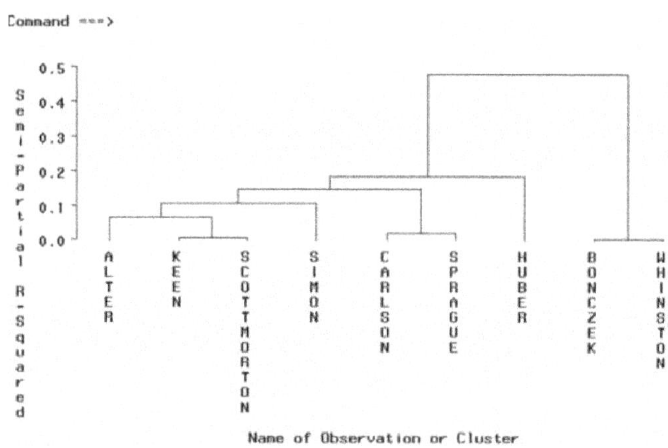

CL 8 = (Bonczek, Whinston)
CL 7 = (Keen, Scott Morton)
CL 6 = (Carlson, Sprague)
CL 5 = (Alter, Keen, Scott Morton)
CL 4 = (Alter, Keen, Scott Morton, Simon)
CL 3 = (Alter, Keen, Scott Morton, Simon, Carlson, Sprague)
CL 2 = (Alter, Keen, Scott Morton, Simon, Carlson, Sprague, Huber)
CL 1 = (Bonczek, Whinston, Alter, Keen, Scott Morton, Simon, Carlson, Sprague, Huber)

other factors. See several previous studies of Eom (1997; , 1998) to see how factor analysis and cluster analysis can be complementary to each other in ACA.

SUMMARY AND CONCLUSION

This chapter describes the distance and cluster procedure of the SAS system. SAS version 9 introduced the proc distance procedure. All previous versions of SAS used two programs (xmacro.sas and distnew.sas) to process a transposed cocitation matrix (input) to produce a distance matrix (output). Specific examples of SAS procedures with dataset, either imported from Excel file or directly entered dataset, are used. The critical inputs to the Proc Distance statement are method = dcorr, var = ratio, and shape=square.

Cluster analysis is a data reduction technique for grouping various entities (individuals, variables, objects) into clusters. By doing so, that the entities in the same cluster have more similarity each other with respect to some predetermined selection criteria. The second part of this chapter focuses on the PROC CLUSTER statement which sets out the CLUSTER procedure steps. ACA research uses the hierarchical agglomerative methods to group authors based on research subspecialties. SAS as well as SPSS alike provide many hierarchical agglomerative procedures such as single linkage, complete linkage, average linkage, Ward method, etc. of these, complete linkage and Ward method are the two methods used by the majority of ACA researchers(McCain, 1990).

REFERENCES

Eom, S. B. (1997). Assessing the Current State of Intellectual Relationships between the Decision Support Systems Area and Academic Disciplines. In K. Kumar & J. I. DeGross (Eds.), *Proceedings of the Eighteenth International Conference on Information Systems* (pp. 167-182). Atlanta, GA: International Conference on Information Systems.

Eom, S. B. (1998). Relationships between the Decision Support System Subspecialties and Reference Disciplines: An Empirical Investigation. *European Journal of Operational Research, 104*(1), 31-45.

Everitt, B. S. (1980). *Cluster Analysis.* London: Heinemann Educational Books Ltd.

Hair Jr., J. F., Anderson, R. E., Tatham, R., & Black, W. C. (1992). *Multivariate Data Analysis with Readings* (3rd ed. ed.). New York: Macmillan Publishing.

McCain, K. W. (1990). Mapping Authors in Intellectual Space: A Technical Overview. *Journal of the American Society for Information Science, 41*(6), 433-443.

SAS Institute Inc. (1988). *Sas/Stat User's Guide, Release 6.03* (3rd ed. ed.). Cary, NC: SAS Institute Inc.

SAS Institute Inc. (2007). *Sas Onlinedoc® 9.1.3.* Cary, NC: SAS Institute Inc.

Stevens, S. S. (1946). On the Theory of Scales of Measurement. *Science, 103*, 677-680.

Chapter X
Mutidimensional Scaling

ABSTRACT

This chapter discusses multidimensional scaling (MDS) procedures. MDS is a class of multivariate statistical techniques/procedures to produce two or three dimensional pictures of data (geometric configuration of points) using proximities among any kind of objects as input. Three SAS procedures (MDS, PLOT, and G3D) are necessary to convert the author cocitation frequency matrix to two or three dimensional pictures of data. The distance matrix produced earlier by using xmacro.sas and distnew.sas programs should be converted to a coordinate matrix, to produce two-dimensional plots, and annotated three-dimensional scatter diagrams. This chapter also discusses how to label data points on a plot. The annotate facility in the SAS system produces figures with the name of the author on each data point. The PROC MDS procedure includes many of the features of the ALSCAL procedure.

INTRODUCTION

This chapter discusses multidimensional scaling (MDS) procedures. MDS is a class of multivariate statistical techniques/procedures to produce two or three dimensional pictures of data (geometric configuration of points) using proximities among any kind of objects as input. The purposes of MDS are to help researchers identify the "hidden structures" in the data and visualize relationships among/within the hidden structures to give clearer explanations of these relationships to others (Hair, Anderson, & Tatham, 1987; Joseph B. Kruskal & Wish, 1990). Three SAS procedures (MDS, PLOT, and G3D) are necessary to convert the author cocitation frequency matrix to two or three dimensional pictures of data.

The distance matrix produced earlier by using xmacro.sas and distnew.sas programs (SAS version 8) or the distance procedures (version 9) should be converted to a coordinate matrix. The coordinate matrix is used to produce two-dimensional plots and annotated three-dimensional scatter diagrams. A distance matrix is the input to the multidimensional scaling procedure, PROC MDS, of the SAS system (version 9.1.3). The PLOT and G3D procedures process the coordinate matrix to visualize the similarity and dissimilarity within each group of an academic discipline as well as the similarity and dissimilarity among the various subspecialties within an academic discipline. In ACA study, 3D scatter plots without labels on data points provide little information for the ACA researchers. This chapter also explains how to label data points on a plot. The annotate facility in the SAS system produces figures with the name of the author on each data point. The PROC MDS procedure includes many of the features of the ALSCAL procedure (F. W. Young, Lewyckyj, & Takane, 1986) and some features of the MLSCALE procedure (Ramsay, 1986) (SAS Institute Inc., 1992).

THE MDS PROCEDURE

Multidimensional scaling is a multivariate statistical analysis tool for examining proximity data among any kind of object. Proximity data consist of one or more square symmetric or asymmetric matrices of similarities or dissimilarities between *objects* or *stimuli* (Joseph B. Kruskal & Wish, 1978, pp. 7-11). The MDS outputs consist of a spatial representation of data which shows underlying relationships on a two or three dimensional map. The MDS map helps visualize relationships more clearly using the ratio of distances on a map to corresponding data values such as a map of a country showing cities. The magnitude of the number indicates how similar/dissimilar two objects are.

Similarity/Proximity Measures

How should the inter-object similarity be measured? Numerous ways of measuring inter-object similarity exist. The non-metric data measures the distance by directly ranking the objects from most preferred to least preferred (preference data) and using the pairwise comparison (similarity data) to determine which items are most similar/dissimilar to each other (all pairs of these objects can be compared).

To measure proximities among authors, the correlations among authors are used most frequently. Correlations are used as proximities by MDS procedures (Joseph B. Kruskal & Wish, 1978). The author cocitation frequency is metric data. As in the PROC Cluster procedure, the cocitation frequency matrix must be converted into a ordinary Euclidian distance data matrix using METHOD= DCORR, This method transformed correlations to Euclidean distance using square root of (1-CORR). Figure 2 shows MDS SAS procedure statements.

Proc MDS Statement

The MDS procedure produces only the iteration history. The procedure does not produce any graphical outputs. It is necessary for ACA analysts to specify several options. The minimum essential options include data= , dimension= , and out= .

The PROC MDS statement is required. All other statements are optional.

PROC MDS < options > ;
VAR variables ;
INVAR variables ;
ID | **OBJECT** variable ;
MATRIX | **SUBJECT** variable ;
WEIGHT variables ;
BY variables ;

Since the MDS procedure produces only the printed iteration history, specifying options are necessary to produce other results listed below.

Figure 1. MDS/plot/g3d procedures in ACA

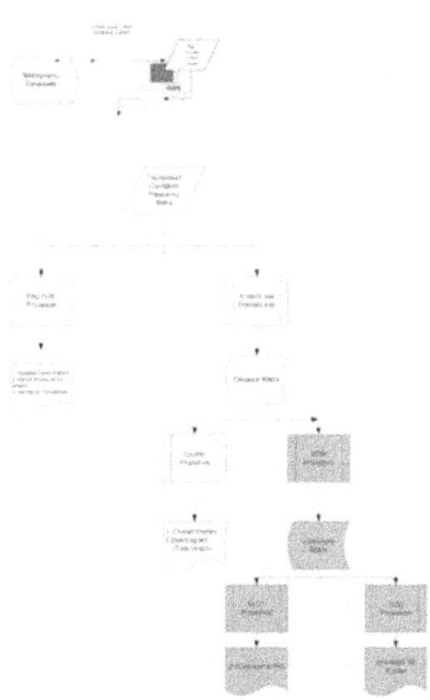

Figure 2. Proc MDS with an imported data set from an Excel file

```
/* ACA data in Excel file is imported to the SAS system */
PROC IMPORT OUT= WORK.ACA
        DATAFILE= "C:\Documents and Settings\sbeom\My Documents\idea
        aca\spss\bookdata.xls"
        DBMS=EXCEL REPLACE;
   SHEET="sasdata$";
   GETNAMES=YES;
   MIXED=NO;
   SCANTEXT=YES;
   USEDATE=YES;
   SCANTIME=YES;
RUN;

/* the dataset is printed to verify the accuracy of the data */
proc print data=aca;
run;

/* Principal component analysis of the data*/
proc factor data=aca method=principal mineigen=1 rotate = promax;
run;

/* Creating the distance matrix using ratio variable and the method = dcorr
option */
Proc distance data=aca out=dist method=dcorr SHAPE=SQUARE;
Var ratio(Alter--Whinston);
ID author;
run;
proc print data=dist;

/*Cluster analysis with the Ward option*/
proc cluster method=ward data=dist pseudo;
ID author;
proc tree space=2;
run;

/*Perform non-metric MDS to produce the configuration*/
PROC MDS
        data=dist
        dimension=3
        condition=un
        level=ordinal
        pconfig
        out=coordordinal;
        ID author;
title 'ordinal option';
RUN;
PROC PRINT data=coordordinal (keep=author dim1 dim2 dim3);
VAR author dim1 dim2 dim3;
RUN;
options ps=40;
/* end */
```

Necessary Options that Must be Specified other than Default Values

Specify the Input Data Set

DATA=*SAS-data-set*

Specify the Out Data Set

OUT=*SAS-data-set*

This option creates a SAS data set containing the estimates of all the parameters of the MDS model and the values of the badness-of-fit criterion.

OUTRES=SAS-data-set

This option creates a SAS data set with one observation for each non-missing datum from the DATA=data set. Each observation contains the original datum, the estimated distance computed from the MDS model, the transformed data and distances, and the residual.

Specify the Measurement Conditionality of the Data

The measurement conditionality is a type of restraints placed on the measurement transformation. There are several types of the conditionality of data depending on the nature of data we want to process. The most important attribute of the data is whether each individual data is measured by an identical scale or not. Young used the following example to show conditional measurement (Forrest W. Young, 1987, p.63).

Let's assume that an experimenter is interested in the perception of smells, but knows that subjects satiate after a very few trials, and must therefore run experiment over a series of days. However, each day equipment used to generate the stimulus smells must be recalibrated. If the experimenter is not confident that the equipment is calibrated precisely the same each day, she should view her data as being portioned into subsets, there being partition for each day. Since the measurement characteristics are conditional on the day of the experiment we also call these data general conditional.

When all observations are comparable, it is called unconditional. Each cocitation frequency count in the cocitation frequency matrix is comparable and measured by same unit of measurement. Therefore, ACA data are unconditional data. The cocitation frequency matrix cannot be subdivided further. There is only one partition of data. We specify the conditionality of the data. ACA analysis puts the entire data set into a single partition. Since ACA analysis use only one data matrix, two options (CONDITION=UN and CONDITION=MATRIX) will produce the same result. Therefore, no CONDITION option is necessary.

CONDITION=UN

With the specification of the CONDITION=UN or CONDITION=MATRIX option and without the SHAPE= option, only one triangle is stored. Readers are referred to (Forrest W. Young, 1987) for a detailed discussion on the measurement of conditionality including matrix conditional, row-conditional, column conditional, and general conditional data (pp. 60-63).

Specify the Number of Dimensions

DIMENSION = n

This option specifies the number of dimensions to use in the MDS model. The default is DIMENSION=2. To plot a 3D graph, it is necessary to specify the number of dimensions.

ID Statement

The ID statement specifies a variable in the DATA= data set that contains descriptive labels (author's name) for each author. The name of authors used in the output are copied to the OUT= data set, instead of non-descriptive variable names such as var1, dist1, obs1, etc.

Options that May Use Default Values

Specify the Measurement Level of the Data and the Type of Transformations

LEVEL= ordinal

LEVEL=ORDINAL specifies a nonmetric analysis, while all other LEVEL= options specify metric analyses. The default is LEVEL=ORDINAL.

Perhaps, the specification of the measurement level of the data could be a difficult task for novice ACA researchers. The SAS system allows 6 different alternative values (absolute, ratio, interval, ordinal, loginterval, ordinal). This is certainly a confusing topic in the study of MDS and ACA. Young wrote the following (Forrest W. Young, 1987, p.68):

Please do not confuse the usage of "metric" here [in discussing metric spaces in his book] with the terms "metric" and nonmetric" when they are used to define the nature of a multi-dimensional scaling algorithm. .... However, "metric" and "non-metric" as commonly used in Psychometrics and multidimensional scaling literature have an entirely different meaning: A metric algorithm assumes they are less than the interval level. All (or at least nearly all) multidimensional scaling models are metric in the mathematical sense, whereas some multidimensional scaling algorithms are , in the psychometric sense, metric and some are nonmetric.

Other ways of distinguishing metric and non-metric MDS is based on the input data values. Metric MDS uses the actual values of the similarity/dissimilarities. Non-metric MDS uses the proximities simultaneously with the configuration which are estimated by an optimal monotone transformation of the metric distance between two variables(J. B. Kruskal, 1964; Shepard, 1962). McCain also suggested use a non-metric scaling and observed that most ACA researchers used a non-metric scaling approach by an ordinal level of measurement (level=ordinal)(McCain, 1990).

Specify the Type of the Dimension Coefficients

COEF= **Identity**

This option specifies the method of computing the dimension coefficients of each variable (author). The coefficients can be Euclidean distances (Identity) or weighted Euclidean distances, in which each subject is allowed differential weights for the dimensions. ACA analysis converts cocitation frequencies into un-weighted Euclidean distances to produce two or three dimensional maps.

The default value is COEF=Identity. Since we do not want to produce weighted Euclidian distance, we will not specify this option.

Specify a Predetermined Transformation

FIT = {**Distance**, Squared, Log,}

This option specifies a predetermined (not estimated) transformation to apply to both sides of the MDS model before the error term is added. The options are

(1)FIT=LOG or FIT=0.
(2)FIT=DISTANCE or FIT=1.
(3)FIT=SQUARED or FIT=2

The default is FIT=DISTANCE or, equivalently, FIT=1, which fits data to distances. ACA analysis uses the default value.

Specify a Stress Formula

FORMULA | FOR=*0* | OLS | O
FORMULA | FOR=*1* | USS | U
FORMULA | FOR=*2* | CSS | C

determines how the badness-of-fit criterion is standardized in correspondence with stress formulas 1 and 2 (Kruskal and Wish 1978, pp. 24-26). The default is FORMULA=1

"Stress" is a measure of fit. The stress is the square root of a normalized "residual sum of squares". The larger values mean a worse fit. For that reason, some call stress a badness-of-fit measure (J. B. Kruskal & Carroll, 1969). With the FIT=DISTANCE and LEVEL=ORDINAL options, this is equivalent to Kruskal's stress formula 1, which is sometimes called Stress 1. The dimensions in the solution are closely related to the level of the stress. McCain suggests that ACA may use a little higher stress value of less than .2 as an acceptable criterion for 2 or 3 dimensional solutions with the higher R-square value, due to the "noisy" nature of cocitation frequency data. In some other areas, the rule of thumb is that a stress value under 0.1 is considered to be excellent and a value over 0.15 may be unacceptable.

Miscellaneous

CUTOFF = n

This option specifies that data less than n is to be replaced by missing values. The default value is CUTOFF =0. Since ACA data has only numbers greater than 0, we will not specify the cutoff option value.

Control displayed output

According to the SAS/STAT user's guide of Online SAS Manual Version 9, displayed output is the results of many interacting options.

Displayed output is controlled by the interaction of the PCONFIG, PCOEF, PTRANS, PFIT, and PFITROW options with the PININ, PINIT, PITER, and PFINAL options. The PCONFIG, PCOEF, PTRANS, PFIT, and PFITROW options specify which estimates and fit statistics are to be displayed. The PININ, PINIT, PITER, and PFINAL options specify when the estimates and fit statistics are to be displayed. If you specify at least one of the PCONFIG, PCOEF, PTRANS, PFIT and PFITROW options but none of the PININ, PINIT, PITER, and PFINAL options, the final results (PFINAL) are displayed. If you specify at least one of the PININ, PINIT, PITER, and PFINAL options but none of the PCONFIG, PCOEF, PTRANS, PFIT and PFITROW options, all estimates (PCONFIG, PCOEF, PTRANS) and the fit statistics for each matrix and for the entire sample (PFIT) are displayed. If you do not specify any of these nine options, no estimates or fit statistics are displayed (except the badness-of-fit criterion in the iteration history).

PCOEF prints the estimated dimension coefficients. The dimension coefficient is the square roots of the *subject weights* (Joseph B. Kruskal & Wish, 1978). The dimension coefficient for each data matrix is "the coefficients that multiply each coordinate of the *common* or *group* weighted Euclidean space to yield the *individual* unweighted Euclidean space" (SAS Institute Inc., 2000).

PCONFIG prints the estimated coordinates of the objects in the configuration.

PFINAL prints final estimates.

MDS PROCEDURE OUTPUT

The major outputs from the PROC MDS procedure are iteration history, convergence status, and configuration under "solution in 3 dimensions" in Table 1.

Iteration History and Convergence Status

The current solution is obtained after 14 iterations. PROC MDS displays the iteration history containing

- Iteration number (0 through 14)

Table 1. MDS procedure outputs with ordinal option

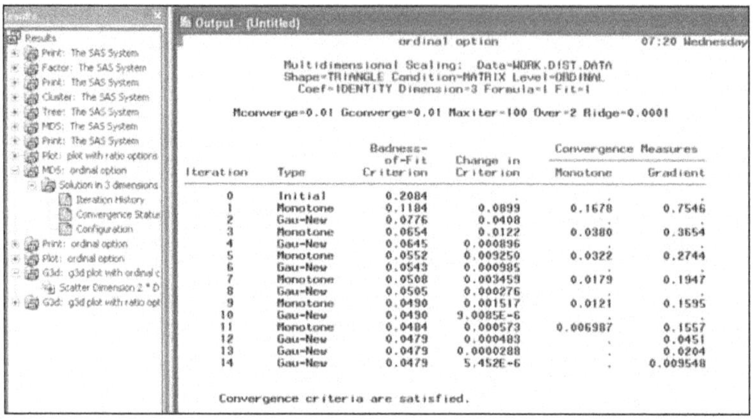

ordinal option 07:20 Wednesday

Multidimensional Scaling: Data=WORK.DIST.DATA
Shape=TRIANGLE Condition=MATRIX Level=ORDINAL
Coef=IDENTITY Dimension=3 Formula=1 Fit=1

Mconverge=0.01 Gconverge=0.01 Maxiter=100 Over=2 Ridge=0.0001

Iteration	Type	Badness-of-Fit Criterion	Change in Criterion	Convergence Measures Monotone	Convergence Measures Gradient
0	Initial	0.2084	.	.	.
1	Monotone	0.1184	0.0899	0.1678	0.7546
2	Gau-New	0.0776	0.0408	.	.
3	Monotone	0.0654	0.0122	0.0380	0.3654
4	Gau-New	0.0645	0.000896	.	.
5	Monotone	0.0552	0.009250	0.0322	0.2744
6	Gau-New	0.0543	0.000985	.	.
7	Monotone	0.0508	0.003459	0.0179	0.1947
8	Gau-New	0.0505	0.000276	.	.
9	Monotone	0.0490	0.001517	0.0121	0.1595
10	Gau-New	0.0490	9.0085E-6	.	.
11	Monotone	0.0484	0.000573	0.006987	0.1557
12	Gau-New	0.0479	0.000483	.	0.0451
13	Gau-New	0.0479	0.0000288	.	0.0204
14	Gau-New	0.0479	5.452E-6	.	0.009548

Convergence criteria are satisfied.

Table 2. MDS procedure outputs with ratio option

Iteration	Type	Badness-of-Fit Criterion	Change in Criterion	Convergence Measure
0	Initial	0.2072	.	0.6970
1	Lev-Mar	0.1381	0.0691	0.2185
2	Gau-New	0.1340	0.004051	0.0961
3	Gau-New	0.1332	0.000858	0.0469
4	Gau-New	0.1329	0.000241	0.0320
5	Gau-New	0.1328	0.000119	0.0249
6	Gau-New	0.1328	0.0000756	0.0216
7	Gau-New	0.1327	0.0000600	0.0209
8	Gau-New	0.1326	0.0000594	0.0224
9	Gau-New	0.1326	0.0000705	0.0254
10	Gau-New	0.1325	0.0000920	0.0292
11	Gau-New	0.1324	0.000121	0.0330
12	Gau-New	0.1322	0.000151	0.0358
13	Gau-New	0.1320	0.000174	0.0370
14	Gau-New	0.1318	0.000182	0.0367
15	Gau-New	0.1317	0.000175	0.0351
16	Gau-New	0.1315	0.000157	0.0327
17	Gau-New	0.1314	0.000135	0.0299
18	Gau-New	0.1313	0.000112	0.0271
19	Gau-New	0.1312	0.0000920	0.0246
20	Gau-New	0.1311	0.0000754	0.0223
21	Gau-New	0.1310	0.0000626	0.0205
22	Gau-New	0.1310	0.0000532	0.0191
23	Gau-New	0.1309	0.0000463	0.0180
24	Gau-New	0.1309	0.0000413	0.0171
25	Gau-New	0.1309	0.0000375	0.0164
26	Gau-New	0.1308	0.0000345	0.0158
27	Gau-New	0.1308	0.0000318	0.0151

- Type of iteration: Initial (initial configuration), Monotone (monotone transformation), and Gau-New (Gauss-Newton step).

  According to Young(Forrest W. Young, 1987), badness-of-fit in non-metric MDS is the discrepancy in distance from the best possible monotone function for a given configuration. Then, a conventional method of gradients (steepest descent) is applied to alter the configuration to reduce badness-of-fit. See page 169 of (Forrest W. Young, 1987).

- Badness-of-Fit Criterion: Using formula=1.

The badness of fit represents as the square root of a normalized "residual sum of squares". The residual sum of squares is equivalent to unexplained variability. Fit can be measured by the square root of multiple correlation (R^2) or square root of 1- R^2. The former is often called goodness-of-fit, and the latter is called badness-of-fit. A stress value of .0479 fits the MDS model very well with less than 5% of unexplained variability, considering the suggested threshold value less than .2 in ACA studies by McCain. Changes in criterion refer to the difference of badness-of -fit criterion between previous alteration and current alteration. For example, the value, .0899, is the difference between the badness-of-fit value (.2084) at iteration 0 and (.1184) that of iteration 1.

Configuration

Table 3 shows a configuration of all of the authors on a three-dimensional space. The configuration of points is the coordinates of each object in a Euclidean space of one or more dimensions (Joseph B. Kruskal & Wish, 1978; Forrest W. Young, 1987). The distances between data points are constant over any similarity transformation such as rotation, permutation, reflection, translation, and dilation of the dimensions. This is the relative location of all the points which are unaffected by any similarity transformation. The configuration is to be used to produce two or three dimensional plots.

Relationships Among Scaling Methods, Stress, and Configuration

In ACA research, producing the MDS results with an acceptable level of stress is critically important. Stress level is very dependent on scaling methods. Choosing metric scaling methods, instead of non-metric scaling, always increases stress. Tables 2 and 3 contrast the differences in Stress 1 values associated with metric and non-metric MDS. Table 2 displays the iteration history with stress 1 (badness-of-fit) value of .0479 with level=ordinal option in PROC MDS statement (non-metric MDS). Table 3 displays the iteration history with stress 1 (badness-of-fit) value of .1308 with level=ratio option in PROC MDS statement (metric MDS). Despite the great difference in stress values, scaling methods produced little differences on the configurations.

Table 3. Configuration with ordinal option

ordinal option 07:20 Wednes

Multidimensional Scaling: Data=WORK.DIST.DATA
Shape=TRIANGLE Condition=MATRIX Level=ORDINAL
Coef=IDENTITY Dimension=3 Formula=1 Fit=1

Configuration

	Dim1	Dim2	Dim3
ALTER	-0.32	-1.28	0.15
BENBASAT	1.77	0.39	0.42
BONCZEK	-2.15	0.30	-0.05
CARLSON	-0.70	-0.66	0.37
DESANCTIS	0.85	1.94	0.45
DICKSON	1.71	1.04	0.49
GINZBERG	-0.09	-1.19	1.08
GORRY	0.43	-1.64	-0.42
HENDERSON	-1.30	0.39	-0.73
HOLSAPPLE	-2.13	0.27	-0.18
HUBER	0.07	1.20	0.71
KEEN	0.06	-0.68	0.22
KONSYNSKI	-1.17	1.64	-0.43
LUCAS	1.83	-0.25	0.88
MASON_R	1.83	-0.12	-1.30
MITROFF	1.82	0.05	-1.43
SCOTT_MORTON	0.14	-0.76	0.07
SIMON	0.12	-0.37	-0.43
SPRAGUE	-0.67	-0.44	0.23
WHINSTON	-2.11	0.16	-0.09

THE PLOT PROCEDURE

As mentioned, the MDS procedure produces only the printed iteration history and coordinate matrix. PROC MDS does not produce any plots. Additional plot procedures are necessary to produce two or three dimensional graphs. This section describes the production of two and three dimensional scatter plots to show the relationships between two or more dimensions. There are many different types of two dimensional plots such as line, high-low, bubble, and scatter. Scatter plots best represent the coordinate matrix. Table 4 shows the coordinate matrix produced by the MDS procedure, which is the input to the plot procedure. The PLOT procedure plots the coordinates of each author in a Euclidean space of two or three dimensions.

PROC PLOT Statement

The basic syntax of producing a two dimensional scatter plot consists of PROC PLOT statement with options, followed by another PLOT statement with options. The following example shows the PROC PLOT statement to produce three two dimensional plots (Figures 2, 3, and 4).

```
PROC PLOT DATA=coordordinal;
PLOT diml*dim2='*' $author;
PLOT (diml dim2)* dim3 = '*' $author;
TITLE 'ordinal option';
RUN;
```

The PROC PLOT statement uses an option, data=data-set name. The syntax of PLOT statement is: **PLOT** < *yvariable*xvariable* >< *=symbol* >. The statement, PLOT dim1*dim2 = '*' $ author, tells the SAS system that dimension 1 variables and dimension 2 variables in Figure 2 be plotted on the vertical (y coordinate) and horizontal (x coordinate) axes, respectively. It also shows that the symbol with single quotation (*) will be used to represent each variable on the plot and the character label of author name ($) will be on the plot.

PROC PLOT *<option(s)>*;
PLOT *plot-request(s) </ option(s)>*;
Plot-request(s) specifies the vertical and horizontal variables to plot and the plotting symbol to use to mark the points on the plot.

To produce plots that can be more meaningful and interpretable, ACA analysts may need to use options that control the appearance and size of plots. Table 4 lists several options to do so. For example, if we want to print all three Figures 2, 3, and 4 on one page (Figure 5), one optional statement (hpercent=33) should be added after the data statement.

```
proc plot data=coordordinal hpercent=33;
plot dim1*dim2='*' $author;
plot (dim1 dim2)* dim3 = '*' $author;

title 'ordinal option';
run;
```

HPERCENT=percent(s)

HPERCENT= enables you to put multiple plots on one page.

hpercent=33 (each plot uses 33% of a page wide and therefore, three plots are printed on one page horizontally.)

hpercent=50 25 25 (the first plot uses 50% of a page wide, the second
and
third occupy 25% of a page wide respectively.

hpercent=300 (this plot uses three pages.)

Figure 3. Plot of dim1 dim2$author*

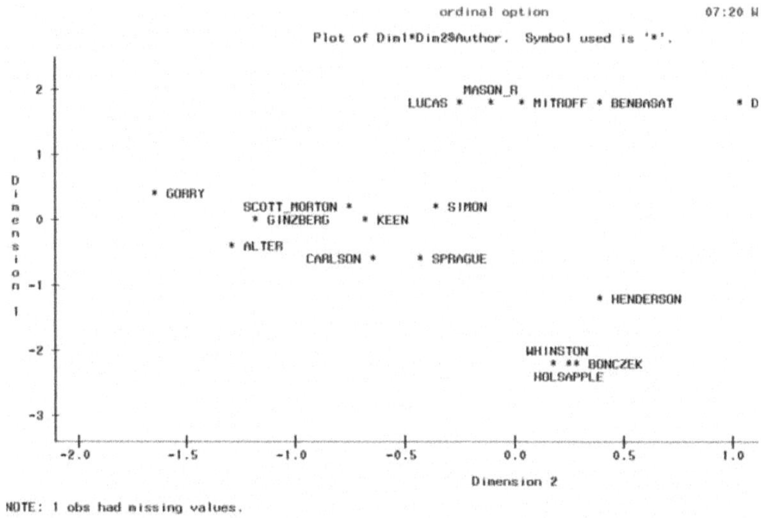

*Figure 4. Plot of dim1*dim3$Author*

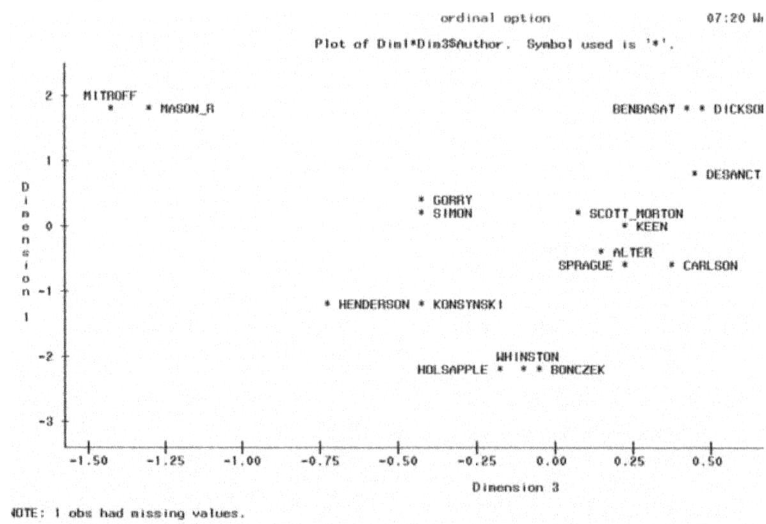

*Figure 5. Plot of dim2*dim3$Author*

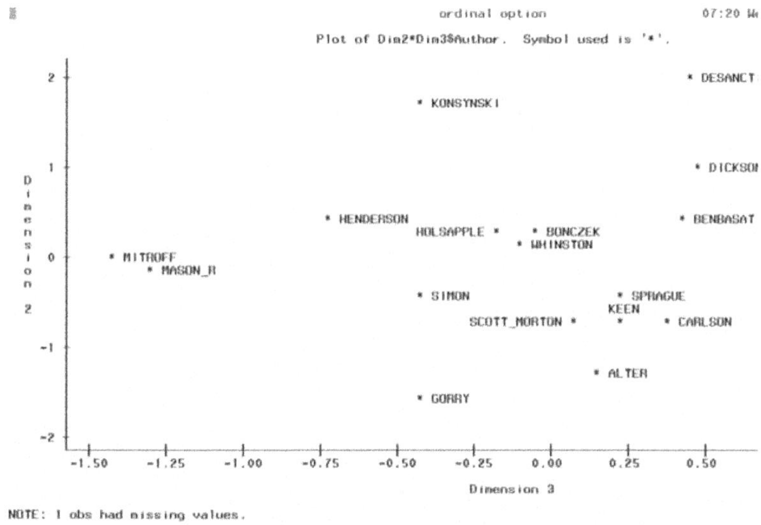

Figure 6. Printing three plots per page with hpercent=33

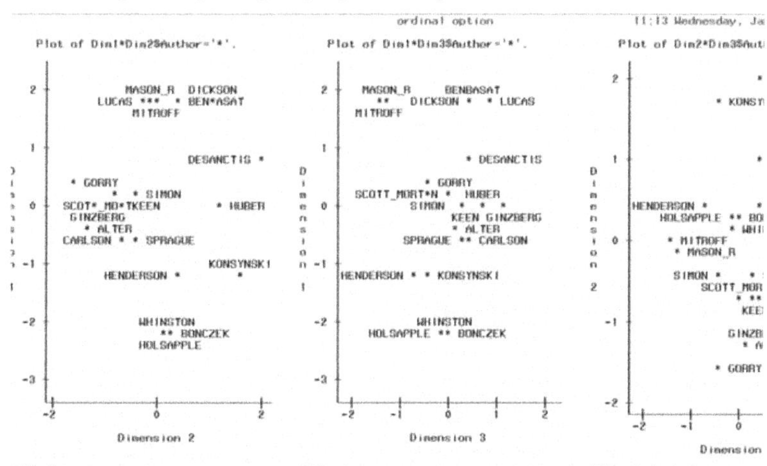

Table 4. PROC PLOT **<option(s)>;**

Task		Option
Specify the input data set		DATA=
Control the appearance of the plot		
	Specify the aspect ratio of the characters on the output device	VTOH=
Control the size of the plot		
	Specify the percentage of the available horizontal space for each plot	HPERCENT=
	Specify the percentage of the available vertical space for each plot	VPERCENT=
Source: SAS Version 9.1 online documentation at: http://support.sas.com/onlinedoc/913/docMainpage.jsp		

VPERCENT=percent(s)

specifies one or more percentages of the available vertical space to use for each plot. For example, using Vpercent=33, all three plots are printed one page. Each plot has one page wide, but height is one third of a page.

VTOH=aspect-ratio

VTOH stands for **V**ertical **TO** **H**orizontal. This option specifies the aspect ratio (vertical to horizontal) of the characters on output device. ***Aspect-ratio*** is a positive real number.

VTOH=*aspect-ratio*

PLOT plot-request(s) </ option(s)>

The PLOT statement followed by the PROC PLOT statement requests the plots to be produced by PROC PLOT. Most of The PLOT statement options are listed in Table 6.6, taken from (SAS Institute Inc., 2000). The following options are used in the example used in ACA plot procedures.

Control Axes by Specifying the Tick-Mark Values

HAXIS = specifies major tick mark values for horizontal axis of box plot.

Table 4. PLOT plot-request(s) </ option(s)>;

		Use this option
Control the axes		
	Specify the tick-mark values	HAXIS=, VAXIS=
	Expand the axis	HEXPAND, VEXPAND
	Specify the number of print positions between tick marks	HSPACE=, VSPACE=
Specify reference lines		
	Draw a line perpendicular to the specified values on the axis	HREF= and VREF=
Put a box around the plot		BOX

VAXIS = specifies major tick mark values for vertical axis of box plot.

The AXIS Statement controls the location, values, and appearance of the axes of the plots and charts. This statement specifies the characteristics of an axis such as:

1. the way the axis is scaled
2. how the data values are ordered
3. the location and appearance of the axis on the tick mark
4. the text and appearance of the axis label and major tick mark values.

TYPE= 'CONFIG' specifies that the output of this PLOT procedure is a spatial arrangement on a two/three dimensional space.

Use a box around the plot

The BOX option is used to draw a box around the plot. Without the box option, the plot procedure produces a plot without a box. See Figure 6.2.

Example: PLOT (diml dim2) * dim3 = '*' $ author / BOX

THE G3D PROCEDURE

The G3D procedure in the SAS system presents solutions in three dimensions using scatter plots, surface plots, and contour plots. The data can be represented as surfaces (surface plots), or as points (scatter plots). The three dimensional numeric variables can also be represented in two dimensions (contour plots). Of these types, the ACA study results can be best displayed by scatter plots. The input to the G3D procedure is the coordinate matrix produced by the MDS procedure.

Syntax of the G3D Procedure

PLOT *plot-request </option(s)>*;

Generating a Simple Scatter Plot

To generate a simple scatter diagram, the following proc g3d statement taken from SAS OnlineDoc Version 8 can be used as a starting point.

Figure 7. PROC PLOT statement with the HAXIS, YAXIS options

```
proc plot data=coordordinal vtoh=2;
plot dim1*dim2='*' $author
     / haxis = by .5
       vaxis= by .5 ;
         where _ type _ = 'CONFIG';
plot (dim1 dim2)* dim3 = '*' $author
      / haxis = by .5
        vaxis= by .5 ;
          where _ type _ = 'CONFIG';

title 'ordinal option';
run;
```

Figure 8. PROC PLOT statement with the BOX, VTOH options

```
PROC PLOT data=coordordinal vtoh=2;
   PLOT dim1*dim2 = '*' $ author
    / BOX
      HAXIS=BY .5
      VAXIS=by .5;
      where _type_='CONFIG';
    PLOT (dim1 dim2) * dim3 = '*' $ author
     / BOX
       HAXIS=BY .5
       VAXIS=by .5;
       where _type_= 'CONFIG';
RUN;
```

*Figure 9. Plot of Dim1*Dim2$Author with HAXIS, YAXIS, VTOH options*

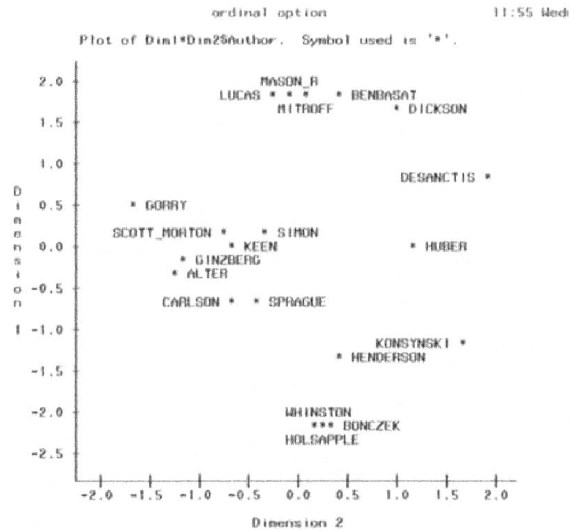


```
proc g3d data=reflib.iris;
  scatter petallen*petalwid=sepallen;
run;
```

The first line, proc g3d, specifies the data set name and starts the procedure. The second line contains the scatter statement for producing three-dimensional scatter plots using values of three numeric variables from the input data set. Using the data imported from external sources, the following PROC G3D statements create a simple 3D scatter plot.

Modifying Plots with SCATTER Options

Plots can be modified with many SCATTER options. To show some of the options, the next PROC G3D statement adds options to add a grid and change the text of the axis labels.

```
proc g3d
data=coordordinal;
scatter dim2*dim1=dim3/grid caxis=black anno=anno shape='pyramid';
label dim1='foundations'
dim2='model management'
```

Figure 10. Proc g3d procedures

```
/* importing data from external sources to the sas system */
PROC IMPORT OUT= WORK.ACA
        DATAFILE= "C:\Documents and Settings\sbeom\My Documents\idea
        aca\spss\bookdata.xls"
        DBMS=EXCEL REPLACE;
   SHEET="sasdata$";
   GETNAMES=YES;
   MIXED=NO;
   SCANTEXT=YES;
   USEDATE=YES;
   SCANTIME=YES;
RUN;
proc print data=aca;
run;
/* principal component analysis */
proc factor data=aca method=principal mineigen=1 rotate = promax;
run;
/* creating distance matrix for cluster and non-metric mds analyses */
Proc distance data=aca out=dist method=dcorr SHAPE=SQUARE;
Var ratio(Alter--Whinston);
ID author;
run;
proc print data=dist;
/* principal component analysis */
proc cluster method=ward data=dist pseudo;
ID author;
proc tree space=2;
run;
/* non-metric mds analysis to produce configuration */
PROC MDS
         data=dist
     dimension=3
         condition=un
     level=ordinal
         pconfig
         out=coordordinal;
     ID author;
title 'non-mteric mds with level=ordinal option';
RUN;
PROC PRINT data=coord (keep=author dim1 dim2 dim3);
VAR author dim1 dim2 dim3;
RUN;
options ps=40;
/* creating two-dimensional plots */
proc plot data=coord vtoh=2;
plot dim1*dim2='*' $author
    / haxis = by .5
     vaxis= by .5 ;
      where _type_ = 'CONFIG';
```

continued on following page

Figure 10. continued

```
plot (dim1 dim2)* dim3 = '*' $author
    / haxis = by .5
      vaxis= by .5 ;
        where _type_ = 'CONFIG';

title 'two dimensional plots of configuration';
run;
/* creating the annotate data set for g3d plot */
data anno;
set coord (keep=dim1 dim2 dim3 author);
length function $8;
RETAIN xsys ysys zsys '2'
function 'label'
 size .7
 position '2'
 color 'blue'
 style 'zapf';
 x=dim1;
 y=dim2;
 z=dim3;
 text=author;
/* Creating three-dimensional scatter plots using G3d Procedure */
proc g3d
 data=coordordinal;
 scatter dim2*dim1=dim3/grid caxis=black anno=anno shape='pyramid';
 label dim1='foundations'
     dim2='model management'
     dim3='user-interface';
 title 'three dimensional scatter plots with labels and caxis';
run;
quit;
```

Figure 11. A simple G3D scatter plot

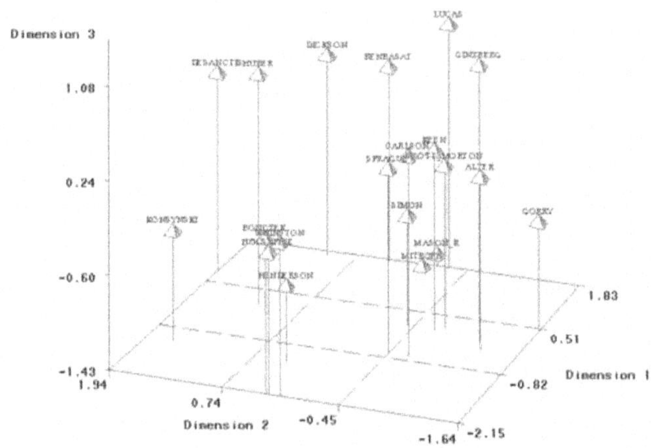


```
dim3='user-interface';
title 'g3d plot with labels and caxis';
run;
```

CREATING THE ANNOTATE DATA SET USING THE DATA STEP

In an ACA study, 3D scatter plots without labels on data points provide little useful information for the ACA researchers. This section discusses how to label data points on a plot. The annotate facility in the SAS system produce a Figure to show the name of the author on each data point.

SAS Program to Create the Annotate Data Set

The following program (Createanno.sas) creates the annotate data set, acaanno (Table 5).

Understanding the SAS Program for Creating Annotate Data Set

Using the SAS program (Createanno.sas) and the annotate data (mysave.anno), this section discusses what each section of the program does by comparing it with the

Figure 12. Scatter plot with labels and caxis

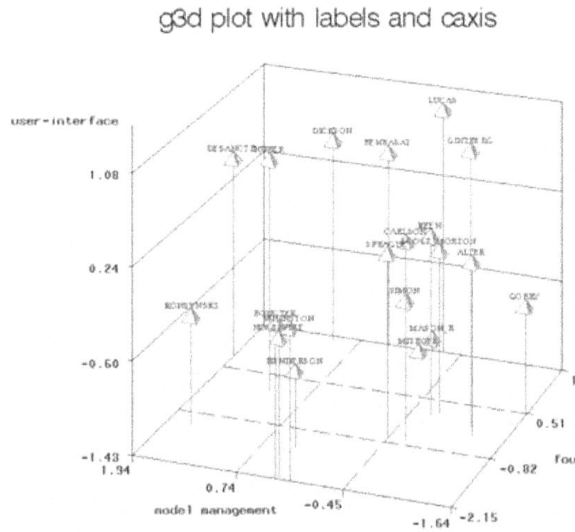

output (mysave.anno). The program produces the annotate data set which contains the observations to generate the output by using a DATA step.

The Data step begins with Data <Data Set Name>

Data <Data Set Name>

/* Data Set Name */
DATA anno;

Set

The set statement reads an observation from one or more SAS data sets. Alternatively the data can be directly entered from the keyboard or the data may come from an infile format. Here, the set statement is one way of reading the data from the previous step, the MDS procedure in Figure 1.

SET<*SAS-data-set(s)* <(*data-set-options(s)*)>>
<*options*>;

The symbol <> should not be included in a SAS statement.

Example
set mysave.coord; or
set mysave.coord (keep=dim1 dim2 dim3 author);

Figure 13. Creating an annotate data set

```
DATA testl;
INPUT Author $ 1-12 @13 x1-x9;
CARDS;
ALTER        121  25   66  23   92   83   38   67   27
BONCZEK       25 103   53  23   46   42   33   59   93
CARLSON       66  53  173  34  112  101   54  133   54
HUBER         23  23   34  68   47   39   37   49   24
KEEN          92  46  112  47  206  174   82  126   50
SCOTTMORTON   83  42  101  39  174  190   82  105   45
SIMON         38  33   54  37   82   82  111   58   37
SPRAGUE       67  59  133  49  126  105   58  182   61
WHINSTON      27  93   54  24   50   45   37   61  104
;
/* SAS version 8 uses xmacro.sas and distnew.sas to create a distance matrix
*/
%include 'c:\program files\SAS Institute\SAS\V8\stat\sample\xmacro.sas';
%include 'c:\program files\SAS Institute\SAS\V8\stat\sample\distnew.sas';
%distance ( data=testl,
            id=author,
            options=print,
            method=dcorr,
            var=x1-x9,
            out=dist);
proc print data=dist;
run;
/* SAS version 9 uses proc distance to create a distance matrix */
/* creating distance matrix for cluster and non-metric
mds analyses */
Proc distance data=aca out=dist method=dcorr
SHAPE=SQUARE;
Var ratio(Alter--Whinston);
ID author;
run;
/**** MDS procedures to produce profiles****/
proc mds
     data=dist
     level=ordinal
     dimension=3
     fit=1
       out=coord
```

continued on following page

Figure 13. continued

```
        outres=res;
    pcoef
        pconfig
        pfinal
            ID author;
;
title1 'MDS Output';
run;
proc print
data=coord (keep=author dim1 dim2 dim3);
var author dim1 dim2 dim3;
run;
options ps=40;
proc plot
data=coord vtoh=.5;
plot dim1*dim2 = '*' $ author / BOX
        HAXIS=BY .4 VAXIS=by 1.3;
        where _type_='CONFIG';
        plot (dim1 dim2) * dim3 = '*' $ author / BOX
        HAXIS=BY .35 VAXIS=by .35;
        where _type_= 'CONFIG';
       TITLE2 'PLOT Of CONFIGURATION';
data anno;
set coord (keep=dim1 dim2 dim3 author);
length function $8;
retain xsys ysys zsys '2'
function 'label'
        size .4
        position '2'
        color 'blue'
        style 'zapf';
        x=dim1;
        y=dim2;
        z=dim3;
        text=author;
proc g3d
data=coord;
scatter dim2*dim1=dim3 /anno=anno shape='pyramid';
run;
```

Table 5. Annotate data set

Obs	_DIMENS_	_MATRIX_	_TYPE_	Author	_NAME_	Dim1
1	3	.	CRITERION			0.11468
2	3	.	CONFIG	ALTER	ALTER	1.04522
3	3	.	CONFIG	BONCZEK	BONCZEK	-2.31542
4	3	.	CONFIG	CARLSON	CARLSON	0.17294
5	3	.	CONFIG	HUBER	HUBER	0.47546
6	3	.	CONFIG	KEEN	KEEN	1.01034
7	3	.	CONFIG	SCOTTMORTON	SCOTTMORTON	1.00292
8	3	.	CONFIG	SIMON	SIMON	0.71084
9	3	.	CONFIG	SPRAGUE	SPRAGUE	0.17918
10	3	.	CONFIG	WHINSTON	WHINSTON	-2.28148
11	3	.	SLOPE			0.34472

Obs	Dim2	Dim3	function	xsys	ysys	zsys	size	position
1	.	.	label	2	2	2	0.4	2
2	1.24929	-0.34354	label	2	2	2	0.4	2
3	0.07483	-0.26530	label	2	2	2	0.4	2
4	0.71272	1.04272	label	2	2	2	0.4	2
5	-1.93306	0.88987	label	2	2	2	0.4	2
6	0.25904	-0.30261	label	2	2	2	0.4	2
7	0.22087	-0.58030	label	2	2	2	0.4	2
8	-1.00424	-1.25772	label	2	2	2	0.4	2
9	0.34683	1.13255	label	2	2	2	0.4	2
10	0.07373	-0.31568	label	2	2	2	0.4	2
11	.	.	label	2	2	2	0.4	2

Obs	color	style	x	y	z	text
1	blue	zapf	0.11468	.	.	
2	blue	zapf	1.04522	1.24929	-0.34354	ALTER
3	blue	zapf	-2.31542	0.07483	-0.26530	BONCZEK
4	blue	zapf	0.17294	0.71272	1.04272	CARLSON
5	blue	zapf	0.47546	-1.93306	0.88987	HUBER
6	blue	zapf	1.01034	0.25904	-0.30261	KEEN
7	blue	zapf	1.00292	0.22087	-0.58030	SCOTTMORTON
8	blue	zapf	0.71084	-1.00424	-1.25772	SIMON
9	blue	zapf	0.17918	0.34683	1.13255	SPRAGUE
10	blue	zapf	-2.28148	0.07373	-0.31568	WHINSTON
11	blue	zapf	0.34472	.	.	

To process the data set, mysave.coord, the keep= option may not be necessary. Table 5 shows that the data set only contains 5 columns of data with an observation number. If there are numerous variables in the data set, only those variables that are listed after the KEEP= data set option are available for processing. With the (keep=dim1 dim2 dim3 author) option, the annotated data set would exclude the four columns of data (**_DIMENS_**, **_MATRIX_**,

TYPE, _NAME_).

Length

The LENGTH statement creates a variable and sets the length of the variable, as in the following example:

```
length function $8;
```

The length statement creates a character variable and function, with a maximum of 8 characters. The length can be set at any number of characters to hold the function name, LABEL. Therefore, it could be permissible to have $5 instead of $8. If the length is set to have a value of less than $5, the annotate facility will not produce annotations.

Retain

A retain statement lists all the variables that retain the same values for all observations in a annotated data set. Notice that in Figure 13, the eight variables listed in the following retain statement hold the same values specified with single quotation.

```
RETAIN     xsys ysys zsys '2'
           function 'label'
                 size .4
                 position '2'
                 color 'blue'
                 style 'zapf';
```

COLOR **'color'** (the color of the text).

POSITION *'text-position'* | '0'; (the placement of the text string (author's name) in relation to the pyramid position of each author.

Text-position can be one of the following:

- Numbers 1 through 9,
- Characters A through F,
- Symbols <, +, or >.

Position '2' indicates that the name of each author is to be printed at the location of one cell above the pyramid shape and it will be centered.

SIZE **height** specifies the height of the text string. The SIZE variable units are based on the value of the HSYS variable. In the above example, HSYS (the coordinate system) value is not specified. The default value is 4, which means the coordinate system unit is a cell in graphics output area. Size .4 means that the height of authors' name is 40% of the height of a cell in graphic area.

Style 'font style' the font style of the TEXT

Function

The purpose of the function statement is to specify a graphics command function for the Annotate facility to perform with the syntax, FUNCTION 'function-name'; The graphics command function includes BAR, CNTL2TXT, COMMENT, DEBUG, DRAW, DRAW2TXT, FRAME, LABEL, MOVE, etc. The default value is LABEL.

Coordinate-System

To produce a three dimensional graph, the value of the coordinate system (xsys, ysys, zsys) must be specified as '2'. The value of 'coordinate-system' are specified in Table 6.

Assignment

Assignment statements in a DATA step evaluate the expression on the right side of the equal sign and store the result in the variable that is specified on the left side of the equal sign.

X=**horizontal-coordinate** is used to represent dimension 1
Y=**vertical-coordinate** is used to represent dimension 2
Z=**depth-coordinate** is used to represent dimension 3
Text=author (Text string values are from ID author names)

PROC G3D with the Annotate =Option

PROC G3D data=coord;
SCATTER dim2*dim1=dim3 /anno=mysave.acaanno grid

Table 6. Value of the coordinate system

Absolute Systems	Relative Systems	Coordinate System Units
1	7	percentage of data area
2	8	data values
3	9	percentage of graphics output area
4	A	cell in graphics output area
5	B	percentage of procedure output area
6	C	cell in procedure output area

Source: SAS Institute (2007)

```
caxis=black shape='pyramid';
RUN;
```

CONCLUSION

This chapter discusses multidimensional scaling (MDS) procedures. MDS is a class of multivariate statistical techniques/procedures to produce two or three dimensional pictures of data (geometric configuration of points) using proximities among any kind of objects as input. Three SAS procedures (MDS, PLOT, and G3D) are necessary to convert the author cocitation frequency matrix to two or three dimensional pictures of data. The distance matrix produced earlier by using xmacro.sas and distnew.sas programs (version 8) or proc distance procedure (version 9) should be converted to a coordinate matrix to produce two-dimensional plots and annotated three-dimensional scatter diagrams. In this process, it is very important to understand the relationships between scaling methods and stress. Stress level is very dependent on scaling methods. Choosing metric scaling methods, instead of non-metric scaling, always increases stress. Therefore, ACA researchers should always use the "level=ordinal" option to perform non-metric scaling. Despite the great difference in stress values, scaling methods produced little differences on the configurations. This chapter also discusses how to label data points on a plot. The annotate facility in the SAS system produces figures with the name of the author on each data point. The PROC MDS procedure includes many of the features of the ALSCAL procedure.

REFERENCES

Hair, Jr., J. F., Anderson, R. E., & Tatham, R. L. (1987). *Multivariate Data Analysis with Readings* (2nd ed.). New York: Macmillan Publishing Company.

Kruskal, J. B. (1964). Multidimensional Scaling by Optimizing Goodness of Fit to a Nonmetric Hypothesis. *Psychometrika, 29*, 1-27.

Kruskal, J. B., & Carroll, J. D. (1969). Geometric Models and Badness-of-Fit Functions. In P. R. Krisnaiah (Ed.), *Multivariate Analysis* (pp. 639-670). New York: Academic Press.

Kruskal, J. B., & Wish, M. (1978). *Multidimensional Scaling.* Beverly Hills, CA: Sage Publications.

Kruskal, J. B., & Wish, M. (1990). *Multidimensional Scaling.* Newbury Park, CA: Sage Publications.

McCain, K. W. (1990). Mapping Authors in Intellectual Space: A Technical Overview. *Journal of the American Society for Information Science, 41*(6), 433-443.

Ramsay, J. O. (1986). *The Mlscale Procedure*. Cary, NC: SAS Institute.

SAS Institute Inc. (1992). *SAS Technical Report P-229, SAS/Stat Software: Changes and Enhancements (Release 6.07).* Cary, NC: SAS Institute Inc.

SAS Institute Inc. (2000). *SAS Onlinedoc® Version 8, February 2000, Pdf Format.* Cary, NC: SAS Institute Inc.

SAS Institute Inc. (2007). *SAS Onlinedoc® 9.1.3.* . Cary, NC: SAS Institute Inc. .

Shepard, R. N. (1962). The Analysis of Proximities: Multidimensional Scaling with an Unknown Distance Function I and Ii *Psychometrika, 27*, 125-139 and 219-246.

Young, F. W. (1987). *Multidimensional Scaling: History, Theory, and Applications.* Hillsdale, New Jersey: Lawrence Erlbaum Associates, Publishers.

Young, F. W., Lewyckyj, R., & Takane, Y. (1986). *The Alscal Procedure.* Cary, NC: SAS Institute.

Chapter XI
ACA Analysis with the 15.0.0 Version of SPSS for Windows

ABSTRACT

This chapter briefly introduces the use of SPSS version 15.0 to conduct ACA analysis. The SPSS accepts datafiles in many different formats including spreadsheets, database files, tab-delimited, and other types of ASCII text files. Assuming that cocitation frequency counts are stored in a spreadsheet file in Excel, we demonstrate each step of ACA analysis to produce outputs using factor, cluster, and multi-dimensional scaling analyses.

INTRODUCTION

The previous three chapters (VIII, IX, and X) discussed how to use the SAS system to conduct ACA using the factor, cluster, and multi-dimensional scaling procedures. This chapter briefly introduces the use of SPSS version 15.0 to conduct ACA analysis. The SPSS accepts data files in many different formats including spreadsheets, database files, and tab-delimited and other types of ASCII text files. Assuming that cocitation frequency counts are stored in a spreadsheet file in Excel form, this chapter starts with how to import the data file in Excel format.

OPENING A DATA FILE

The very first step is to go to window Start menu → all programs → SPSS for Windows → SPSS 15.0 for Windows. The first Window pop-up dialog box in Figure 1, select an existing data source.

The open data dialog box in Figure 2 by default shows all SPSS data format (.sav) files. To read an Excel file you created, first, click "v" in the first "look in" dropdown menu box to change to the directory in which the Excel data file is stored.

Table 1. Sample cocitation frequency counts

alter	bonczek	carlson	huber	keen	smorton	Simon	sprague	Whinston
121	25	66	23	92	83	38	67	27
25	103	53	23	46	42	33	59	93
66	53	173	34	112	101	54	133	54
23	23	34	68	47	39	37	49	24
92	46	112	47	206	174	82	126	50
83	42	101	39	174	190	82	105	45
38	33	54	37	82	82	111	58	37
67	59	133	49	126	105	58	182	61
27	93	54	24	50	45	37	61	104

Figure 1. Windows "What would you like to do?" dialog box

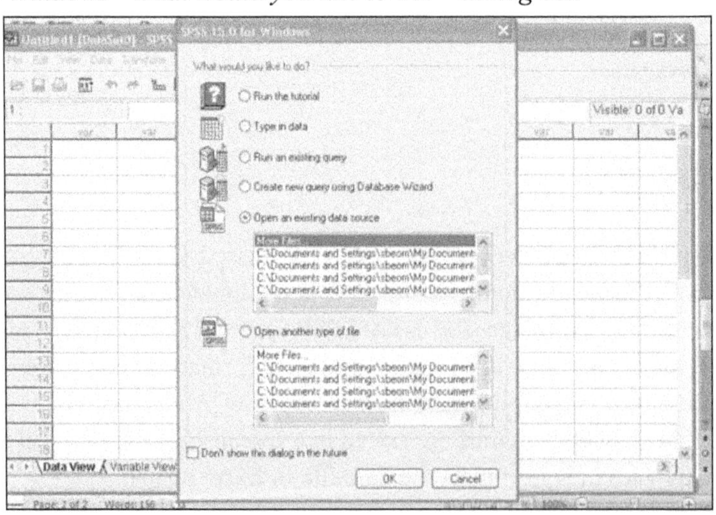

Figure 2. The "Open data" dialog box

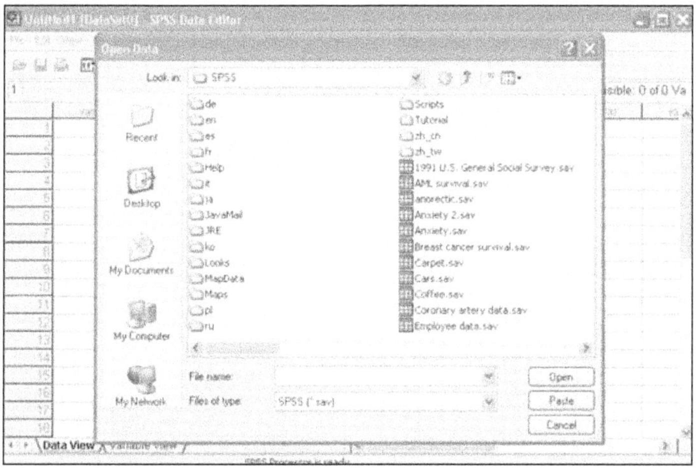

Second, click "v" in the second "Files of types:" dropdown menu box to change to file type to Excel: (*.xls).

Third, click the file name from the list of excel files listed in the Open data dialog box. Finally, Click "open" at the bottom right.

If your data file contains more than one worksheet, you may be able to select one of them. Click "OK".

Figure 5 is Data view of the data file. The data file is rectangular. The columns represent variable names (alter .. whinston). The rows represent cases. In SAS, cases are called observations.

Figure 6 illustrates the variable view of the input file. The variable view describes the data structure. The level of measurement can be specified as scale (numeric data on an interval or ratio scale), ordinal or nominal. The labels provide descriptive information and they are displayed in output. The variable names can be 8 characters long, and the labels can be up to 256 charters long.

FACTOR ANALYSIS

After the dataset is imported and saved as SPSS data format, several steps are necessary to conduct factor analysis. As Figure 12.7 shows, select the following steps.

Analyze
Data Reduction
Factor

Figure 3. Select the excel data file (9vardata.xls)

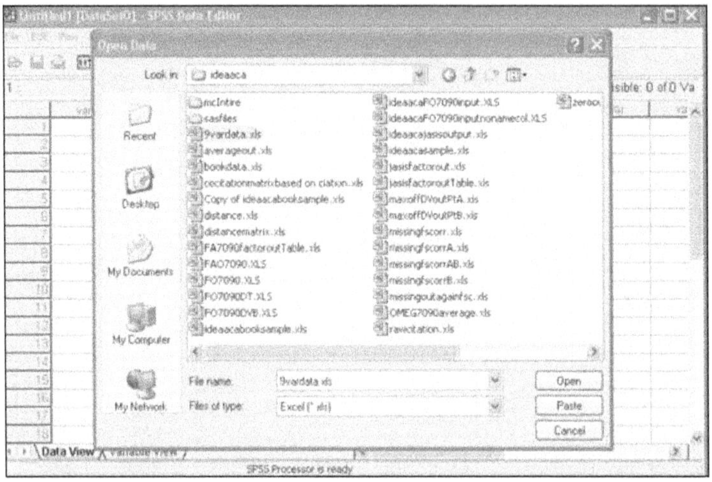

Figure 4. Opening Excel data source

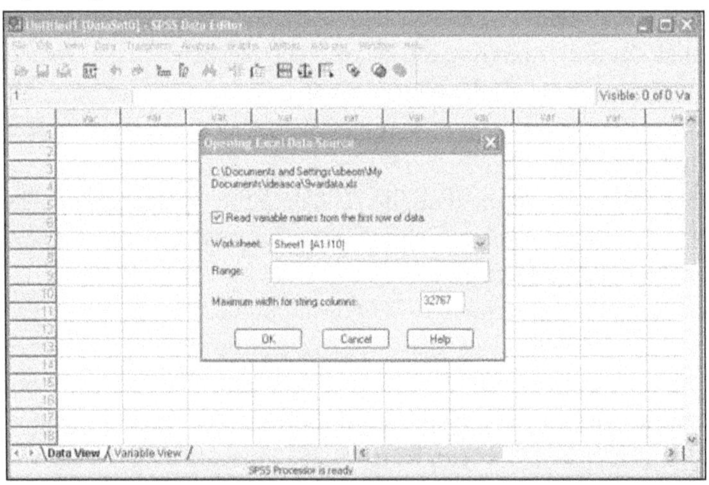

Figure 7 shows the steps to conduct a factor analysis. From SPSS data editor menu, choose Analyze → Data Reduction → Factor. Once you click the factor submenu, the factor analysis dialog box appears (Figure 8).

Select the variables for the factor analysis. Double clicking on each variable moves it to the variables box at right. Or, while pressing the "shift" key on the keyboard, left-click all variables one by one and move them to the Variables box by clicking the button at the left hand side of the variables box. Next, click the

Figure 5. Data view

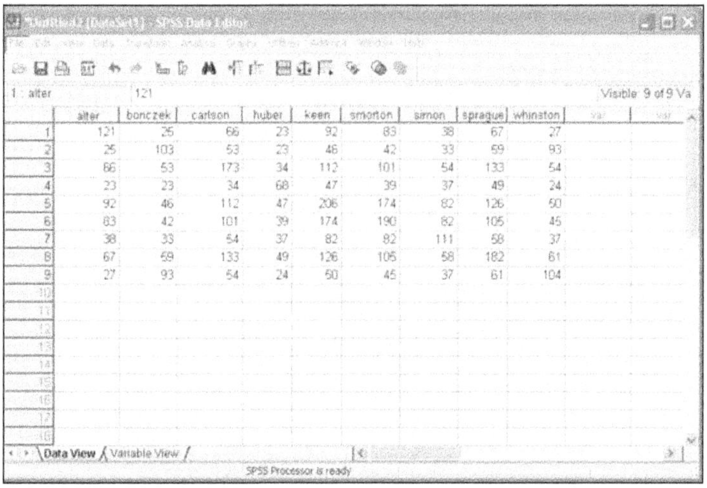

	alter	bonczek	carlson	huber	keen	smorton	simon	sprague	whinston
1	121	25	66	23	92	83	38	67	27
2	25	103	53	23	46	42	33	59	93
3	66	53	173	34	112	101	54	133	54
4	23	23	34	68	47	39	37	49	24
5	92	46	112	47	206	174	82	126	50
6	83	42	101	39	174	190	82	105	45
7	38	33	54	37	82	82	111	58	37
8	67	59	133	49	126	105	58	182	61
9	27	93	54	24	50	45	37	61	104

Figure 6. Variable view

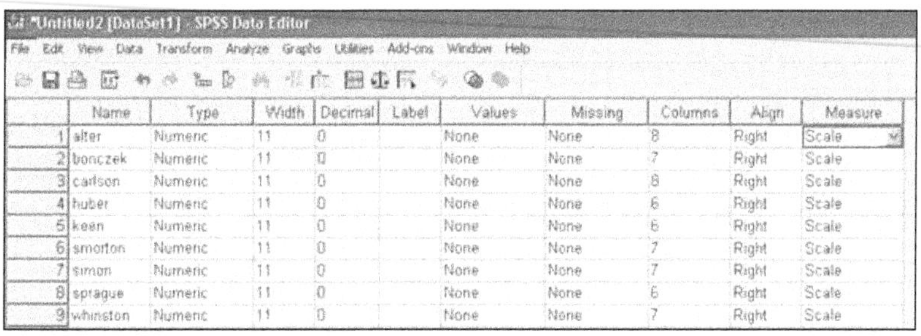

	Name	Type	Width	Decimal	Label	Values	Missing	Columns	Align	Measure
1	alter	Numeric	11	0		None	None	8	Right	Scale
2	bonczek	Numeric	11	0		None	None	7	Right	Scale
3	carlson	Numeric	11	0		None	None	8	Right	Scale
4	huber	Numeric	11	0		None	None	6	Right	Scale
5	keen	Numeric	11	0		None	None	6	Right	Scale
6	smorton	Numeric	11	0		None	None	7	Right	Scale
7	simon	Numeric	11	0		None	None	7	Right	Scale
8	sprague	Numeric	11	0		None	None	6	Right	Scale
9	whinston	Numeric	11	0		None	None	7	Right	Scale

"Descriptives" button. At the bottom of the factor analysis dialog box, there are four more buttons that activate other dialog boxes: Descriptives, Extraction, Rotation, Scores, and Options.

FACTOR ANALYSIS DESCRIPTIVES DIALOG BOX

The factor analysis descriptives dialog box (Figure 10) allows the user specify what kind of information to be included in the factor analysis output.

Checking only statistics (univariate descriptive and initial solution), the following outputs will be produced. Univariate statistics include the mean, standard deviation, and number of observations/cases for all variables (Figure 11). Initial solution output (Figure 12) includes initial communalities, eigen values, and the percentage of variance accounted for by each factor. A communality is defined as the percent of variance in a variable that is accounted for by the retained components (or factors). Eigen values are defined as the amounts of variance that is accounted for by a given factor or component.

Figure 7. Steps to conduct factor analysis

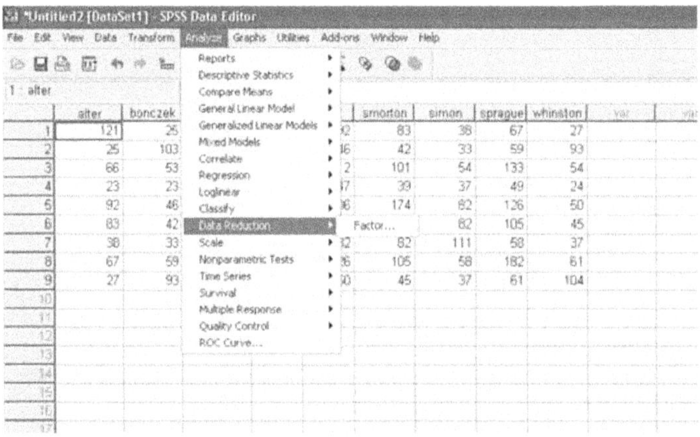

Figure 8. Factor Analysis dialog box

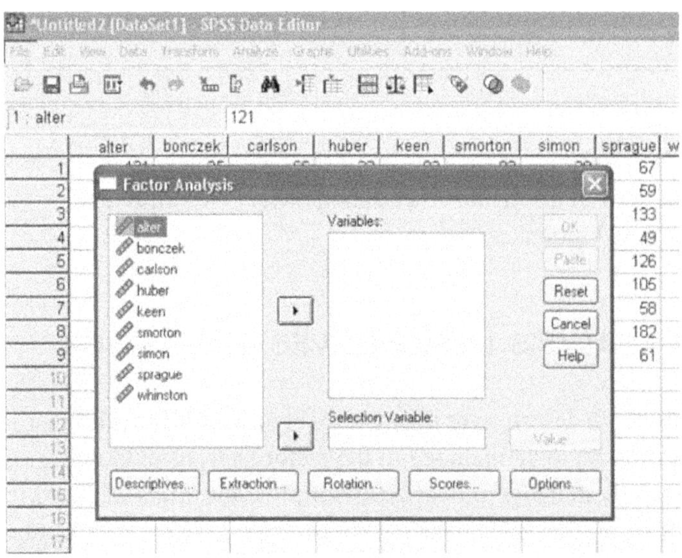

FACTOR ANALYSIS: EXTRACTION DIALOG BOX

The extraction dialog box allows the user to specify the extraction method. The term "extraction" refers to the process of extracting a maller number of factors from a large set of variables. The extraction dialog box offers several choices of factor extraction techniques for the user. As Figure 14 shows, the methods include principal components, unweighted least squares, generalized least squares, maximum likelihood, principal axis factoring, alpha factoring, and image factoring. For more information about these extraction methods, see (Tabachnick & Fidell, 1996). The most widely used extraction method is principal component analysis (Mertler & Vannatta, 2005).

Choose principal components under method
Choose correlation matrix under analyze
Choose both "Unrotated factor solution" and "Scree Plot" under display
Choose "eigen value over 1 under extract.

FACTOR ANALYSIS: ROTATION DIALOG BOX

Rotation is "a process of manipulating the reference axes." (Child, 1970) Rotation is a process by which a factor solution is made more interpretable without altering the underlying mathematical structure"(Mertler & Vannatta, 2005).

Figure 9. The variables are selected for factor analysis

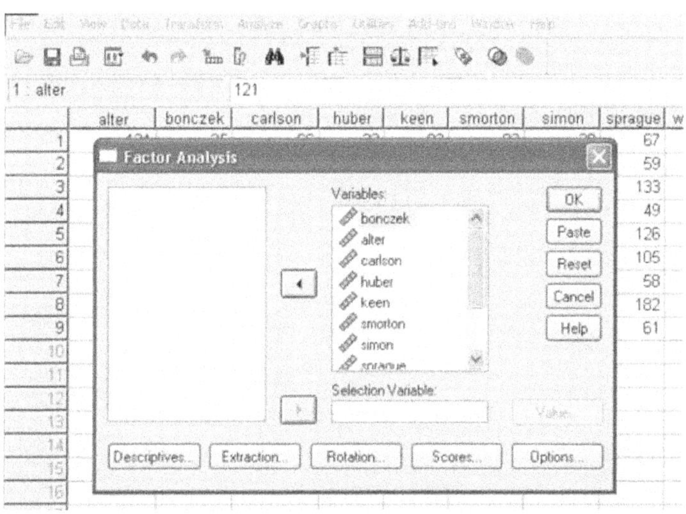

Figure 10. Factor analysis descriptives (only statistics checked)

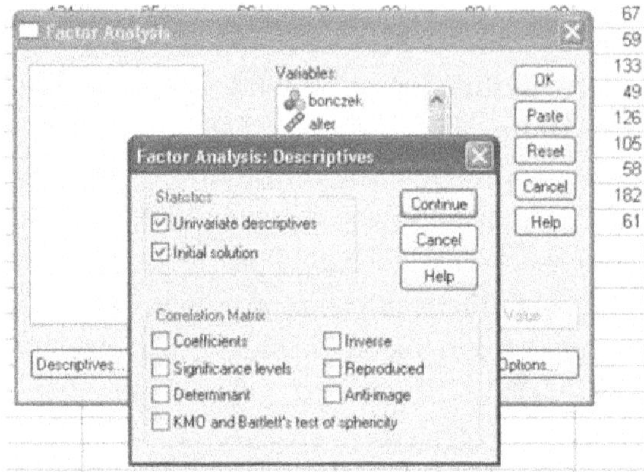

Figure 11. Descriptive statistics

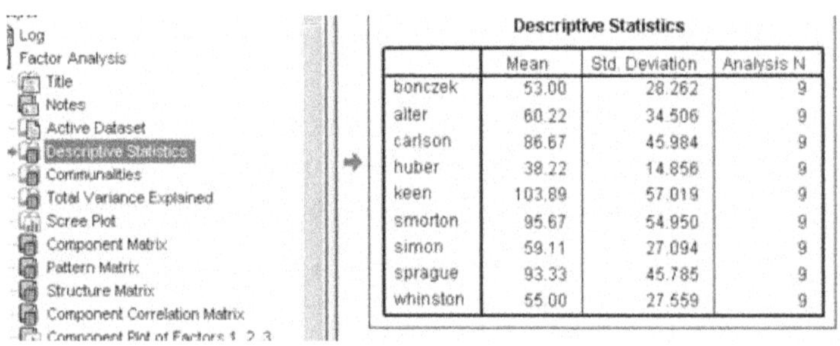

Descriptive Statistics

	Mean	Std. Deviation	Analysis N
bonczek	53.00	28.262	9
alter	60.22	34.506	9
carlson	86.67	45.984	9
huber	38.22	14.856	9
keen	103.89	57.019	9
smorton	95.67	54.950	9
simon	59.11	27.094	9
sprague	93.33	45.785	9
whinston	55.00	27.559	9

Figure 12. Initial solution

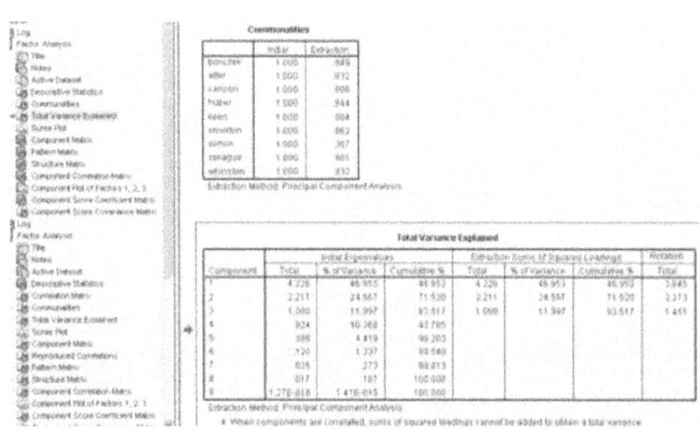

Figure 13. Factor analysis descriptives (correlations matrix)

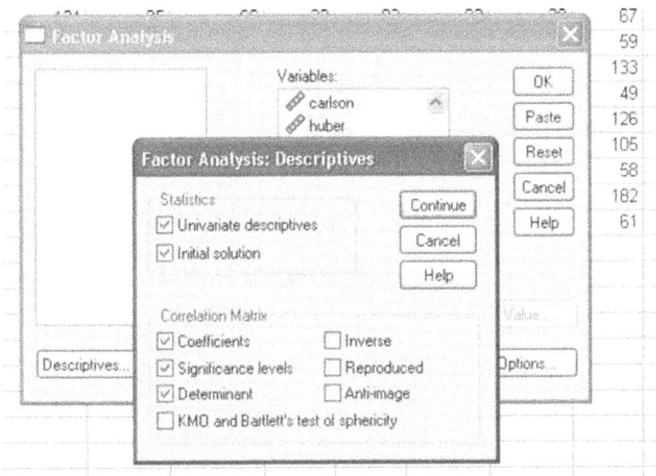

An unrotated component matrix is the result of direct solution. The primary purpose of principal components analysis is to cluster a large number of variables into a smaller number of unobservable factor/components. The rotation of factors produces derived solutions (rotated factor pattern and factor structure correlations) to make the solution more interpretable by simplifying the rows or columns. This means making as many values in each row and column as close as possible to zero (Hair, Anderson, & Tatham, 1987).

Figure 15 shows several different rotation methods. They are Quartimax, Varimax, Equamax, Direct oblimin, and Promax. Promax and direct oblimin are oblique methods. The other three are orthogonal methods. The orthogonal factor rotation is applied when there are no correlations between factors. Oblique rotation is preferred when there are high degrees of correlation among factors. It is a reasonable statement that factors (research subspecialties in an academic discipline) in ACA are highly correlated each other. Therefore, most ACA researchers prefer to use oblique rotation method to produce correlated factor structure. The SAS system's proc factor with rotate=promax option produces rotated factor pattern (orthogonal rotation) and factor structure correlation (oblique rotation). In SPSS, you can specify only one rotation option.

Varimax rotation is a orthogonal method that minimizes factor complexity by maximizing variance for each factor.

Promax rotation is a oblique method that produces a correlated structure matrix (SAS factor structure correlations).

Figure 14. Factor analysis: Extraction dialog box

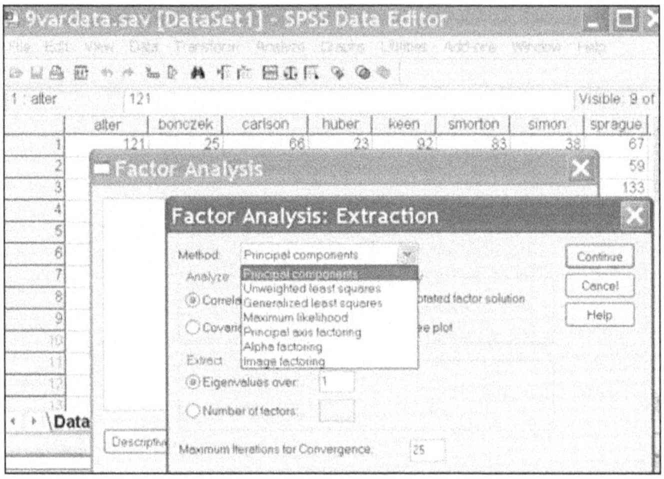

HIERARCHICAL CLUSTER ANALYSIS

Cluster analysis aims to identify and classify variables based on characteristics they possessed. In ACA, our purpose is to create an empirical map of authors in an academic discipline. Cluster analysis shows the process of joining all individual authors into one final cluster. There are two different clustering algorithms: Hierarchical and nonhierarchical. ACA uses agglomerative hierarchical clustering procedure. The hierarchical clustering procedure has two types: agglomerative and divisive.

Input to the Cluster analysis is a correlation matrix. The correlation matrix as a proximity measure has been discussed earlier (McCain, 1990). As shown in the factor analysis section in this chapter, correlation matrix can be created as part of the factor analysis procedure. Or, the matrix can be created by choosing ANALYZE→CORRELATE →DISTANCE procedure. Let's start with original dataset (Figure 5). The dataset is the cocitation frequency count matrix. This matrix needs to be converted to a correlations matrix. The correlations matrix must be converted to distance matrix.

The following SPSS procedures shows the step-by-step procedure from the cocitation frequency data file to correlations matrix and to distance (proximity) matrix. These three steps can be performed by taking two successive steps.

First, open your dataset (9vardata.sav). Convert the cocitation frequency count matrix to correlations matrix by choosing ANALYZE→CORRELATE →DISTANCE procedure.

```
GET FILE='C:\Documents and Settings\sbeom\My Documents\ideaaca\spss\
9vardata.sav'.

PROXIMITIES
  alter bonczek carlson huber keen smorton simon sprague whinston
  /MATRIX OUT (*)
  /VIEW=VARIABLE
  /MEASURE= CORRELATION
  /STANDARDIZE= NONE .

PROXIMITIES  alter bonczek carlson huber keen smorton simon sprague
whinston
  /MATRIX OUT  (*)
```

Figure 15. Factor analysis: Rotation dialog box

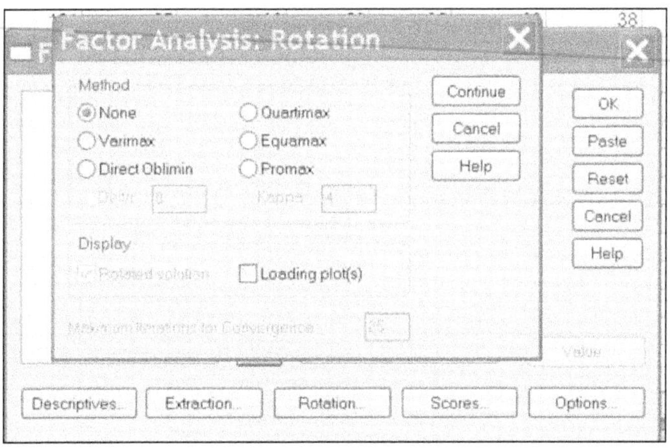

Figure 16. Correlation matrix

Correlation Matrix[a,b]

		bonczek	alter	carlson	huber	keen	smorton	simon	sprague
Correlation	bonczek	1.000	-.470	-.023	-.529	-.305	-.315	-.390	-.006
	alter	-.470	1.000	.445	-.159	.683	.656	.167	.416
	carlson	-.023	.445	1.000	.024	.619	.563	.193	.877
	huber	-.529	-.159	.024	1.000	.194	.138	.168	.219
	keen	-.305	.683	.619	.194	1.000	.972	.563	.660
	smorton	-.315	.656	.563	.138	.972	1.000	.608	.572
	simon	-.390	.167	.193	.168	.563	.608	1.000	.204
	sprague	-.006	.416	.877	.219	.660	.572	.204	1.000
	whinston	.981	-.456	-.616	-.529	-.273	-.285	-.348	.007

a. Determinant = .000

b. This matrix is not positive definite.

Figure 17. Reproduced correlations

		Reproduced Correlations						
		bonczek	alter	carlson	huber	keen	smorton	simon
Reproduced Correlation	bonczek	.949[b]	-.418	.059	-.555	-.363	-.382	-.477
	alter	-.418	.832[b]	.401	-.190	.723	.744	.457
	carlson	.059	.401	.806[b]	.093	.728	.676	.299
	huber	-.555	-.190	.093	.944[b]	.137	.091	.246
	keen	-.363	.723	.728	.137	.904[b]	.879	.518
	smorton	-.382	.744	.676	.091	.879	.862[b]	.513
	simon	-.477	.457	.299	.246	.518	.513	.387[b]
	sprague	.037	.324	.839	.237	.727	.664	.306
	whinston	.940	-.395	.079	-.557	-.338	-.357	-.463
Residual[a]	bonczek		-.051	-.082	.025	.059	.067	.087
	alter	-.051		.044	.030	-.040	-.088	-.289
	carlson	-.082	.044		-.069	-.109	-.113	-.106
	huber	.025	.030	-.069		.058	.046	-.078
	keen	.059	-.040	-.109	.058		.093	.045
	smorton	.067	-.088	-.113	.046	.093		.095
	simon	.087	-.289	-.106	-.078	.045	.095	
	sprague	-.043	.092	.038	-.018	-.068	-.092	-.102
	whinston	.041	-.060	-.094	.027	.065	.072	.115

Extraction Method: Principal Component Analysis.

a. Residuals are computed between observed and reproduced correlations. There are 24 (66.0%) nonredundant residuals with ab greater than 0.05.

b. Reproduced communalities.

Figure 18. Component matrix

					eigen values and communalities				
comp	1	comp 2	comp 3		=b2*b2	=c2*c2	=d2*d2	=sum(f2:h2)	
bonczek	-0.584	0.775	0.084		0.341056	0.600625	0.007056	0.948737	94.87%
alter	0.736	0.073	-0.534		0.541696	0.005329	0.285156	0.832181	83.22%
carlson	0.666	0.551	0.242		0.443556	0.303601	0.058564	0.805721	80.57%
huber	0.327	-0.548	0.732		0.106929	0.300304	0.535824	0.943057	94.31%
keen	0.921	0.231	-0.052		0.848241	0.053361	0.002704	0.904306	90.43%
smorton	0.898	0.198	-0.129		0.806404	0.039204	0.016641	0.862249	86.22%
simon	0.599	-0.159	-0.051		0.358801	0.025281	0.002601	0.386683	38.67%
sprague	0.683	0.518	0.407		0.466489	0.268324	0.165649	0.900462	90.05%
whinston	-0.559	0.783	0.077		0.312481	0.613089	0.005929	0.931499	93.15%
				Eigen	4.225653	2.209118	1.080124	communality	7.514895
					=sum(f2:f10)				
				% Variance	46.95%	24.55%	12.00%	7.514895	

```
/VIEW= VARIABLE
/MEASURE= SEUCLID
/PRINT NONE
    /STANDARDIZE= NONE .
```

The dendrogram shows the process of starting with nine observation (authors) the first step in this procedure is to find the two observations/variables with the shortest distance and they are combined into a new aggregate cluster. Now, there

Figure 19. Comparison of SPSS and SAS outputs

SPSS and SAS use different terminologies. This sections compare outputs from SPSS and SAS.

	SAS output section	SPSS output section
Eigen values	Initial Factor Method: Principal component section	Descriptive statistics (total variance explained)
Factor pattern	Initial Factor Method: Principal component section	Component matrix
Rotated factor pattern	Pre-rotation method: Varimax section	Rotated component matrix
Factor structure correlations	Rotation method=promax section: Factor structure	Structure matrix

are a total of 8 clusters/observations (1 cluster and 7 observations). As the process continues, the number of observations and clusters to be joined is reduced by one each time.

CONDUCTING CLUSTER ANALYSIS

After the correlations dataset is imported and saved as SPSS data format, several steps are necessary to conduct cluster analysis. As Figure 12.18 shows, select the following steps.

Analyze
Classify
Hierarchical Cluster

There are two different clustering algorithms: Hierarchical and nonhierarchical (K-means clustering). ACA uses agglomerative hierarchical clustering procedure. The hierarchical clustering procedure has two types: agglomerative and divisive. Choose Hierarchical Cluster.

Do the following.

Move all 9 variables to Variable(s) box.

Check Variables, not Cases under Cluster.

Check both Statistics and Plots. Checking "Statistics" results in producing a proximity matrix and agglomeration schedule. Checking "Plots" generates dendrogram and vertical icicle.

Since the dataset contains 9 authors (variables), checking the range of solutions from minimum 2 to maximum 8 will produce a cluster membership table (Figure 24). You must click either agglomeration schedule or proximity matrix or both. If you leave these two boxes unchecked, you cannot continue. Cluster membership table is produced by using range of solution on the statistics subdialog box.

The cluster membership table provides ACA researchers with complementary information that can be used with thedendrogram. The dendrogram shows the joining sequence to the final clusterl. Starting from the left to right, it shows how all authors/variables are joining to form the final cluster. What is not clear is that which of the three author sets (bonczek, whinston), (carlson, sprague), (keen, scott morton, alter) forms the first cluster. The cluster membership table (8 clusters column) shows that bonczek and whinston formed the first cluster. The 7 cluster column of the same table shows that keen and scott morton joined next to form a new cluster.

Checking proximity matrix in Statistics Subdialog box in Figure 23 will generate a proximity matrix as shown below. The proximity matrix is an input to multidimensional scaling to be discussed later.

The plots subdialog box prompts you to indicate whether the dendrogram (Figure 27) should be included as part of output. A dendrogram is also called a tree graph. Next, it asks about the icicle plot. This name is used because icicle plot resembles a row of icicles hanging from an eave. The icicle plot summarizes the steps in forming clusters. It shows the number of clusters at each step and the members of each cluster.

This vertical icicle diagram shows the case (author variables) horizontally at the top and the number of clusters vertically at the number of clusters column. You will notice that between each case (author variable) column, there is a blank column. Therefore, each row begins with cluster number, followed by 9 case (author variable) and 8 blank columns. The easiest way of understanding the vertical icicle plot is to begin from the bottom (number of cluster = 8). Agglomerate hierarchical clustering begins with all 9 variables as observations/clusters.

After joining two observations to form the first cluster, the number of clusters becomes 8. The vertical icicle plot shows that at the very bottom, there are 8 cluster members (7 case/authors and one joined cluster (whinstone and bonczek). The "X" under each case(author) denotes membership status to each cluster, and the "X" under the blank between two cases/authors indicates the joined status. Similarly, row 3 can be interpreted as follows. When there are three clusters, the cluster members are one individual case (Huber) and two joined clusters [(bonczek, whinstone) and (Sprague, Carlson, Simon, Scott Morton, Keen, Alter). I noticed that the SAS output distinguishes the differences between cluster and observation. In SAS, a cluster is formed by joining two observations, or joining a cluster with an observation, or a

Figure 20. First step in cluster analysis

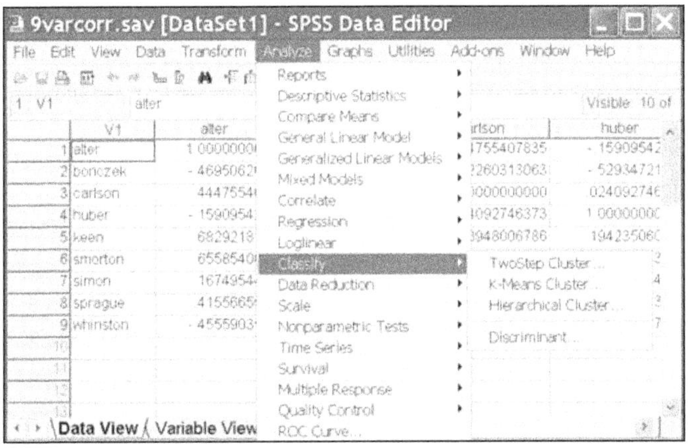

Figure 21. Hierarchical cluster analysis dialog box before selecting variables

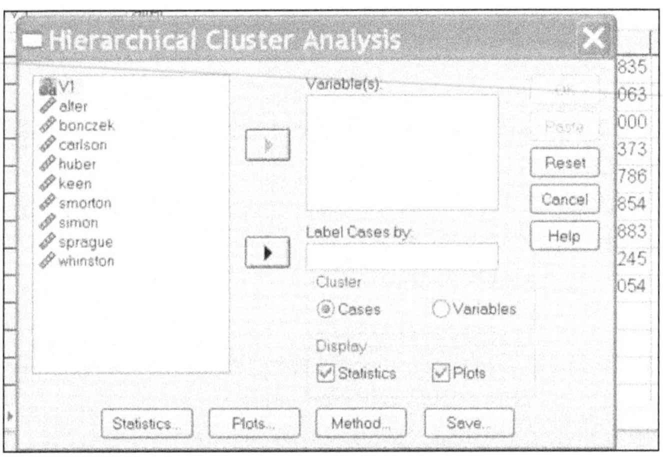

Clicking "Hierarchical Cluster" activates Hierarchical cluster analysis dialog box.

cluster with another cluster. Each variable before joining is called an observation. In SPSS, the cluster is not used by the same way used in SAS output. In other words, each variable, before or after joining, is a cluster.

The method subdialog box is a critically important part of inputs, which is needed to get correct outputs. Cluster method box offers 7 different cluster methods. Choose Ward's method. Readers are referred to Chapter IX in which a section to discuss different cluster methods is included. McCain suggests using either Ward's method or complete linkage [McCain, 1990 #8409]. In SPSS, complete linkage is equivalent to between-group linkage.

The method subdialog box is a critically important part of inputs to get correct outputs. The cluster method box offers. No value transformation is necessary.

Saving Command Syntax

As the user follows a step-by-step procedure, the SPSS system records each and every step the user took. Rather than repeating the same procedure again and again, you may want to create a command file containing all commands. To do so, click the "paste" button on Hierarchical Cluster Analysis Dialog Box (Figure 32). You will be directed to the SPSS syntax editor. The editor window contains the following command syntax. You may save it for later use as "Clustersyntax". This command is recorded after opening the input file. You may add the command to get the input file from the output.

```
GET FILE='C:\Documents and Settings\sbeom\My Documents\ideaaca\spss\
9vardata.sav'.
PROXIMITIES
  alter bonczek carlson huber keen smorton simon sprague whinston
  /MATRIX OUT (*)
  /VIEW=VARIABLE
  /MEASURE= CORRELATION
```

Figure 22. Hierarchical cluster analysis dialog box after selecting variables

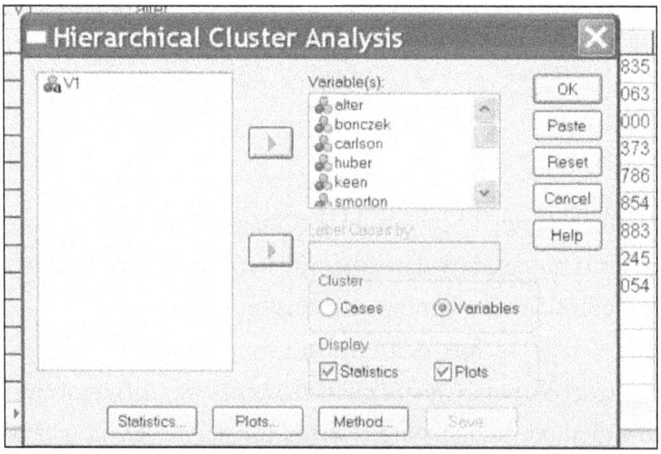

Clicking the "statistics" button will activate the statistics dialog box.

Figure 23. Statistics subdialog box

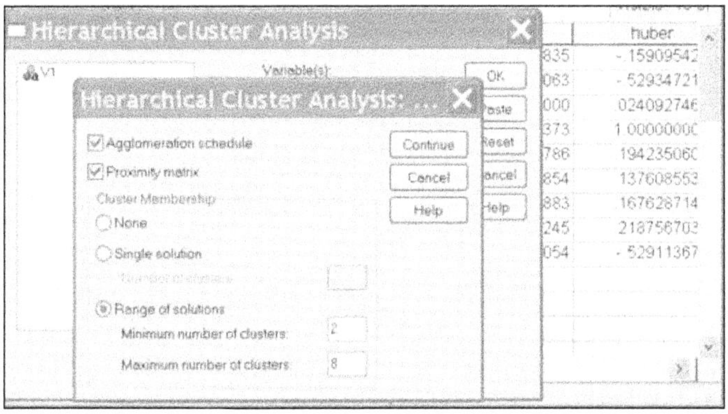

Figure 24. Cluster membership table

Cluster Membership

Case	8 Clusters	7 Clusters	6 Clusters	5 Clusters	4 Clusters	3 Clusters	2 Clusters
alter	1	1	1	1	1	1	1
bonczek	2	2	2	2	2	2	2
carlson	3	3	3	3	3	1	1
huber	4	4	4	4	4	3	1
keen	5	5	5	1	1	1	1
smorton	6	5	5	1	1	1	1
simon	7	6	6	5	1	1	1
sprague	8	7	3	3	3	1	1
whinston	2	2	2	2	2	2	2

Figure 25. Proximity matrix

Proximity Matrix

	Matrix File Input								
Case	alter	bonczek	carlson	huber	keen	smorton	simon	sprague	whinston
alter	.000	9.146	1.270	3.419	.732	.695	1.636	1.451	8.923
bonczek	9.146	.000	6.971	7.873	10.249	10.077	8.220	7.195	.005
carlson	1.270	6.971	.000	3.585	.873	.988	2.104	.072	6.752
huber	3.419	7.873	3.585	.000	3.524	3.472	1.931	3.246	7.756
keen	.732	10.249	.873	3.524	.000	.019	1.183	.827	9.954
smorton	.695	10.077	.988	3.472	.019	.000	.996	.971	9.781
simon	1.636	8.220	2.104	1.931	1.183	.996	.000	2.083	7.968
sprague	1.451	7.195	.072	3.246	.827	.971	2.083	.000	6.971
whinston	8.923	.005	6.752	7.756	9.954	9.781	7.968	6.971	.000

Figure 26. Plots subdialog box

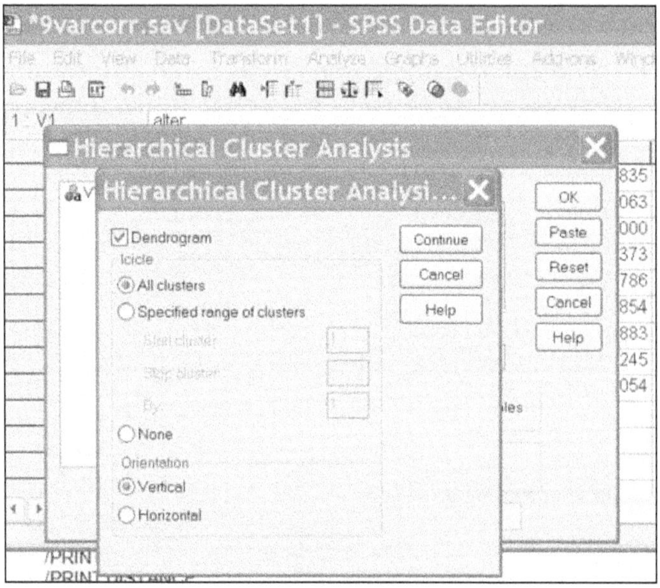

Figure 27. Dendrogram

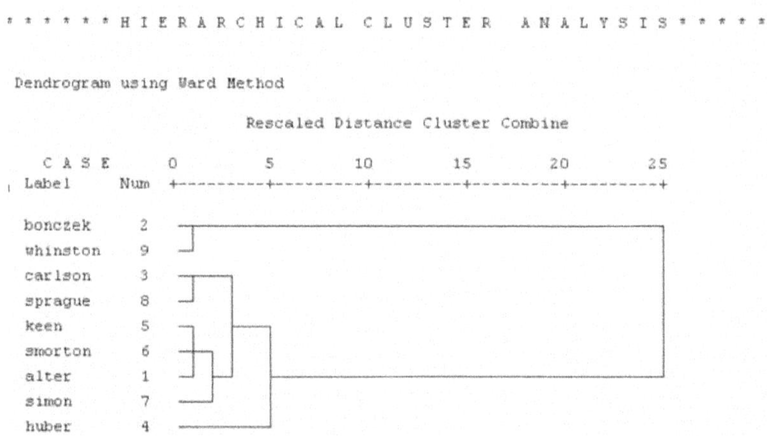

/STANDARDIZE= NONE .
PROXIMITIES alter bonczek carlson huber keen smorton simon sprague whinston
/MATRIX OUT (*)
/VIEW= VARIABLE

Figure 28. Horizontal icicle

Horizontal Icicle

	Number of clusters							
Case	1	2	3	4	5	6	7	8
whinston	X	X	X	X	X	X	X	X
	X	X	X	X	X	X	X	X
bonczek	X	X	X	X	X	X	X	X
	X							
huber	X	X	X	X	X	X	X	X
	X	X						
sprague	X	X	X	X	X	X	X	X
	X	X	X	X	X	X		
carlson	X	X	X	X	X	X	X	X
	X	X	X					
simon	X	X	X	X	X	X	X	X
	X	X	X	X				
smorton	X	X	X	X	X	X	X	X
	X	X	X	X	X	X	X	
keen	X	X	X	X	X	X	X	X
	X	X	X	X	X			
alter	X	X	X	X	X	X	X	X

Figure 29. Vertical icicle

Vertical Icicle

	Case																
Number of clusters	whinston		bonczek		huber		sprague		carlson		simon		smorton		keen		alter
1	X	X	X	X	X	X	X	X	X	X	X	X	X	X	X	X	X
2	X	X	X		X	X	X	X	X	X	X	X	X	X	X	X	X
3	X	X	X		X		X	X	X	X	X	X	X	X	X	X	X
4	X	X	X		X		X	X	X		X	X	X	X	X	X	X
5	X	X	X		X		X	X	X		X		X	X	X	X	X
6	X	X	X		X		X	X	X		X		X	X	X		X
7	X	X	X		X		X		X		X		X	X	X		X
8	X	X	X		X		X		X		X		X		X		X

```
 /MEASURE= SEUCLID
 /PRINT  NONE
 /STANDARDIZE= NONE .
CLUSTER
 /MATRIX IN  (*)
 /METHOD WARD
 /PRINT SCHEDULE CLUSTER(2,8)
 /PRINT DISTANCE
 /PLOT DENDROGRAM VICICLE.
```

Executing Saved Syntax Files

From SPSS data editor, Choose file→Open→ Syntax and the syntax file you saved and open the file.

File
Open
Syntax

MULTIDIMENSIONAL SCALING

Multidimensional scaling (MDS) aims to construct a geometric representation of data in a Euclidian space. In ACA, our purpose is to create an empirical map of authors in an academic discipline. While cluster analysis shows the process of joining all individual authors into one final cluster, MDS outputs focus on the final geometrical configuration on a two or three dimensional space. On a MDS map, authors whose works are cocited together frequently will be mapped closely. The higher co-citation frequency is translated into shorter distance between the co-cited authors.

Figure 30. Method subdialog box

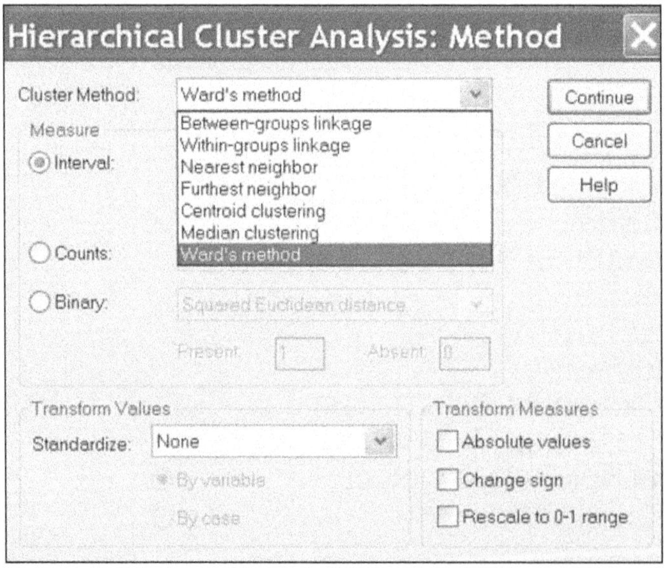

Figure 31. Method subdialog box specifying variable measure

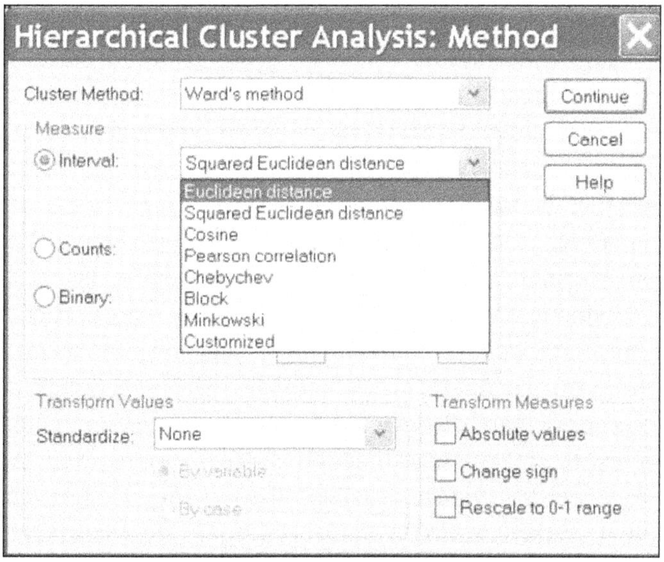

Figure 32. Method subdialog box specifying variable measure

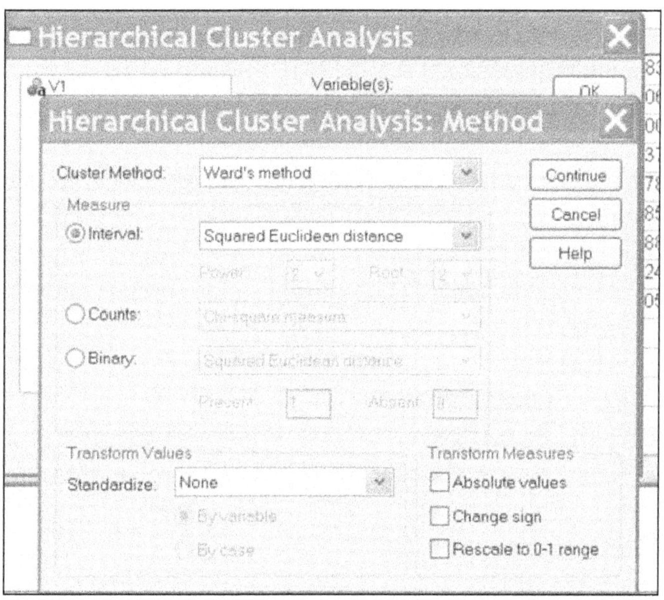

Input to MDS is a proximity matrix. The proximity matrix is already created in the process of conducting hierarchical clustering procedures (Figure 12.23). We begin MDS by opening the proximity matrix . Then Choose Analyze → Scale → Multidimensional Scaling.

After choosing multidimensional scaling (ALSCAL), the multidimensional scaling dialog box (Figure 34) appears. As shown, move all variables to the variable box. Figure 34 shows that the variables are already moved. Since the proximity matrix is used as an input file, choose "Data are distance" as shown above. Next click the "Model" button. The model subdialog box is shown in Figure 35.

The model subdialog box needs three important user inputs. First, the level of measurement must be specified. The proximity matrix we are using as an input is treated as "ordinal". Therefore, check "ordinal". Since the proximity matrix is a single matrix, check "matrix" under conditionality. Enter 2 and 3 to minimum and maximum box to create a two dimensional MDS map as well as a three dimensional map. Check "Euclidean distance" under scaling model. After clicking "Continue" button, you will see MDS dialogbox (Figure 35). Click "Option" button to see the option subdialog box (Figure 36). Readers are referred to Chapter X in regard to this issue.

Check all items under "Display" and all other inputs under "Criteria" are default values. After clicking the "Continue" button, you will see MDS dialog box (Figure 35) again. The MDS outputs will be produced after clicking "OK". The major outputs are summarized below.

To perform multidimensional scaling (MDS) using ALSCAL on a similarities matrix, command syntax is used.

```
GET FILE='C:\Documents and Settings\sbeom\My Documents\ideaaca\spss\
9vardata.sav'.
PROXIMITIES
  alter bonczek carlson huber keen smorton simon sprague whinston
  /MATRIX OUT (*)
  /VIEW=VARIABLE
  /MEASURE= CORRELATION
  /STANDARDIZE= NONE .
PROXIMITIES  alter bonczek carlson huber keen smorton simon sprague
whinston
  /MATRIX OUT  (*)
  /VIEW= VARIABLE
  /MEASURE= SEUCLID
  /PRINT  NONE
  /STANDARDIZE= NONE .
```

```
ALSCAL
 /MATRIX= IN (*)
 /SHAPE=SYMMETRIC
 /LEVEL=ORDINAL
 /CONDITION=MATRIX
 /MODEL=EUCLID
 /CRITERIA=CONVERGE(.001) STRESSMIN(.005) ITER(30) CUTOFF(0)
  DIMENS(2,3)
 /PLOT=DEFAULT ALL
 /PRINT=DATA HEADER.
```

MDS PROCEDURE OUTPUTS

The following is a selected portion of MDS(ALSCAL) outputs.

```
GET
  FILE='C:\Documents and Settings\sbeom\My
Documents\ideaaca\spss\9vardata.sav'.
PROXIMITIES
    alter bonczek carlson huber keen smorton simon sprague
  whinston
  /MATRIX OUT (*)
  /VIEW=VARIABLE
  /MEASURE= CORRELATION
  /STANDARDIZE= NONE .
```

Figure 33. Conducting MDS procedures

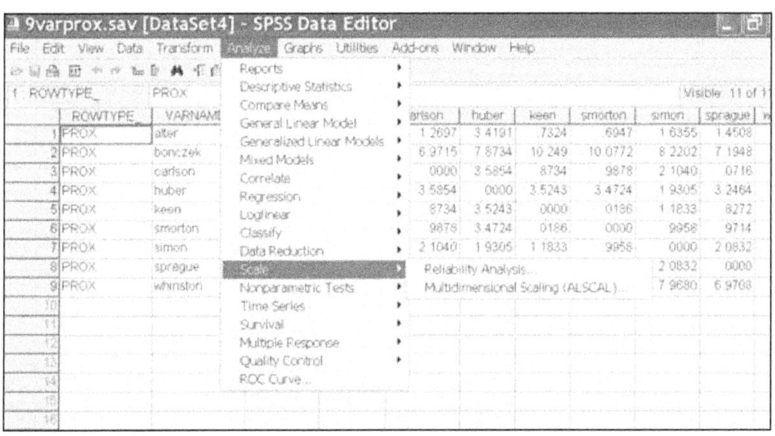

PROXIMITIES

Notes		
Output Created		09-JUN-2007 12:10:17
Comments		
Input	**Data**	C:\Documents and Settings\sbeom\My Documents\ideaaca\spss\9vardata.sav
	Filter	<none>
	Weight	<none>
	Split File	<none>
Missing Value Handling	**Definition of Missing**	User-defined missing values are treated as missing.
	Cases Used	Statistics are based on cases with no missing values for any variable used.
Syntax		PROXIMITIES alter bonczek carlson huber keen smorton simon sprague whinston /MATRIX OUT (*) /VIEW=VARIABLE /MEASURE= CORRELATION /STANDARDIZE= NONE .
Resources	**Elapsed Time**	0:00:00.78
	Workspace Bytes	544
	Processor Time	0:00:00.45

Figure 34. MDS dialog box

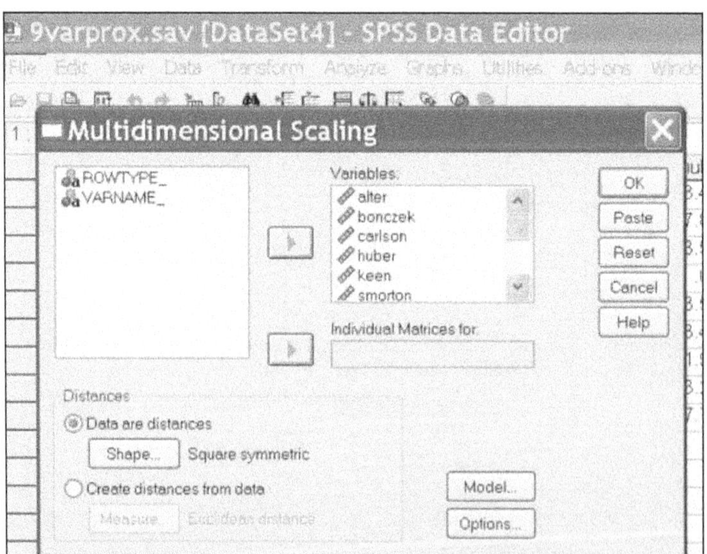

Figure 35. MDS model subdialog box

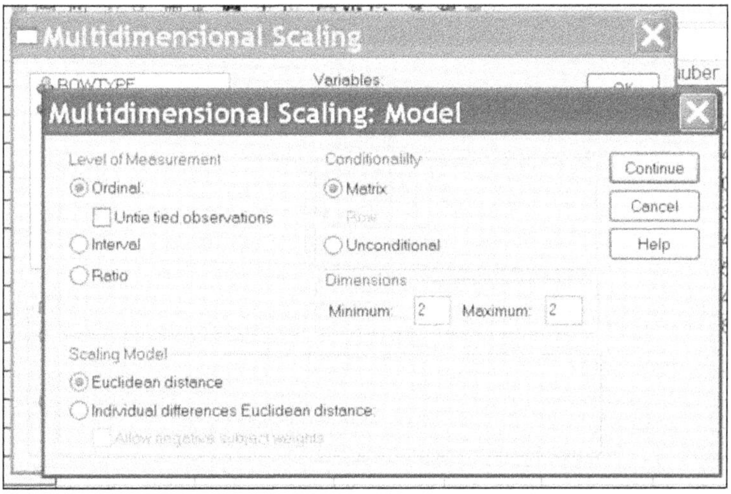

Figure 36. MDS Option subdialog box

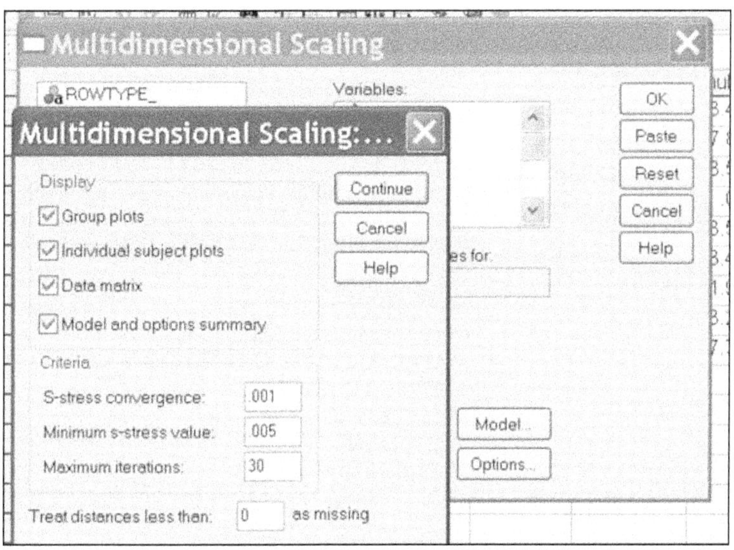

Files Saved	**Matrix File**	working data file

```
C:\Documents and Settings\sbeom\My Documents\ideaaca\spss\9vardata.
sav
```

Case Processing Summary					
Cases					
Valid		Missing		Total	
N	Percent	N	Percent	N	Percent
9	100.0%	0	.0%	9	100.0%

```
PROXIMITIES alter bonczek carlson huber keen smorton simon sprague whinston
 /MATRIX OUT  (*)
 /VIEW= VARIABLE
 /MEASURE= SEUCLID
 /PRINT  NONE
 /STANDARDIZE= NONE .
```

PROXIMITIES

Notes		
Output Created		09-JUN-2007 12:10:18
Comments		
Input	**File Label**	MATRIX FILE
	Filter	<none>
	Weight	<none>
	Split File	<none>
Missing Value Handling	**Definition of Missing**	User-defined missing values are treated as missing.
	Cases Used	Statistics are based on cases with no missing values for any variable used.
Syntax		PROXIMITIES alter bonczek carlson huber keen smorton simon sprague whinston /MATRIX OUT (*) /VIEW= VARIABLE /MEASURE= SEUCLID /PRINT NONE /STANDARDIZE= NONE .
Resources	**Elapsed Time**	0:00:00.36
	Workspace Bytes	472
	Processor Time	0:00:00.26
Files Saved	**Matrix File**	working data file

Case Processing Summary(a)					
Cases					
Valid		Missing		Total	
N	Percent	N	Percent	N	Percent
9	100.0%	0	.0%	9	100.0%
a Squared Euclidean Distance used					

```
ALSCAL
  /MATRIX= IN (*)
  /SHAPE=SYMMETRIC
  /LEVEL=ORDINAL
  /CONDITION=MATRIX
  /MODEL=EUCLID
  /CRITERIA=CONVERGE(.001) STRESSMIN(.005) ITER(30)
  CUTOFF(0) DIMENS(2,3)
  /PLOT=DEFAULT ALL
  /PRINT=DATA HEADER.
```

REFERENCES

Child, D. (1970). *The Essentials of Factor Analysis*. London: Holt, Rinehart and Winston, Inc.

Hair, Jr., J. F., Anderson, R. E., & Tatham, R. L. (1987). *Multivariate Data Analysis with Readings* (2nd ed.). New York: Macmillan Publishing Company.

McCain, K. W. (1990). Mapping Authors in Intellectual Space: A Technical Overview. *Journal of the American Society for Information Science, 41*(6), 433-443.

Mertler, C. A., & Vannatta, R. A. (2005). *Advanced and Multivariate Statistical Methods: Practical Application and Interpretaion* (Third ed.). Glendale, CA: Pyrczak Publishing.

Tabachnick, B. G., & Fidell, L. S. (1996). *Using Multivariate Statistics* (Third ed.). New York: Harper Collins.

Proximity Matrix									
	Correlation between Vectors of Values								
	alter	**bonczek**	**carlson**	**huber**	**keen**	**smorton**	**simon**	**sprague**	**whinston**
alter	1.000	-.470	.445	-.159	.683	.656	.167	.416	-.456
bonczek	-.470	1.000	-.023	-.529	-.305	-.315	-.390	-.006	.981
carlson	.445	-.023	1.000	.024	.619	.563	.193	.877	-.016
huber	-.159	-.529	.024	1.000	.194	.138	.168	.219	-.529
keen	.683	-.305	.619	.194	1.000	.972	.563	.660	-.273
smorton	.656	-.315	.563	.138	.972	1.000	.608	.572	-.285
simon	.167	-.390	.193	.168	.563	.608	1.000	.204	-.348
sprague	.416	-.006	.877	.219	.660	.572	.204	1.000	.007
whinston	-.456	.981	-.016	-.529	-.273	-.285	-.348	.007	1.000
This is a similarity matrix									

Section V
Applications in the MIS Area

Chapter XII
The Intellectual Structure of Decision Support Systems Research (1969–1989)

ABSTRACT

This is the capstone chapter that shows how the concepts, tools, and techniques discussed in each of the previous chapters can be applied in conducting author cocitation analysis using a real data in the DSS area. The step-by-step procedures are shown in detail from the preparation of data file in Excel format and importing the file to the SAS system for multivariate statistical analysis. This chapter also guides the readers through the process of analyzing the results of principal components analysis, cluster analysis, and multi-dimensional scaling. The chapter also shows how to apply different criteria to select the optimal number of factor solutions, cluster solutions, and evaluating the acceptability of multi-dimensional scaling outputs. This chapter reports part of the intellectual structure of the DSS field by means of an empirical assessment of the DSS literature over the period 1969 through 1989. Three multivariate data analysis tools (factor analysis, multidimensional scaling, and cluster analysis) are applied to an author cocitation frequency matrix derived from a large database file of comprehensive DSS literature over the same period. Four informal clusters of decision support systems (DSS) research subspecialties and a reference discipline were uncovered. Four of these represent DSS research subspecialties—foundations, model/data management, user-interface/individual differences, and group support systems. One other conceptual grouping defines a reference discipline of DS—organizational science.

INTRODUCTION

All previous discussions in the book provide the readers with detailed descriptions of concepts, issues, processes, and tools of author cocitation analysis. This chapter demonstrates how the concepts, tools, and techniques discussed in each of the previous chapters can be applied in conducting author cocitation analysis, using real data in the MIS area. The step-by-step procedures will show the readers what to do and what not to do and how to do it in a most efficient and effective way. The application section, section V, contains two chapters. The first chapter in the section emphasizes the technical dimension of ACA, while the last chapter put more emphasis on the non-technical aspects of ACA.

Within the managerial and organizational context, it was in the early 1950s when one of the first computers began to process payroll data. Since then, the study of computers and information systems has evolved continuously. Since the early 1970s, scholars in the management information systems (MIS)/decision support systems (DSS) areas have recognized the important roles computer-based information systems play in supporting managers in their semi-structured or unstructured decision making activities. For example, Gorry and Scott Morton (1971, p. 57) made the controversial claim that, "information systems should exist only to support decisions." Since then, there has been a growing amount of research in the area of DSS (Elam, Huber, & Hurt, 1986; Hyun B. Eom & Sang M. Lee, 1990; Farhoomand, 1987; Teng & Galletta, 1990). As Keen (1980) indicated in the early 1980s, it is necessary for information systems research to clarify reference disciplines and to build a cumulative tradition, in order to become a coherent and substantive field. This is necessary for DSS research as well. In the DSS area, Eom, Lee, and Kim (1993) conducted an initial study to identify two areas of contributing disciplines (management science and multiple criteria decision making) and five subspecialties of DSS research (foundations, group DSS, database management systems, multiple-criteria DSS, marketing DSS, and routing DSS). Due to the restrictive nature of their data set (specific DSS applications only), their study failed to provide a comprehensive picture of DSS research subspecialties.

Decision support systems are a relatively young field of study, compared to others, such as economics, organizational behavior, etc. As an area continues to grow and becomes a coherent field, the study of the intellectual development of the field is important. A number of prior studies have been conducted to assess the extent of progress within these stages in the decision support systems area (Eom, 2007).

The decision support systems (DSS) area originated from the late 1960's. It continues to be a coherent academic discipline. This chapter investigates the intellectual development of DSS during 1969 and 1989 using ACA. Cocitation counts data is to be analyzed to assess the progress that has been made during this time.

DATA

The primary data for this study were gathered from a total of 632 citing articles in the DSS area over the past 20 years (1969-1989). The details of the nature of these citing articles can be found in (Eom, 1995; Eom & Farris, 1996). These citing articles consist of two categories—specific DSS application articles based on the same definition and collection criteria of Eom and Lee (1990) and non-application articles.

The majority of published ACA studies use commercial databases as introduced in Chapter II. However, this chapter is based on the custom database and cocitaion matrix generation systems we developed using Foxbase database management systems. Custom database inherently requires a huge amount of labor to build and manage. Regardless of the database types (custom vs. on-line), all cocitation counts data files must be entered to the processing systems. The SAS system allows you to import external data into the system. Alternatively, it can be manually entered or the spreadsheet file such as xls extension Microsoft Excel file can be manually copied and pasted into the SAS system. The most efficient way is to import it from Excel file as discussed in Chapter VIII. To do so, the cocitation file should have the specific format (See Table 5 in Chapter VII). The row 1 of the Table 5 contains the ID variable and is followed by variable names. Cell A1 contains label, "author" and cell A2 contains first author name, ALTER, followed by BENBSAT and others. The matrix in Figure 1 consists of 59 rows and 59 columns matrix. Column 1 lists all 58 author names selected along with ID variable Author. This ID variable is used in the SAS statement later to display author names in principal component analysis, cluster analysis and multidimensional scaling and 2D and G3D plot analyses. Cocitation counts are retrieved from the custom database of the decision support systems area by the cocitation generator we developed as introduced in Chapter V.

Two strong advantages of our cocitation count generator over commercial cocitation retrieval system are the ability to compute cocitaion counts of non-primary authors and the treatment of name homographs. SSCI only indexes author's last name and initials. Consequently, citation records by an author of the same last name and initials may have not been authored by the same author. The user should be aware the possible consequences of inexact result of query due to name-homographs (e.g., Smith, G., Smith, GA, Smith, GN, Smith, GD, Smith, GR, etc.). It is practically not possible for the indexers to differentiate among authors bearing the same first and middle initials.

Our system uses only the last name without the first or middle name information. It was very difficult to design a system that could use a combination of last names and first names. The question now is how to distinguish Johnson, Andrew and Johnson, John in the database if we use only last names. The only way we could handle this situation was adding the first initial to the last name. Records of Johnson, John are changed to JohnsonJ, whether he is the primary or non-primary author. However, in a given academic field, the database builders are familiar with most of authors in the filed. In our database of the decision support systems area, there are authors of same last name with different first names such as Davis, G. and Davis, R. The whole databases records should be changed from Davis, G to DavisG, G, and from Davis, R. to DavisR, R. Figure 1 includes several last names with concatenated their initials of first and middle names such as AntonyRN, CarlsonED, DavisGB, DavisR, etc.

Authors for this study were selected by a two-stage process. The first step is the loose screening based on the minimum citation of 60 or more times during the period of 1969-1989, based on the prior work of Culnan (1986). This step is an intuitive first step of loosely screening the final author set using the custom database. ACA researchers recommend applying cocitation counts data to finalize the author set for multivariate analysis. The average number of reference articles in our custom database is about 30 cited references per citing article. We can estimate the total number of records to be 18960 cited records from 632 citing articles. From the 18960 records, it will be an extremely difficult task to identify all the names of authors and to analyze and generate cocitation frequency counts. For that reason, citation based loose screening is necessary to cut down the number of authors as an intermediate step. This stage yields a list of about 90 names. With this list of 90 authors, we finalize the 58 author set after evaluating the 90 author cocitation matrix generation system, as discussed below.

COMPILATION OF COCITATION FREQUENCY MATRIX

To overcome a standard problem with the Institute for Scientific Information (ISI) databases which code only the first author of a cited work, a Fox-Base based matrix generation system was developed to compute a cocitation frequency between any pair of authors. The cocitation matrix generation system we developed gives access to cited coauthors as well as first authors.

In the next stage, the final author set was determined. Due to the possible instability of small cocitation counts, author cocitation analysis researchers introduced several ad hoc criteria for further screening a large pool of candidate authors to finalize a list of authors. The criteria include a *mean cocitation rate* above a certain lower limit per author in each time period (e.g., nine for 10 years of Social Scisearch data), cocitation with at least one-third of the entire author set, or restricting the final author set to the 20% receiving the highest number of citations and cocitations in initial retrieval trials. For further details on several different approaches to

Figure 1. The data files (period 69-79, triangle shape)

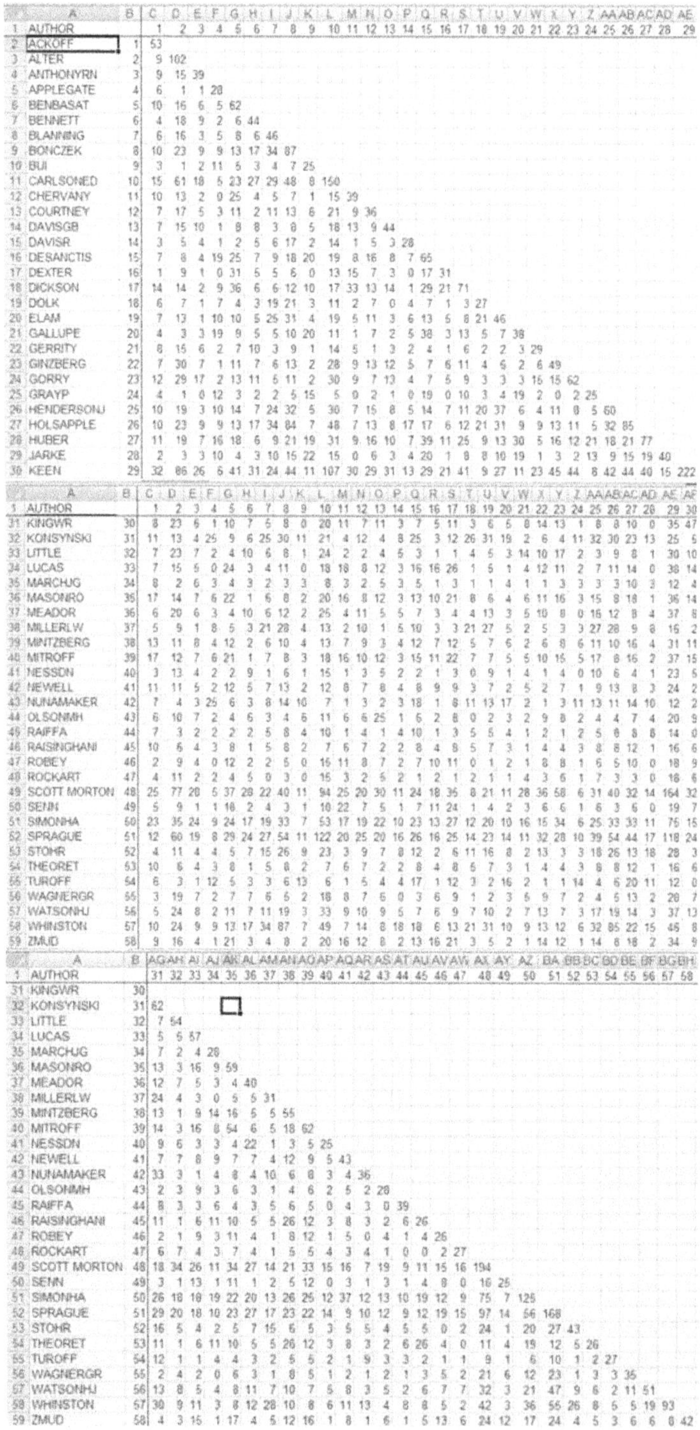

compiling a predetermined list of authors, see McCain (1990). However, all these criteria were suggested to be applied to the commercial on-line databases such as SCISEARCH and SOCIAL SCISEARCH.

Our databases are significantly different from those commercial databases in terms of size of records. Besides, the cocitation matrix generation system we developed gives access to cited coauthors as well as first authors. Due to these differences, we could not follow the suggested criteria e.g., (McCain, 1990) such as "a mean cocitation rate of 'x' or more cocitations in each time period." Rather, we had to invent a new criterion through the method of trial and error. We experimented with the sensitivity of changing the cocitation threshold on the final outcomes (number of meaningful factors to accurately represent DSS research subspecialties). With our databases, we conclude that the number of cocitations of an author with himself/herself can be a better criterion to determine the final author set due to the simplicity. Applying the mean cocitation criteria to our Lotus worksheet file (the output from the cocitation matrix generation system) involves too many computations. Whenever we delete/add an author to the final author set, we have to compute the mean cocitation rate of each author again. Using the cocitation rate of 25 with himself/herself in the period (1969-1989), the initial set of 90 authors was reduced to 58 as the final author set for further analysis. Later, we experimented by lowering the threshold from the 25 cocitation rate to a 20 cocitation rate. Although the number of authors increased slightly, the number of meaningful factors remained the same, and the major conclusions reported here did not change. Also, it is important to point out that cocitation thresholds themselves, as sole connection criteria, are suspect in a highly multidisciplinary area. One should look at the overall connectedness and the focused cocitation counts as well. Figure 2 contains all SAS statements that perform all the necessary statistical analyses to produce the outputs for our study.

RESEARCH METHODOLOGY

This study is based on the assumptions that "bibliographic citations are an acceptable surrogate for the actual influence of various information sources" (McCain, 1984) and that the cocitation analysis of a field yields a valid representation of the intellectual structure of the field (Bellardo, 1980; McCain, 1984, 1986; Smith, 1981).

Three multivariate data analysis tools (factor analysis, multidimensional scaling, and cluster analysis) were applied to an author cocitation frequency matrix derived from a large database file of comprehensive DSS literature over the period of 1969 through 1989. The primary research method for this study, Author Cocitation Analysis, consists of the following steps (Figure 2).

Figure 2. SAS statements

```
   /* importing data from external sources to the sas system
*/
   PROC IMPORT OUT= WORK.aca
                DATAFILE= "C:\Documents and Settings\sbeom\My
Documents\idea
               aca\spss\book7089.xls"
               DBMS=EXCEL REPLACE;
         SHEET="aca7089$";
         GETNAMES=YES;
         MIXED=NO;
         SCANTEXT=YES;
         USEDATE=YES;
         SCANTIME=YES;
   RUN;
   proc print data=aca;
   run;
   /* principal component analysis */
   proc factor data=aca method=principal mineigen=1 rotate =
promax;
   run;
   /* creating distance matrix for cluster and non-metric mds
analyses */
   Proc distance data=aca out=dist method=dcorr SHAPE=SQUARE;
   Var ratio(Ackoff--zmud);
   ID author;
   run;
   proc print data=dist;
   /* principal component analysis */
   proc cluster method=ward data=dist pseudo;
   ID author;
   proc tree space=2;
   run;
   /* non-metric mds analysis to produce configuration */
   PROC MDS
      data=dist
               dimension=3
      condition=un
               level=ordinal
      pconfig
      out=coord;
               ID author;
   title 'non-mteric mds with level=ordinal option';
   RUN;
   PROC PRINT data=coord (keep=author dim1 dim2 dim3);
   VAR author dim1 dim2 dim3;
   RUN;
```

continued on following page

Figure 2. continued

```
   options ps=40;
   /* creating two-dimensional plots */
   proc plot data=coord vtoh=2;
   plot dim1*dim2='*' $author
        / haxis = by .5
          vaxis= by .5 ;
     where _ type _ = 'CONFIG';
   plot (dim1 dim2)* dim3 = '*' $author
         / haxis = by .5
           vaxis= by .5 ;
      where _ type _ = 'CONFIG';

   title 'two dimensional plots of configuration';
   run;
   /* creating the annotate data set for g3d plot */
   data anno;
   set coord (keep=dim1 dim2 dim3 author);
   length function $8;
   RETAIN xsys ysys zsys '2'
   function 'label'
     size .7
     position '2'
     color 'blue'
     style 'zapf';
     x=dim1;
     y=dim2;
     z=dim3;
     text=author;
   /* Creating three-dimensional scatter plots using G3d Pro-
cedure */
   proc g3d
     data=coord;
        scatter  dim2*dim1=dim3/grid  caxis=black  anno=anno
shape='pyramid';
     label dim1='foundations'
           dim2='model management'
           dim3='user-interface';
      title 'three dimensional scatter plots with labels and
caxis';
   run;
   quit;
```

PRINCIPAL COMPONENT ANALYSIS

The raw cocitation matrix of 58 authors was analyzed by the principal components analysis. Selecting the optimal number of factors is an important decision to be made by an analyst using the latent root criterion (eigenvalue 1 criterion), scree test, etc. The eigenvalue criterion suggests the initial solution of 9 factors (Figure 3). The scree test is another approach that is often suggested. Figure 4 shows that the eigenvalues (latent roots) are plotted against the number of factors (1 through 12) in their order of extraction. Notice that as the number of factor increases, the slope of the line connecting two successive factors almost becomes flattened out. When looking at the point above factor 6, you will see that the curve beginning with factor 6 almost becomes a horizontal line. Thus, the scree test suggests that the 6 factor solution is an optimal solution. Finally, exercising expert judgment in regard to the interpretability of each factor could be the most important criterion.

Table 1 presents the nine factors extracted from the correlation matrix using principal components analysis. Care must be exercised when interpreting the substantiality of factors 6, 7, 8, and 9. Factor 6 has only two authors. Factors 7, 8, and 9 consist of a small number of authors with very low factor loadings. Furthermore, all these authors in factors 6 through 9 loaded on other factors with the same or higher absolute values. Although it is a usual convention to report all authors with loadings greater than .4, only those authors with loadings greater than .7 should be used to define factors. Consequently, I suggests that the five factor solution is more meaningful than the other two solutions suggested by the eigenvalue and scree test.

The uncovered DSS research subspecialties seem to be foundations (factor 1), model management (factor 2), user interface/individual differences (factor 3), group decision support systems (factor 4), and organization science (factor 5). Readers who are interested in the contributions of each author that appeared in the factor loadings are referred to (Eom & Farris, 1996).

The SAS outputs include also factor structure correlations from the oblique factor rotation (Table 2). The PROMAX rotation specification provides both orthogonal and oblique rotations with only one invocation of PROC FACTOR. Compared to an orthogonal rotation method, the oblique factor rotation is "more desirable because it is theoretically and empirically more realistic" (Hair, Anderson, & Tatham, 1987, p.245). It allows a more natural rotation without the imposition of orthogonal factors. Moreover, it generates additional information about the correlations between the factors (Table 3). Results from factor structure correlations (Table 2) suggest that the six factor solution may be more reasonable than the five factor solution only if we can interpret the results of factor six. I will not discuss the issue here.

Figure 3. Eigenvalues of the correlations matrix

Eigenvalues of the Correlation Matrix: Total = 58 Average = 1

	Eigenvalue	Difference	Proportion	Cumulative
1	21.0403292	11.2105452	0.3628	0.3628
2	9.8297840	3.8173298	0.1695	0.5322
3	6.0124542	2.1474359	0.1037	0.6359
4	3.8650183	0.6214201	0.0666	0.7025
5	3.2435981	1.5851264	0.0559	0.7585
6	1.6584717	0.2677114	0.0286	0.7871
7	1.3907603	0.1978882	0.0240	0.8110
8	1.1928721	0.1108442	0.0206	0.8316
9	1.0820279	0.2767053	0.0187	0.8503
10	0.8053227	0.0327492	0.0139	0.8641
11	0.7725734	0.1042077	0.0133	0.8775
12	0.6683658	0.1039644	0.0115	0.8890
13	0.5644013	0.0420454	0.0097	0.8987

Figure 4. Scree test

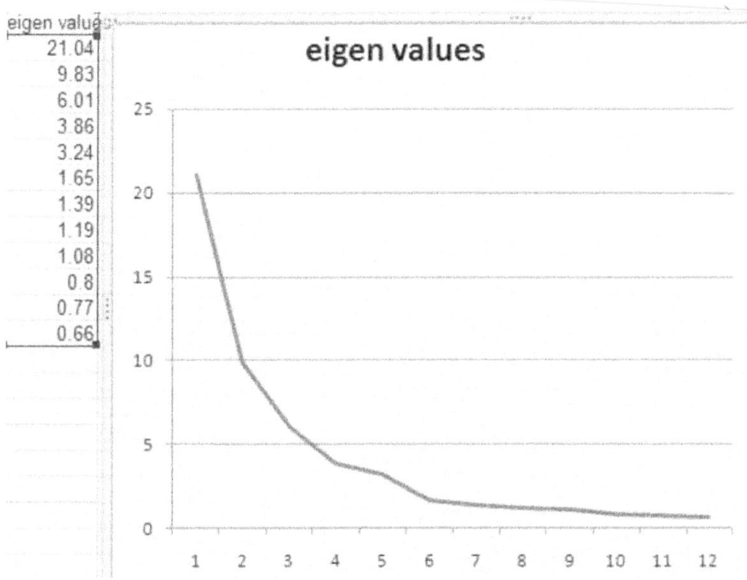

Table 1. Rotated factor pattern (1969-1989) rotation method = promax

Factor 1		Factor 2		Factor 3	
LITTLE	1.09	MILLERLW	1.03	DEXTER	1.11
GERRITY	1.03	DOLK	0.99	BENBASAT	1.02
ROCKART	0.88	BLANNING	0.97	SENN	1.00
ALTER	0.86	BONCZEK	0.95	DICKSON	0.99
GORRY	0.86	HOLSAPPLE	0.95	CHERVANY	0.95
SCOTT MORTON	0.80	WHINSTON	0.94	LUCAS	0.82
BENNETT	0.78	ELAM	0.90	ZMUD	0.76
WAGNERGR	0.75	HENDERSONJ	0.77	ROBEY	0.63
CARLSONED	0.73	STOHR	0.77	DESANCTIS	0.47
KEEN	0.73	KONSYNSKI	0.71	MITROFF	0.41
ANTHONYRN	0.72	DAVISR	0.67		
KINGWR	0.66				
SPRAGUE	0.66				
GINZBERG	0.60				
WATSONHJ	0.59				
MEADOR	0.55				
NESSDN	0.54				
SIMONHA	0.44				
Variance	21.04		9.83		6.01
%Variance	36.28%		16.95%		10.37%
Factor 4		Factor 5		Factor 6	
GALLUPE	0.99119528	RAISINGHANI	1.0521	OLSONMH	1.0284
GRAYP	0.97221291	THEORET	1.0521	DAVISGB	0.8868
TUROFF	0.96351253	MINTZBERG	0.9408		
BUI	0.95444498	MARCHJG	0.9057		
DESANCTIS	0.8420684	RAIFFA	0.5073		
JARKE	0.7598758	NEWELL	0.4144		
APPLEGATE	0.75560365				
HUBER	0.72123859				

continued on following page

Table 1. continued

NUNAMAKER	0.60193003				
Variance	3.865		3.243		1.66
%Variance	6.66%		5.59%		2.86%

Factor 7		Factor 8		Factor 9	
MASONRO	0.66271069	NESSDN	0.813	NEWELL	0.6657
MITROFF	0.64490607	MEADOR	0.6853		
ACKOFF	0.45317693				
NUNAMAKER	0.41160753				
Variance	1.39		1.19		1.08
%Variance	2.40%		2.06%		1.87%

Table 2. Factor structure correlations (1969-1989)

Factor 1		Factor 2		Factor 3	
SCOTT MORTON	0.94619	BLANNING	0.92265	BENBASAT	0.9367
KEEN	0.93061	ELAM	0.91375	CHERVANY	0.90878
ALTER	0.89236	WHINSTON	0.91153	DICKSON	0.90865
CARLSONED	0.87097	BONCZEK	0.91063	DEXTER	0.89841
SPRAGUE	0.84866	HOLSAPPLE	0.91043	ZMUD	0.89268
GORRY	0.84647	MILLERLW	0.90733	LUCAS	0.87835
LITTLE	0.84109	HENDERSONJ	0.87632	SENN	0.86518
GERRITY	0.80743	STOHR	0.86383	ROBEY	0.83517
SIMONHA	0.79706	DOLK	0.85067	KEEN	0.64918
ANTHONYRN	0.79581	KONSYNSKI	0.83304	MASONRO	0.64028
BENNETT	0.79525	DAVISR	0.73337	MITROFF	0.63848
GINZBERG	0.78107	SPRAGUE	0.60862	COURTNEY	0.61424
KINGWR	0.76044	CARLSONED	0.57269	SCOTT MORTON	0.61021
WAGNERGR	0.74216	JARKE	0.57125	KINGWR	0.58916
WATSONHJ	0.73106	COURTNEY	0.55549	GINZBERG	0.56362
ROCKART	0.72947	WATSONHJ	0.52404	WAGNERGR	0.5539
MEADOR	0.67788	RAIFFA	0.50206	SPRAGUE	0.54517
DAVISGB	0.66144	SIMONHA	0.49977	DAVISGB	0.54449
NESSDN	0.5843	MEADOR	0.45615	SIMONHA	0.53767
ROBEY	0.56199	NUNAMAKER	0.45009	ALTER	0.50668

continued on following page

Table 2. continued

ACKOFF	0.5571	HUBER	0.44489	CARLSONED	0.49654
ZMUD	0.55484	BENNETT	0.44176	HUBER	0.49346
COURTNEY	0.5491	KEEN	0.44087	GORRY	0.47181
LUCAS	0.5389	SCOTT MORTON	0.41073	MINTZBERG	0.46983
OLSONMH	0.53013			ACKOFF	0.43935
BENBASAT	0.47979			WATSONHJ	0.43331
DAVISR	0.47892			DESANCTIS	0.42542
STOHR	0.47071			NEWELL	0.421
HUBER	0.4603			OLSONMH	0.40818
NEWELL	0.4416				
CHERVANY	0.44009				
WHINSTON	0.43721				
HOLSAPPLE	0.42623				
MASONRO	0.42308				
HENDERSONJ	0.42007				
BONCZEK	0.41944				
Variance	17.21		13.01		13.41
Variance %	29.67%		22.43%		23.12%

Factor4		Factor5		Factor 6	
GALLUPE	0.94958	MINTZBERG	0.94637	DAVISGB	0.94079
BUI	0.91561	RAISINGHANI	0.88986	OLSONMH	0.92891
GRAYP	0.90546	THEORET	0.88986	KEEN	0.65226
TUROFF	0.89636	MARCHJG	0.79534	SCOTT MORTON	0.64885
DESANCTIS	0.87331	SIMONHA	0.65879	GINZBERG	0.63198
JARKE	0.85564	ACKOFF	0.58942	ANTHONYRN	0.60589
APPLEGATE	0.79947	NEWELL	0.56075	SIMONHA	0.60008
HUBER	0.7597	MITROFF	0.5293	KINGWR	0.58346
NUNAMAKER	0.71404	KEEN	0.52667	LUCAS	0.57028
KONSYNSKI	0.58347	COURTNEY	0.52529	GORRY	0.5618
		RAIFFA	0.52274	ROBEY	0.55342
		HUBER	0.51156	COURTNEY	0.54878
		SPRAGUE	0.50163	ALTER	0.53934
		MASONRO	0.4992	ZMUD	0.53902

continued on following page

Table 2. continued

		ROBEY	0.49462	SPRAGUE	0.53395
		SCOTT MORTON	0.47598	CARLSONED	0.52758
		ZMUD	0.46771	CHERVANY	0.48358
		WATSONHJ	0.44068	WAGNERGR	0.45295
		CARLSONED	0.43634	DICKSON	0.44566
		ANTHONYRN	0.43159	ACKOFF	0.44107
		LUCAS	0.41544	WATSONHJ	0.4354
		KINGWR	0.40772	BENNETT	0.41962
		BENBASAT	0.40701	MASONRO	0.41858
				MITROFF	0.41331
				MEADOR	0.40218
Variance	8.14		10.05		10.08
Variance %	14.03%		17.33%		17.38%
	Factor 7		Factor 8		Factor 9
MITROFF	0.56257	NESSDN	0.72801	NEWELL	0.42106
MASONRO	0.55636	MEADOR	0.69479	COURTNEY	-0.40148
NUNAMAKER	0.44532	HENDERSONJ	0.59955		
		ELAM	0.56011		
		KONSYNSKI	0.53439		
Variance	2.26		3.92		1.93
Variance %	3.90%		6.76%		3.33%

CLUSTER ANALYSIS

The cluster analysis resulted in the dendrogram (Figure 3), which illustrates the hierarchical clustering of several groups of DSS related researchers. Figure 3 shows the cluster structure. Figure 4 illustrates the joining sequence to show how each of the authors in the study is combined into a new aggregate cluster until all authors are grouped into the final one cluster. In addition, Figure 4 provides the reader with the pseudo F statistic and the pseudo T^2 statistic labeled as PST2.

PROC CLUSTER output (Figure 4) displays

- the Number of Clusters (NCL). Starting with 58 author variables, NCL starts with 57, meaning that two authors (Mason and Mitroff) joined together to

Table 3. Inter-factor correlations

	Fact 1	Fact 2	Fact 3	Fact 4	Fact 5	Fact 6	Fact 7	Fact 8	Fact 9
Factor1	1.00								
Factor2	0.34	1.00							
Factor3	0.50	0.14	1.00						
Factor4	0.01	0.33	0.02	1.00					
Factor5	0.41	0.27	0.45	0.11	1.00				
Factor6	0.61	0.13	0.51	0.01	0.40	1.00			
Factor7	-0.17	-0.09	-0.01	0.01	0.02	-0.17	1.00		
Factor8	0.12	0.36	0.07	0.12	0.16	0.02	0.21	1.00	
Factor9	-0.05	-0.07	-0.25	0.01	-0.14	-0.13	0.06	-0.25	1.00

form the first cluster and the remaining each of 56 authors make 56 separate clusters. Thus, the total number of clusters are 57 (1+56). The joining process continues until all observations are clustered into one whole cluster (agglomerative hierarchical clustering).

- the Clusters Joined. It shows which observations are joined to form a cluster under NCL column. The observations are identified by either the value of the ID variable or the observation number (if the ID variable is not used).
- Frequency of New Cluster or FREQ. It shows the total number of observations in the cluster. For example, CL44 has FREQ of 3. The members of CL44 are Dolk and cluster 48 (Elam and MillerLW).
- SPRSQ (Semipartial R-Squared). It shows the decrease in the proportion of variance accounted for resulting from joining the two clusters. It equals the between-cluster sum of squares divided by the corrected total sum of squares.
- RSQ (R-Squared). R^2 is the proportion of variance accounted for by the clusters.
- Pseudo F or PSF, the pseudo F statistic measuring the separation among all the clusters at the current level.
- Pseudo t^2 or PST2, the pseudo t^2 statistic measuring the separation between the two clusters most recently joined.

Selecting the Number of Clusters

In Chapter IX, we introduced how the PSF and PST2 statistic can be used to find an optimal number of clusters.

Figure 3. Dendrogram

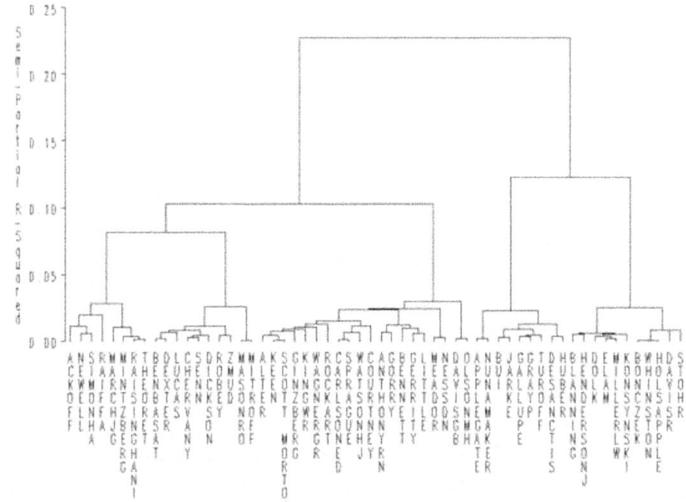

Figure 4. Cluster history

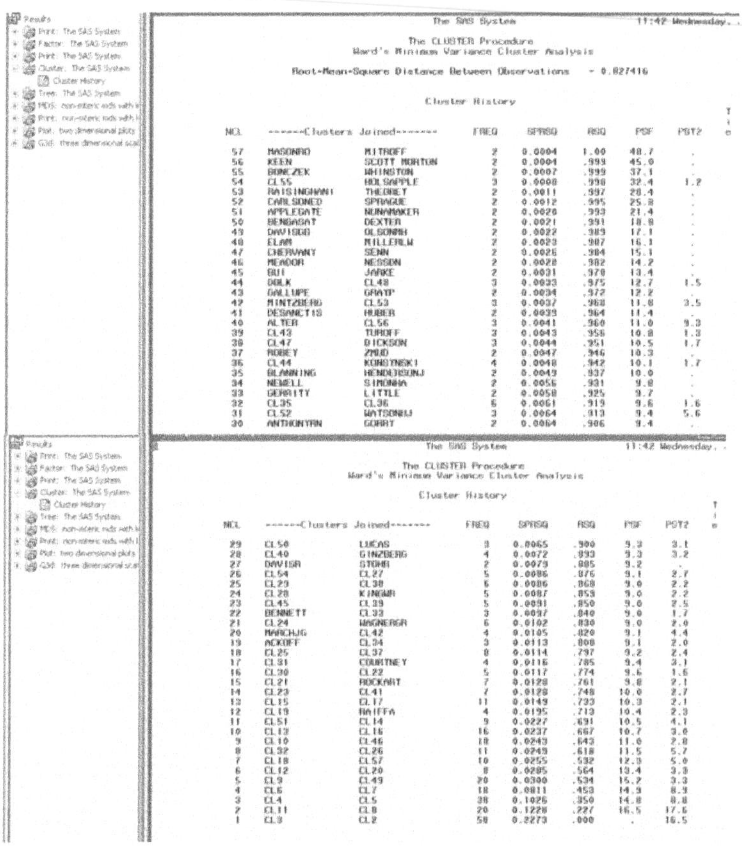

The SAS System 11:42 Wednesday, .

The CLUSTER Procedure
Ward's Minimum Variance Cluster Analysis

Root-Mean-Square Distance Between Observations = 0.827416

Cluster History

NCL	------Clusters Joined------		FREQ	SPRSQ	RSQ	PSF	PST2
57	MASONRO	MITROFF	2	0.0004	1.00	48.7	.
56	KEEN	SCOTT MORTON	2	0.0004	.999	45.0	.
55	BONCZEK	WHINSTON	2	0.0007	.999	37.1	.
54	CL55	HOLSAPPLE	3	0.0008	.998	32.4	1.2
53	RAISINGHANI	THEORET	2	0.0011	.997	28.4	.
52	CARLSONED	SPRAGUE	2	0.0012	.995	25.8	.
51	APPLEGATE	NUNAMAKER	2	0.0020	.993	21.4	.
50	BENBASAT	DEXTER	2	0.0021	.991	18.9	.
49	DAVISGB	OLSONMH	2	0.0022	.989	17.1	.
48	ELAM	MILLERLW	2	0.0023	.987	16.1	.
47	CHERVANY	SENN	2	0.0026	.984	15.1	.
46	MEADOR	NESSON	2	0.0028	.982	14.2	.
45	BUI	JARKE	2	0.0031	.978	13.4	.
44	DOLK	CL48	3	0.0033	.975	12.7	1.5
43	GALLUPE	GRAYP	2	0.0034	.972	12.2	.
42	MINTZBERG	CL53	3	0.0037	.968	11.8	3.5
41	DESANCTIS	HUBER	2	0.0039	.964	11.4	.
40	ALTER	CL56	3	0.0041	.960	11.0	9.3
39	CL43	TUROFF	3	0.0043	.956	10.8	1.3
38	CL47	DICKSON	3	0.0044	.951	10.5	1.7
37	ROBEY	ZMUD	2	0.0047	.946	10.3	.
36	CL44	KONSYNSKI	4	0.0048	.942	10.1	1.7
35	BLANNING	HENDERSONJ	2	0.0049	.937	10.0	.
34	NEWELL	SIMONHA	2	0.0056	.931	9.8	.
33	GERRITY	LITTLE	2	0.0058	.925	9.7	.
32	CL35	CL36	6	0.0061	.919	9.6	1.6
31	CL52	WATSONHJ	3	0.0064	.913	9.4	5.6
30	ANTHONYRN	GORRY	2	0.0064	.906	9.4	.
29	CL50	LUCAS	3	0.0065	.900	9.3	3.1
28	CL40	GINZBERG	4	0.0072	.893	9.3	3.2
27	DAVISGR	STOHR	2	0.0079	.885	9.2	.
26	CL54	CL27	5	0.0086	.876	9.1	2.7
25	CL29	CL38	6	0.0086	.868	9.0	2.2
24	CL28	KINGWR	5	0.0087	.859	9.0	2.2
23	CL45	CL39	5	0.0091	.850	9.0	2.5
22	BENNETT	CL33	3	0.0097	.840	9.0	1.7
21	CL24	WAGNERGR	6	0.0102	.830	9.0	2.0
20	MARCHJG	CL42	4	0.0105	.820	9.1	4.4
19	ACKOFF	CL34	3	0.0113	.808	9.1	2.0
18	CL25	CL37	8	0.0114	.797	9.2	2.4
17	CL31	COURTNEY	4	0.0116	.785	9.4	3.1
16	CL30	CL22	5	0.0117	.774	9.6	1.6
15	CL21	ROCKART	7	0.0128	.761	9.8	2.1
14	CL23	CL41	7	0.0128	.748	10.0	2.7
13	CL15	CL17	11	0.0149	.733	10.3	2.1
12	CL19	RAIFFA	4	0.0195	.713	10.4	2.3
11	CL51	CL14	9	0.0227	.691	10.5	4.1
10	CL13	CL16	16	0.0237	.667	10.7	3.0
9	CL10	CL46	18	0.0243	.643	11.0	2.8
8	CL32	CL26	11	0.0249	.618	11.5	5.7
7	CL18	CL57	10	0.0255	.592	12.3	5.0
6	CL12	CL20	8	0.0285	.564	13.4	3.3
5	CL9	CL49	20	0.0300	.534	15.2	3.3
4	CL6	CL7	18	0.0811	.453	14.9	8.9
3	CL4	CL5	38	0.1026	.350	14.8	8.8
2	CL11	CL8	20	0.1228	.227	16.5	17.6
1	CL3	CL2	58	0.2273	.000	.	16.5

Plot of the Pseudo F Statistic Against Number of Clusters

Plots of the pseudo F statistic against the number of clusters are useful in selecting the number of clusters (Figure 5). The peaks of pseudo F values in the plot can be used to decide on the number of clusters. Depending on the number of variables, ACA analysts may ignore the pseudo F values of the last 2 or 3 clusters (e.g., clusters 8, 9 and 10). Figure 5 has peaks at 2 and 5 clusters. Ignoring the last cluster, this criterion suggests only two clusters to be used. In ACA, the important criteria that cannot be ignored is the interpretability of clusters. When looking at the dendrogram (Figure 3), a five cluster solution appears to be the optimal solution.

Plot of the Pseudo t2 Statistic Against Number of Clusters

Plots of the pseudo t^2 statistic against the number of clusters are also useful to select the number of clusters (Figure 6). The valleys of pseudo t^2 statistic values in the plot can be used to decide the number of clusters. Depending on the number of variables, ACA analysts may ignore the values of the last 2 or 3 clusters. Figure 6 has two valleys at 5 and 6 clusters.

Considering the pseudo t^2 statistic and the pseudo F statistic simultaneously narrows down the number of clusters for the final selection. We can select the five cluster solution as the optimal solution. The interpretability of cluster solution is the decisive criteria in selecting the number of clusters. ACA analysts should be able to name each cluster based on contents of each author under each cluster.

Results of Cluster Analysis

Careful examination of cluster history, dendrogram, and inter-factor correlations should provide the ACA researchers with useful and essential information to assess the intellectual structure and trace the process of how each area has been developed to reach the current state. See (Eom & Farris, 1996) to study an example of how to use cluster history and dendrogram to analyze the process of how each authors contributed in the development and evolution of the decision support systems field. The dendrogram shows a linkage among various clusters, and influence could be inferred based on the close examination of the works of authors in the clusters. The linkage in the dendrogram may not in and of itself demonstrate influence. Comparison of the two solutions from the factor analysis and the cluster analysis may provide some valuable information on the similarities and differences of the two solutions to help us reach a better interpretation of the results of multivariate analysis.

Figure 5. Plot of pseudo F statistic against number of clusters

No of Clusters	1	2	3	4	5	6	7	8	9	10
Psedo F		16.5	14.8	14.9	15.2	13.4	12.3	11.5	11	10.7

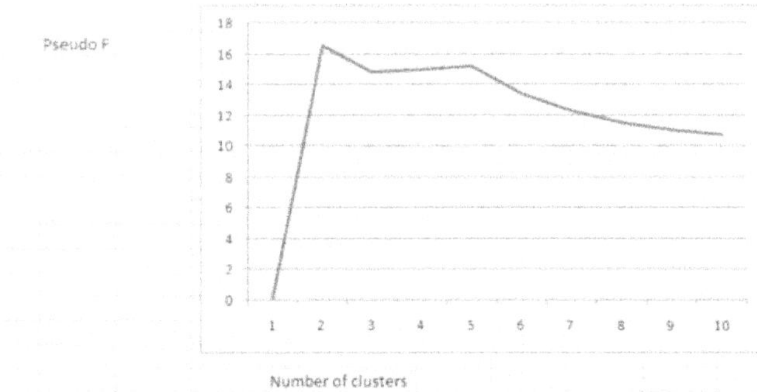

Figure 6. Plot of pseudo f statistic against number of clusters

No of Clusters	1	2	3	4	5	6	7	8	9	10
PST2	16.5	17.6	8.8	8.9	3.3	3.3	5	5.7	2.8	3

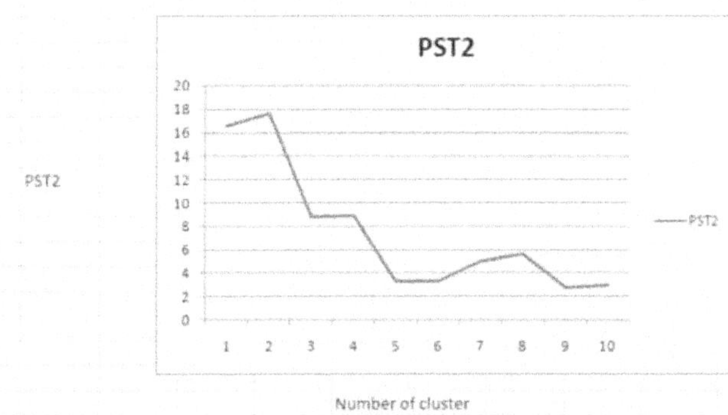

The dendrogram (Figure 3) illustrates the hierarchical clustering of five groups of DSS related researchers. Figure 4 shows the joining sequence to show how each of the authors in the study is combined into a new aggregate cluster until all 58 authors are grouped into the final one cluster (cluster one (CL1).

In cluster analysis, each author belongs to only one cluster, whereas an author can belong to more than one factor. The first cluster in Figure 3, starting from left, consist of Ackoff, Newell, Simon HA, Raiffa, March JG, Mintzbergh, Raisinghani, and Theoret. This cluster roughly corresponds to factor 5 in Table 1. Factor 5 does not include Ackoff and H.A. Simon. MDS output shows that these two authors are close to factor 5 authors as well as factor 1 authors. Next cluster of Benbasat through MItroff seems to represent user interface subspecialty group (Factor 3). In the middle, the cluster with the largest number of authors is equivalent to the foundation factor (Factor 1). Nine authors in the next cluster are identical to those authors in factor 4. The last cluster with 11 authors corresponds to factor 2 which has identical authors.

MULTIDIMENSIONAL SCALING

The distance matrix derived from the author cocitation matrix was used as an input to the multidimensional scaling program PROC MDS of personal computer SAS (version 9.1). This was done to visualize the similarity and dissimilarity within each group of DSS researchers, as well as the similarity and dissimilarity among the various groups of DSS researchers. The distance matrix was generated by the proc distance procedure of the SAS system (version 9). The output data set is used as input to the MDS and CLUSTER procedures (a hierarchical agglomerative clustering program with Ward's trace option).

In Chapter X, we have discussed the badness-of-fit-criterion as to the acceptability of MDS solution. the rule of thumb is that the stress value under 0.1 is considered to be excellent and over 0.15 may be unacceptable in most social science areas. In ACA area, a little higher stress value of less than .2 can be an acceptable criterion for 2 or 3 dimensional solutions with the higher R-square value due to the "noisy" nature of cocitation frequency data (McCain, 1990). This solution is an acceptable one with badness-of-fit value .1502 (2 dimensional solutions) and .0962 (3 dimensional solution).

ACA results reveal that over period of 1969-1989, four research subspecialties emerged along with one reference discipline.

REFERENCE (CONTRIBUTING) DISCIPLINES

There seem to be only one contributing discipline that has influenced the evolution of the state of DSS research subspecialties during the period of 1969-1989. There have been a number of assumed reference disciplines in the DSS area such as psychology, economics, computer science, political science, etc. Nevertheless, this study identified only a weak influence from the organizational sciences on the development of DSS research subspecialties.

Organizational Sciences

Factor 5 appears to represent *Organizational Sciences.* DSS are designed and implemented to support organizational as well as individual decision making. Without a detailed understanding of decision making behavior in organizations, "decision support is close to meaningless as a concept" (Peter G. W. Keen & Scott Morton, 1978, p. 61). Organizational scientists have classified organizational decision making in terms of several schools of thought: 1) the rational model focusing on the selection of the most efficient alternatives, with the assumption of a rational, completely informed, single decision maker; 2) the organizational process model by Cyert and March (1963) stressing the compartmentalization of the various units in any organization; 3) the satisficing model by Simon and his colleagues (Newell & Simon, 1972; Simon, 1969 (1981)) to find an acceptable, good enough solution, reflecting "bounded rationality"; 4) and other models.

DSS RESEARCH SUBSPECIALTIES

In 1980, Keen (1980) stated that MIS research lacked a cumulative tradition. In his view, there was virtually no cumulative research tradition in the MIS area without "continued follow-up on interesting lines of inquiry." Several areas of the DSS research subspecialties that emerged in this study provide us with some evidence as to the existence of a cumulative research tradition.

Foundations

Most authors in factor 1 conducted descriptive research to provide definitions and concepts in the very early stage of DSS development. Some authors clearly pinpointed a need for another type of information system to relieve managers' suffering from an "over abundance of irrelevant information" (Ackoff, 1967). Therefore Gorry and Scott Morton (1971) claimed, "Information systems should exist only to support decisions." Anthony (1965) classified all managerial activities into three categories: strategic planning, management control, and operational control. This taxonomy combined with that of Simon (1960) which classified all decisions into structured, semistructured, and unstructured, provided a simple schema for classifying organizational decisions to be best supported by TPS, MIS, and DSS. Little (1970) suggested a concept of *decision calculus* as "a model-based set of procedures for processing data and judgments to assist a manager in his decision making." Although he did not use the term DSS, he proposed the concept of a decision calculus which has several desirable characteristics of DSS

Figure 7. Three dimensional solution's badness-of-fit criterion

```
non-mteric mds with level=ordinal option          11:42 Wednesd

Multidimensional Scaling: Data=WORK.DIST.DATA
Shape=TRIANGLE Condition=UN Level=ORDINAL
Coef=IDENTITY Dimension=3 Formula=1 Fit=1

Mconverge=0.01 Gconverge=0.01 Maxiter=100 Over=2 Ridge=0.0001 Alternate=ROW=29
```

Iteration	Type	Badness-of-Fit Criterion	Change in Criterion	Convergence Measures	
				Monotone	Gradient
0	Initial	0.2534	.	.	.
1	Monotone	0.1757	0.0778	0.1817	0.6166
2	Gau-New	0.1185	0.0572	.	.
3	Monotone	0.1138	0.004618	0.0280	0.3098
4	Gau-New	0.1095	0.004336	.	.
5	Monotone	0.1053	0.004208	0.0240	0.2760
6	Gau-New	0.1014	0.003844	.	.
7	Monotone	0.1004	0.001059	0.0134	0.2402
8	Gau-New	0.0991	0.001337	.	.
9	Monotone	0.0987	0.000400	0.008255	0.2057
10	Gau-New	0.0981	0.000550	.	0.1686
11	Gau-New	0.0963	0.001795	.	0.0434
12	Gau-New	0.0962	0.0000859	.	0.0140
13	Gau-New	0.0962	9.8241E-6	.	0.006118

Convergence criteria are satisfied.

Figure 8. Two dimensional solution's badness-of-fit criterion

```
non-mteric mds with level=ordinal option          11:42 Wednesd

Multidimensional Scaling: Data=WORK.DIST.DATA
Shape=TRIANGLE Condition=UN Level=ORDINAL
Coef=IDENTITY Dimension=2 Formula=1 Fit=1

Mconverge=0.01 Gconverge=0.01 Maxiter=100 Over=2 Ridge=0.0001 Alternate=ROW=29
```

Iteration	Type	Badness-of-Fit Criterion	Change in Criterion	Convergence Measures	
				Monotone	Gradient
0	Initial	0.3215	.	.	.
1	Monotone	0.2305	0.0910	0.2233	0.5135
2	Gau-New	0.1732	0.0573	.	.
3	Monotone	0.1692	0.004063	0.0347	0.3437
4	Gau-New	0.1614	0.007743	.	.
5	Monotone	0.1557	0.005715	0.0300	0.2142
6	Gau-New	0.1534	0.002304	.	.
7	Monotone	0.1528	0.000569	0.0124	0.1471
8	Gau-New	0.1521	0.000754	.	.
9	Monotone	0.1517	0.000366	0.009508	0.1298
10	Gau-New	0.1511	0.000589	.	0.1021
11	Gau-New	0.1502	0.000947	.	0.0146
12	Gau-New	0.1502	0.0000120	.	0.009420

Convergence criteria are satisfied.

(*simple, robust, easy to control, adaptive, complete on important issues, easy to communicate with*). Sprague and Watson (1975) examined the necessity of including decision models in an integrated MIS. They emphasized that there is a need for a systematic way of embedding decision support models into MIS to support managers' decision making processes.

Keen and Scott Morton (1978) extended these previous works and suggested a widely accepted definition of DSS which implies "the use of computers to: assist managers in their decision processes in semistructured tasks; support, rather than

*Figure 9. Three dimensional MDS map (plot of dim1*dim2)*

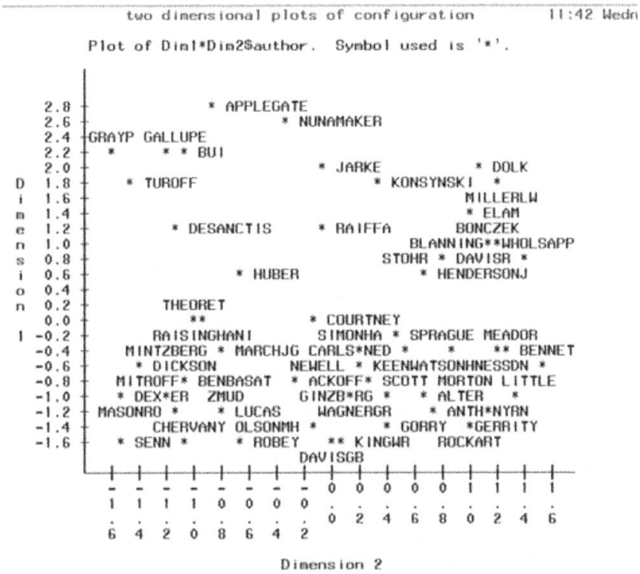

*Figure 9. Three dimensional MDS map (plot of dim2*dim3)*

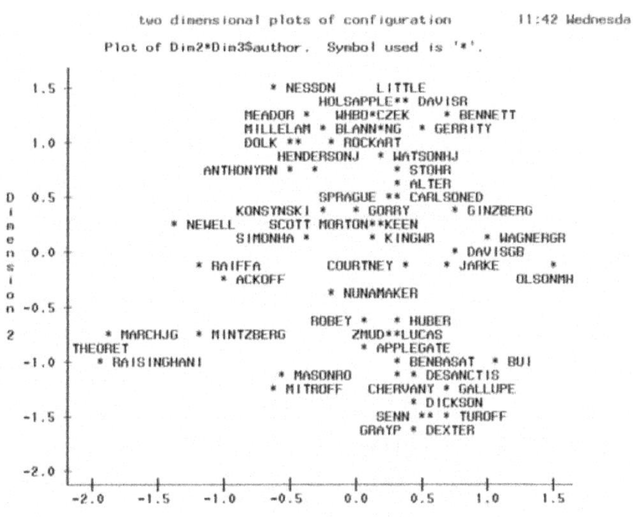

Figure 10. Three dimensional MDS map

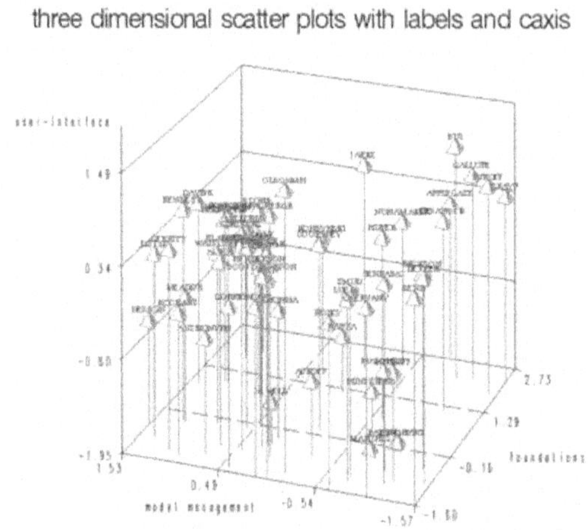

replace managerial judgment; and improve the effectiveness of decision making rather than its efficiency. Keen and Scott Morton (1978) suggested three important areas of DSS research from an organizational perspective: design, implementation, and evaluation of DSS. Sprague and Carlson (1982) added several important research areas —- data, model, dialogue, and decision makers, which can be termed DSS architecture. In addition, Sprague (1980) suggested an important and widely accepted definition and concept, which was termed specific decision support systems.

Through the analysis of 56 implemented specific DSS, Alter (1975; , 1977; , 1980) classified all DSS into seven distinct types and added several folders into the implementation drawers: patterns, risk factors, and strategies of DSS implementations. In the early 1980s, Wagner (1981, p. 77) maintained, "If the DSS concept has a valid core, it must be secured against adulteration and overburdening by *evidence drawn from actual practice.*" Also, the valid core of the DSS concept is to provide "interactive support for the thought processes of one or more executives in their principal function of making decisions." Others suggested theoretical models. William King and Rodriguez (1978) suggested an evaluation process model for evaluating MIS and DSS, which measures attitude, value perception, information usage, and decision performance in every stage of the system development life cycle in a simulated environment.

Several of these authors began to conduct empirical studies. Among them, Ginzberg's earlier work (1981), based on an empirical test of the level-of-adoption hypothesis, suggested that if full benefit is to be realized, DSS must be used as a catalyst for changes in the definition of the manager's role. Also, DSS should be viewed in the broader context of organization change, and therefore the design of DSS is likely to be more successful if it incorporates (1) user participation, (2) normative system modeling, and (3) evolutionary or iterative design. Sanders and Courtney (1985) reported the results of a field study of organizational factors to ascertain the influence of success factors (the decision context, task interdependence, and task constraints) of DSS implementations.

Model/Data Management

Data/Model Management Systems have emerged as the third major research area (factor 3) in the DSS field. Since 1975, model/data management has been researched to encompass several central topics such as model base structure and representation, model base processing, and the application of artificial intelligence to model integration, construction, and interpretation (Chang, Holsapple, & Whinston, 1993). In the model structure and representation area, the structured modeling approach by Geoffrion (1987) has significantly advanced the model representation area of model management, which is a significant extension of the entity-relationship data model and a necessary step for advancing to the next stage of model management (model manipulation). Dolk and Konsynski (1984) developed the model abstraction structure for representing models as a feasible basis for developing model management systems. Dolk attempts to connect both artificial intelligence and database management to evolve a theory of model management by model integration that relies heavily upon the relational database theory.

In the model processing area, Blanning (1982) investigated important issues in the design of relational model bases and presented a framework for the development of a relational algebra for the specification of join implementation in model bases. In the area of AI application to model management, Bonczek, Holsapple, and Whinston (1979; , 1980; , 1980; , 1981) suggested the use of artificial techniques for determining how models and data should be integrated in response to a user query. Elam, Henderson, and Miller (1980) introduced the concept of knowledge-based model management systems (MMS) to support a variety of complex decision problems with the use of semantic nets. They contended that the knowledge-based MMS could facilitate the use of the analytical tools in structuring as well as analyzing decision problems. Although model management research has not progressed enough to develop a theory of models, Dolk and Kottemann (Dolk & Kottemann, 1993, p. 51) believe that the emergence of a theory of models is imminent. They also believed that the current model integration research is projected as "the springboard for building a theory of models equivalent in power to relational theory in the database community."

Dolk and Kottemann (1993) further believe that model management needs to see some effective implementations, much like the relational theory needed ORACLE and other commercially viable products. The expense of building such systems is high, however, and it is not clear that there is market support for such a product. It is hoped that someone will achieve a breakthrough in this regard. Comprehensive literature reviews on model management can be found in (Blanning, 1993; Chang, Holsapple, & Whinston, 1993).

User Interface/Individual Differences

The initial investigation of user interface/individual differences (factor 4) started with the earlier works of Mason and Mitroff (Mason & Mitroff, 1973, p. 478), who hypothesized, "What is information for one type will definitely not be information for another. Thus, as designers of MIS, our job is not to get (or force) all types to conform to one, but to give each type the kind of information he is psychologically attuned to and will use most effectively." Bariff and Lusk (1977) presented a model for the useful classification of behavioral variables for attaining successful MIS design. The Bariff and Lusk model proposed that the successful design and implementation of an MIS should explicitly involve the consideration of the system's user cognitive styles. Benbasat and Dexter (1979; , 1982) conducted a series of similar experiments to conclude that "an appropriate information system design can help overcome a mismatch between task environment and psychological type" (Izak Benbasat & Dexter, 1982, p. 8). Despite those positive claims emphasizing the user's cognitive style as an important consideration in the design of management information systems and DSS, Huber (1983, p. 567) concluded, "(1) the currently available literature on cognitive styles is an unsatisfactory basis for deriving operational guidelines for MIS and DSS designs." (2) "Further cognitive style research is unlikely to lead to operational guidelines for MIS and DSS designs."

Other subgroups of researchers in this factor have focused on the evaluation of graphical and color enhanced information presentation and other presentation formats (e.g., Tabular). They include Chervany and Dickson (1974) and Dickson, *et al.* (1977): comparison of the decision impacts of detailed reports with summarized reports; Lusk and Kersnick (1979), Lucas and Nielson (1980), and Lucas (1981): comparison of tabular with graphics; and DeSanctis (1984) comprehensively investigated previous research in this area up to 1984. Despite the numerous previous research reports, the results are confusing and inconclusive: "the extravagant claims favoring graphic presentation formats may be considerably overstated" (Ives, 1982, p. 21), "a picture may *not* be worth a thousand words" (Gerardine DeSanctis, 1984, p. 482), and there are no differences between tabular and graphical reports in terms of decision quality (Izak Benbasat, 1974). Jarvenpaa, Dickson, and DeSanctis

(1985) argued that numerous equivocal findings could be attributable to the various tasks used in these experiments and the match between the task and presentation method and the lack of a sound taxonomy for classifying data extraction tasks. They recommended the development of some type of taxonomy of tasks as a basis of interpreting the impact of the graphical presentation format. Tan and Benbasat (1993) have provided taxonomies for classifying various tasks and for classifying information presentation methods and concluded that "the task and presentation notion of matching provides a way to explain the conflicting results by showing that the information presentation methods can not be evaluated outside the given task context in which they are applied" (Tan & Benbasat, 1993, p. 168).

GDSS

Since the mid-1980's we have witnessed an emerging DSS research theme: group decision support systems (factor 2). Earlier works by Delbecq, Van de Ven, and Gustafson (1975) experimentally compared three alternative methods for group decision making: the conventional interacting (discussion) group, the nominal group technique, and the Delphi technique. Many of these techniques (silent and independent idea generation, presenting each idea in a round-robin procedure, silent independent voting, etc.) were successfully utilized in the development of GDSS in the 1980's. Turoff and Hiltz (1982) conducted two experiments to study the impact of computer-based conferencing systems on group decision making. They concluded that GDSS helped the computer-aided groups reach quality decisions more often than groups unaided by a GDSS. In the early stages of GDSS development, several descriptive research papers have been the cornerstones for subsequent GDSS empirical research. Huber (1984) provided a comprehensive definition and proposed an architecture of GDSS. Further, alternative GDSS design strategies were examined to conclude that an activity-driven design strategy is superior to either a task-driven or techniques-driven strategy. His analysis of group activities led to another important conclusion that textual and relational information (PERT network, or organizational chart) is relatively more important for GDSS than it is for single user DSS. Another landmark paper is the result of the work of DeSanctis and Gallupe (1987); it presents an overview of GDSS, the potential impact of GDSS on group processes and outcomes, and proposes a multidimensional taxonomy of GDSS, based on the four dimensions: group size (smaller, larger); member proximity (dispersed, face-to-face); task type (6 types); and GDSS tool type (levels 1, 2, and 3). Kraemer and King, J.L. (1988) presents a comprehensive assessment of GDSS development and the use in the U.S. by reviewing the current status of GDSS activities. They conceive GDSS as a sociotechnical "package" of (1) hardware, (2) software, (3) organizationware and (4) people; They classified GDSS into the following 6 types: The Electronic Boardroom, The Teleconferencing Facilities, The Information Center, The Decision Conference, The Collaboration Laboratory, and The Group Network.

During the second half of the 1980's, a group of researchers began to conduct empirical GDSS research. There are four comprehensive reviews of major GDSS research (Izak Benbasat, DeSanctis, & Nault, 1993; Dennis & Gallupe, 1993; Dennis, George, Jessup, Nunamaker, & Vogel, 1988; Pinsonneault & Kraemer, 1989). Dennis *et al.* (1988) identified at least four streams of research under the broader label of experimental GDSS research to compare: Local Area Decision Nets (LADNs) to Decision Rooms, LADNs to no computer support, Decision Rooms to no computer support, and two different configurations of the same Decision Room. Gallupe, DeSanctis, and Dickson (1988) added one more value of task type (group problem finding) to the dimension III of the GDSS cabinet and conducted an empirical investigation of group problem finding (smaller group, face-to-face, level 1, problem finding). Jarke, Jelassi, and Bui seem to define an important field of GDSS — multiple criteria decision making (MCDM)-model embedded group decision support systems (Bui & Jarke, 1984). The next subgroup includes Nunamaker (1987), Applegate, George (Dennis, George, Jessup, Nunamaker, & Vogel, 1988), and Konsynski (1984-1985) of the electronic meeting systems research. The taxonomy of EMS environments presented by Dennis *et al.* (1988) added a new time dimension (dimension VI: synchronous and asynchronous meetings) and another value (multiple group sites) to the dimension II of the DeSanctis and Gallupe taxonomy.

CONCLUSION

We have identified several decision support systems research subspecialties based upon factor analysis, multidimensional scaling, and cluster analysis of a massive amount of bibliographical data. Two major conclusions from this study are that decision support systems are *weakly* grounded in only one contributing discipline (organizational science) and that researchers in the DSS research area are in the process of developing their own theories in the areas of foundations, model management, user interface/individual differences, and group support systems. This research provides a piece of evidence of the existence of the cumulative research tradition in the DSS area.

Organizational scientists have made important contributions to the development of DSS research subspecialties such as foundations, user interface/individual differences, and model management. An important implication of this result is that although DSS writers have referred to such parent disciplines as psychology, political science, accounting, computer science, and management science/operations research, these disciplines were not visible in this study and have apparently not played as important a role in the evolution of current DSS research as some previous commentators have opined.

Over the last two decades (1970-1989), DSS research has mainly concentrated on each component (data/model/dialogue/decision maker) of specific DSS. Keen and Scott Morton's DSS frameworks based on organizational perspectives have not been widely adopted over the past two decades. Our results show that none of Keen's suggested areas of DSS study (design, implementation, evaluation of DSS) have been shown to be substantive DSS research subspecialties.

All previous chapters in this book detailed aspects of author cocitation process, along with fundamental issues and recent developments. This is a capstone chapter that shows how the concepts, tools, and techniques discussed in each of the previous chapters can be applied in conducting author cocitation analysis using a real data in the DSS area. The step-by-step procedures are shown in detail from the preparation of data file in Excel format and importing the file to the SAS system for multivariate statistical analysis. This chapter also guides the readers through the process of analyzing the results of principal components analysis, cluster analysis, and multi-dimensional scaling. Especially, the chapter shows how to apply different criteria to select the optimal number of factor solutions, cluster solutions, and evaluating the acceptability of multi-dimensional scaling outputs.

REFERENCES

Ackoff, R. L. (1967). Management Misinformation Systems. *Management Science, 14*(12), B147-B156.

Alavi, M., & Joachimsthaler, E. A. (1992). Revisiting DSS Implementation Research: A Meta-Analysis of the Literature and Suggestions for Researchers. *MIS Quarterly, 16*(1), 95-116.

Alter, S. L. (1975). *A Study of Computer Aided Decision Making in Organizations.* Unpublished Ph. D. Thesis, MIT, Boston, MA.

Alter, S. L. (1977). A Taxonomy of Decision Support Systems. *Sloan Management Review, 19*(1), 39-56.

Alter, S. L. (1980). *Decision Support Systems: Current Practice and Continuing Challenges.* Reading, MA: Addison-Wesley.

Anthony, R. N. (1965). *Planning and Control Systems: A Framework for Analysis.*

Cambridge, MA: Division of Research, Graduate School of Business, Harvard University.

Bariff, M. L., & Lusk, E. J. (1977). Cognitive and Personality Tests for the Design of Management Information Systems. *Management Science, 23*(8), 820-829.

Bellardo, T. (1980). The Use of Co-Citations to Study Science. *Library Research, 2*, 231-237.

Benbasat, I. (1974). *An Experimental Evaluation of the Effects of Information System and Decision Maker Characteristics on Decision Effectiveness.* Unpublished Ph.D. Thesis, University of Minnesota, Minneapolis, MN.

Benbasat, I., DeSanctis, G., & Nault, B. R. (1993). Empirical Research in Managerial Support Systems: A Review and Assessment. In C. W. Holsapple & A. B. Whinston (Eds.), *Recent Development in Decision Support Systems* (Vol. 101, pp. 383-437). Berlin: Springer-Verlag.

Benbasat, I., & Dexter, A. S. (1979). Value and Events Approaches to Accounting: An Experimental Evaluation. *Accounting Review, 54*(4), 735-749.

Benbasat, I., & Dexter, A. S. (1982). Individual Differences in the Use of Decision Support Aids. *Journal of Accounting Research, 20*(1), 1-11.

Blanning, R. W. (1982). A Relational Framework for Model Management in Decision Support Systems. In G. W. Dickson (Ed.), *Decision Support Systems-82 Transactions* (pp. 16-28). San Francisco, CA.

Blanning, R. W. (1993). Model Management Systems: An Overview. *Decision Support Systems, 9*(1), 9-18.

Bonczek, R. H., Holsapple, C. W., & Whinston, A. B. (1979). Computer-Based Support of Organizational Decision Making. *Decision Sciences, 10*(2), 268-291.

Bonczek, R. H., Holsapple, C. W., & Whinston, A. B. (1980). The Evolving Roles of Models in Decision Support Systems. *Decision Sciences, 11*(2), 337-356.

Bonczek, R. H., Holsapple, C. W., & Whinston, A. B. (1980). Future Directions for Developing Decision Support Systems. *Decision Sciences, 11*(4), 616-631.

Bonczek, R. H., Holsapple, C. W., & Whinston, A. B. (1981). *Foundations of Decision Support Systems.* New York: Academic Press.

Bui, T. X., & Jarke, M. A. (1984). A DSS for Cooperative Multiple Criteria Group Decision Making. In *Proceedings of the 5th International Conference on Information Systems* (pp. 101-113). Tucson, AZ.

Chang, A.-M., Holsapple, C. W., & Whinston, A. B. (1993). Model Management Issues and Directions. *Decision Support Systems, 9*(1), 19-37.

Chervany, N. L., & Dickson, G. W. (1974). An Experimental Evaluation of Information Overload in a Production Environment. *Management Science, 20*(10), 1335-1344.

Culnan, M. J. (1986). The Intellectual Development of Management Information Systems, 1972-1982: A Co-Citation Analysis. *Management Science, 32*(2), 156-172.

Cyert, R. M., & March, J. G. (1963). *A Behavioral Theory of the Firm*. Englewood Cliffs, NJ: Prentice Hall.

Delbecq, A. L., Van de Ven, A. H., & Gustafson, D. H. (1975). *Group Techniques for Program Planning: A Guide to Nominal Group and Delphi Processes*. Glenview, IL: Scott, Foresman and Company.

Dennis, A. R., & Gallupe, R. B. (1993). A History of Group Decision Support Systems Empirical Research: Lessons Learned and Future Directions. In L. M. Jessup & J. S. Valacich (Eds.), *Group Support Systems: New Perspectives* (pp. 59-77). New York: Macmillan.

Dennis, A. R., George, J. F., Jessup, L. M., Nunamaker, J. F., Jr., & Vogel, D. R. (1988). Information Technology to Support Electronic Meetings. *MIS Quarterly, 12*(4), 591-624.

DeSanctis, G. (1984). Computer Graphics as Decision Aids: Directions for Research. *Decision Sciences, 15*(4), 463-487.

DeSanctis, G., & Gallupe, B. (1987). A Foundation for the Study of Group Decision Support Systems. *Management Science, 33*(5), 589-609.

Dickson, G. W., Senn, J. A., & Chervany, N. L. (1977). Research in Management Information Systems: The Minnesota Experiments. *Management Science, 23*(9), 913-923.

Dolk, D. R., & Konsynski, B. R. (1984). Knowledge Representation for Model Management Systems. *IEEE Transactions On Software Engineering, SE-10*(6), 619-628.

Dolk, D. R., & Kottemann, J. E. (1993). Model Integration and a Theory of Models. *Decision Support Systems, 9*(1), 51-63.

Elam, J. J., Henderson, J. C., & Miller, L. W. (1980). Model Management Systems: An Approach to Decision Support in Complex Organizations. In E. R. McLean (Ed.), *Proceedings of the First International Conference on Information Systems* (pp. 98-110). Philadelphia, PA.

Elam, J. J., Huber, G. P., & Hurt, M. E. (1986). An Examination of the DSS Literature (1975-1985). In E. R. McLean & H. G. Sol (Eds.), *Decision Support Systems: A Decade in Perspective* (pp. 239-251). Amsterdam, North-Holland: Elsevier Science.

Eom, H. B., & Lee, S. M. (1990). Decision Support Systems Applications Research: A Bibliography (1971-1988). *European Journal of Operational Research, 46*(3), 333-342.

Eom, H. B., & Lee, S. M. (1990). A Survey of Decision Support System Applications (1971-April 1988). *Interfaces, 20*(3), 65-79.

Eom, S. B. (1995). Decision Support Systems Research: Reference Disciplines and a Cumulative Tradition. *Omega: The International Journal of Management Science, 23*(5), 511-523.

Eom, S. B. (2007). *The Development of Decision Support Systems Research: A Bibliometrical Approach.* Lewiston, NY: The Edwin Mellen Press.

Eom, S. B., & Farris, R. (1996). The Contributions of Organizational Science to the Development of Decision Support Systems Research Subspecialties. *Journal of the American Society for Information Science, 47*(12), 941-952.

Eom, S. B., Lee, S. M., & Kim, J. K. (1993). The Intellectual Structure of Decision Support Systems (1971-1989). *Decision Support Systems, 10*(1), 19-35.

Farhoomand, A. F. (1987). Scientific Progress of Management Information Systems. *Data Base, 18*(4), 48-56.

Gallupe, R. B., DeSanctis, G., & Dickson, G. W. (1988). The Impact of Computer Support on Group Problem Finding: An Experimental Approach. *MIS Quarterly, 12*(2), 276-296.

Geoffrion, A. M. (1987). An Introduction to Structured Modeling. *Management Science, 33*(5), 547-588.

Ginzberg, M. J. (1981). Early Diagnosis of MIS Implementation Failure: Promising Results and Unanswered Questions. *Management Science, 27*(4), 459-478.

Gorry, G. A., & Scott Morton, M. S. (1971). A Framework for Management Information Systems. *Sloan Management Review, 13*(1), 55-70.

Hair, J. F., Jr., Anderson, R. E., & Tatham, R. L. (1987). *Multivariate Data Analysis with Readings* (2nd ed.). New York: Macmillan Publishing Company.

Huber, G. P. (1983). Cognitive Style as a Basis for MIS and DSS Design: Much Ado About Nothing? *Management Science, 29*(5), 567-579.

Huber, G. P. (1984). Issues in the Design of Group Decision Support Systems. *MIS Quarterly, 8*(3), 195-204.

Ives, B. (1982). Graphical User Interfaces for Business Information Systems. *MIS Quarterly, 6*(4), 15-46.

Jarvenpaa, S. L., Dickson, G. W., & DeSanctis, G. (1985). Methodological Issues in Experimental Is Research: Experiences and Recommendations. *MIS Quarterly, 9*(2), 141-156.

Keen, P. G. W. (1980). MIS Research: Reference Disciplines and a Cumulative Tradition. In E. R. McLean (Ed.), *Proceedings of the First International Conference on Information Systems* (pp. 9-18). Philadelphia, PA.

Keen, P. G. W., & Scott Morton, M. S. (1978). *Decision Support Systems: An Organizational Perspective*. Reading, MA: Addison-Wesley.

King, W. R., & Rodriguez, J. I. (1978). Evaluating Management Information Systems. *MIS Quarterly, 2*(3), 43-51.

Konsynski, B. R., Kottemann, J. E., Nunamaker, J. F., Jr., & Stott, J. W. (1984-1985). Plexsys-84: An Integrated Development Environment for Information System. *Journal of Management Information Systems, 1*(3), 64-104.

Kraemer, K. L., & King, J. L. (1988). Computer-Based Systems for Cooperative Work and Group Decision Making. *ACM Computing Surveys, 20*(2), 115-146.

Little, J. D. C. (1970). Models and Managers: The Concepts of a Decision Calculus. *Management Science, 16*(8), B466-B485.

Lucas, H. C., Jr. (1981). An Experimental Investigation of the Use of Computer-Based Graphics in Decision Making. *Management Science, 27*(7), 757-768.

Lucas, H. C., Jr., & Nielsen, N. R. (1980). The Impact of the Mode of Information Presentation on Learning and Performance. *Management Science, 26*(10), 982-993.

Lusk, E. J., & Kersnick, M. (1979). The Effect of Cognitive Style and Report Format on Task Performance: The MIS Design Consequences. *Management Science, 25*(8), 787-798.

Mason, R. O., & Mitroff, I. I. (1973). A Program for Research on Management Information Systems. *Management Science, 19*(5), 475-487.

McCain, K. W. (1984). Longitudinal Author Cocitation Mapping: The Changing Structure of Macroeconomics. *Journal of the American Society for Information Science, 35*(6), 351-369.

McCain, K. W. (1986). Cocited Author Mapping as a Valid Representation of Intellectual Structure. *Journal of the American Society for Information Science, 37*(3), 111-122.

McCain, K. W. (1990). Mapping Authors in Intellectual Space: A Technical Overview. *Journal of the American Society for Information Science, 41*(6), 433-443.

Newell, A., & Simon, H. A. (1972). *Human Problem Solving.* Englewood Cliffs, NJ: Prentice Hall.

Nunamaker, J. F., Jr., Applegate, L. M., & Konsynski, B. R. (1987). Facilitating Group Creativity: Experience with a Group Decision Support System. *Journal of Management Information Systems, 3*(4), 5-19.

Pinsonneault, A., & Kraemer, K. L. (1989). The Impact of Technological Support on Groups: An Assessment of the Empirical Research. *Decision Support Systems, 5*(2), 197-216.

Sanders, G. L., & Courtney, J. F. (1985). A Field Study of Organizational Factors Influencing DSS Success. *MIS Quarterly, 9*(1), 77-93.

Simon, H. A. (1960). *The New Science of Management Decision.* New York: Harper & Row.

Simon, H. A. (1969 (1981)). *The Sciences of the Artificial* (2nd ed.). Cambridge, MA: The MIT Press.

Smith, L. C. (1981). Citation Analysis. *Library Trends, 30*(1), 83-106.

Sprague Jr., R. H. (1980). A Framework for the Development of Decision Support Systems. *MIS Quarterly, 4*(4), 1-26.

Sprague Jr., R. H., & Carlson, E. D. (1982). *Building Effective Decision Support Systems.* Englewood Cliffs, NJ: Prentice Hall.

Sprague Jr., R. H., & Watson, H. J. (1975). MIS Concepts: Part Ii. *Journal of Systems Management, 26*(2), 35-40.

Tan, J. K. H., & Benbasat, I. (1993). The Effectiveness of Graphical Presentation for Information Extraction: A Cumulative Experimental Approach. *Decision Sciences, 24*(1), 167-191.

Teng, J. T. C., & Galletta, D. F. (1990). MIS Research Directions: A Survey of Researcher's Views. *Data Base, 21*(3/4), 1-10.

Turoff, M., & Hiltz, S. R. (1982). Computer Support for Group Versus Individual Decisions. *IEEE Transactions On Communications, COM-30*(1), 82-92.

Wagner, G. R. (1981). Decision Support Systems: The Real Substance. *Interfaces, 11*(2), 77-86.

Chapter XIII
The Changing Structure of Decision Support Systems Research:
An Empirical Investigation through Author Cocitation Mapping (1990–1999)

ABSTRACT

This chapter extends an earlier benchmark study (Sean B. Eom, 1995) which examined the intellectual structure, major themes, and reference disciplines of decision support systems (DSS) over the last two decades (1960-1990). Factor analysis of an author cocitation matrix over the period of 1990 through 1999 extracted 10 factors, representing 6 major areas of DSS research: group support systems, DSS design, model management, implementation, and multiple criteria decision support systems and five contributing disciplines: cognitive science, computer supported cooperative work, multiple criteria decision making, organizational science, and social psychology. We have highlighted several notable trends and developments in the DSS research areas over the 1990s.[1]

INTRODUCTION

Earlier studies documented the intellectual development of the decision support systems (DSS) area over the last two decades (1969-1990) in terms of two of the three main needs defined by Keen (1980)— reference disciplines and a cumulative tradition. Eom (1995 p. 517) concluded that "After 20 years of research, the DSS literature does not exhibit an overall DSS research paradigm. Nonetheless, this study convinces the author that DSS is in the active process of solidifying its domain and demarcating its reference disciplines." This paper assesses the ongoing changes in the intellectual development and structure of DSS research, using multivariate analysis of an author cocitation matrix over the period of 1990 through 1999. This study aims at identifying the intellectual structure, reference disciplines, and major themes in DSS research over the past ten years (1990-1999) with a particular emphasis on contrasting the structural changes in the intellectual structures in the DSS area over the period of 1969 through 1990 and the period of 1990 through 1999.

DATA AND RESEARCH METHOD

The data for this study were gathered from a total of 984 articles in the DSS area over the past ten years (1990-1999) using the criteria described in our earlier study. The number of citing articles can be an indicator of vitality of the DSS area. During the past 10 years, DSS researchers have published 984 articles at an average rate of 98.4 articles per year, while the previous two decades (1969-1990) had published a total of 632 articles at an average rate of 31.6 articles per year.

The raw cocitation matrix of 171 authors is analyzed by the factor analysis program of SAS (statistical analysis systems) to ascertain the underlying structure of DSS research subspecialties. Principal component analysis (varimax rotation) with the latent root criterion (eigenvalue 1 criterion) is applied to obtain the initial solution of 15 factors (see Table 1). The scree tail test indicates that only the first twelve factors should be qualified. For further details of author cocitation analysis including the statistical method used, readers are referred to Eom (Sean B. Eom, 2003).

RESULTS

Based on careful examinations and interpretation of these outputs, ten factors resulted, as shown in Appendix 1. The ten extracted factors account for 84.11 percent of the total variances of the data set. Factor analysis extracted ten factors consisting of six major areas of DSS research (group support systems, design, model management, user interface/implementation, evaluation, and multiple criteria decision support systems) and five contributing disciplines (cognitive science, computer supported cooperative works, multiple criteria decision making, organization science, and social psychology). Factor 4 seems to represent both the user interface and implementation factors. Factor 8 also includes authors in multiple criteria decision making (MCDM) and multiple criteria decision support systems (MCDSS).

Through the comparison of the previous studies and this study, we identified that the DSS area has undergone profound structural changes in the intellectual structure over the past ten years (1990-1999). DSS research areas and reference disciplines can be categorized into four different groups —steady, strengthening, emerging, dying, and slowly growing areas. The steady areas include model management which have appeared in the previous study (1995) and this study. In the reference discipline areas, organizational science and multiple criteria decision making have been consistently influenced to the development of DSS research subspecialties over the past 30 years (1969-1999). The GSS area has been strengthened significantly since 1990s. *The emerging areas* are represented by design, implementation, MCDSS, and evaluation in the DSS research area and cognitive science, CSCW, social psychology in the contributing discipline. *The dying area* includes two DSS research subfields (foundation and individual differences) that appeared to be no longer active. Group support systems (GSS) research has been strengthened and model management research has been a continuously central research theme. Organization science has been a steady field of DSS contributing discipline.

Group Support Systems

Factor 1 appears to define ***group support systems (GSS)***. This investigation period include 171 scholars as markers of the DSS intellectual space. Seventy-nine of these scholars (46%) were clustered into factor 1. In terms of percent of variance, the percent of variance of the GSS factor (36.6%) tells the relative importance of the GSS factor.

Appendix A lists 79 authors under Factor 1. Undoubtedly GSS have now become the mainstream research field by many different measures. The unique contributions of ACA analysis is the reconstruction of bird's eye view of each subfield thorough the identification of a group of influential and responsible DSS researchers who represent major forces that have charted and perhaps will chart the future directions for DSS research and redirect DSS research efforts toward a common paradigm. "Any study of paradigm-directed or of paradigm-shattering research must begin by locating the responsible group or groups" (Kuhn, 1970, p.242). Some of the important recent developments in this area can be summarized as follows.

GSS Tool Developments

There have been continuing developments and enhancements of GSS tools to support and augment the existing group DSS and electronic meeting systems such as an idea consolidator (M. Aiken & Carlisle, 1992), an optimization-based group decision tool for combining subjective estimates and extracting the underlying knowledge of group members (Singh, Bennavail, & Chen, 1992), a hypertext and computer-mediated communication systems (Turoff, Hiltz, Bahgat, & Rana, 1993), a group software for modelling and analysing business process re-engineering (Dennis, Daniels, Hayes, & Nunamaker, 1993-1994), an interactive videodisc-based GDSS for directing the pattern, timing, and contents in group decision making (Reisman, Johnson, & Mayes, 1992).

GSS Application Developments

A wide range of GSS/electronic meeting systems/decision conferencing system applications has been reported to support/facilitate a wide rage of decisions (Sean B. Eom, Lee, Kim, & Somarajan, 1998).

Empirical/Laboratory Experimental Studies

A number of empirical/laboratory experimental studies have been conducted to investigate the effects of a variable on the quality of group decisions, level of agreement, subjective satisfaction, etc. For a comprehensive reviews of GSS field studies, See (Fjermestad & Hiltz, 1998-1999), which presents a descriptive evaluation of 54 case and field studies from 79 published papers spanning two decades of group support systems research.

Integrated GSS with Other Technologies

GSS is being integrated with other technologies such as expert systems and case-based reasoning, etc. A prototype system that embedded expert systems into GDSS is developed to make a GSS a more user-friendly and powerful tool for group support by capturing the scarce expertise of human facilitators GSS session management knowledge (M. W. Aiken, Liu Sheng, & Vogel, 1991). The distributed artificial intelligence approach for designing and developing group problem solving systems is being investigated to coordinate organizational activities in a distributed environment through the development of prototype systems comprising a network of expert systems (Shaw & Fox, 1993).

Design of DSS

Factor 2 seems to represent Design of DSS. Over the past 10-year period (1990-1999), design of DSS has emerged as a DSS research subfield. Traditional assumptions in the DSS area are being challenged. One of them includes the role of cognitive styles in the DSS design. A significant development in the 1990s is the demise of Individual differences (cognitive style) research as a basis of DSS design in the DSS literature. After over a decade of cognitive styles and individual difference research, Huber (1983, p. 567) concluded that "the currently available literature on cognitive styles is an unsatisfactory basis for deriving operational guidelines for MIS and DSS designs" and "Further cognitive style research is unlikely to lead to operational guidelines for MIS and DSS designs."

A number of empirical studies conducted to test the existing DSS design frameworks/theory. Remus (1984) investigated a key issue in DSS design, the impact of graphical and tabular data presentation on decision making. Igbaria and Guimaraes (1994) empirically tested the outcomes of user involvement in the DSS development to establish the positive relationship between user involvement and several measures of system success such as DSS usages, user overall satisfaction with the DSS, and user perceived DSS benefits. A contingency model of DSS design methodology is developed by Arinze (1991) to help DSS developers select an appropriate methodology out of several methodologies of data-driven, process driven (Peter G. W. Keen & Scott Morton, 1978; Sprague & Carlson, 1982), decision-driven (Stabell, 1983), and systemic paradigms (Ariav & Ginzberg, 1985).

Model Management

Factor 3 appears to represent Model management. Since 1975, model management has been researched to encompass several central topics such as model base structure and representation, model base processing, and application of artificial intelligence to model integration, construction, and interpretation

Structured Modeling

In the model structure and representation area, the structured modeling approach by Geoffrion (1987) has advanced the model representation area of model management, which is an extension of the entity-relationship data model and a necessary step for advancing to the next stage of model management (model manipulation).

According to Geoffrion, structured modeling is a systematic way of thinking about models and their implementations to overcome perceived shortcomings of modeling systems available in the 1980s. It is based on the idea that every model can be viewed as a collection of distinct elements, each of which has a definition that is either primitive or based on the definition of other elements in the model.

Dolk and Konsynski (Dolk & Konsynski, 1984; Konsynski & Dolk, 1982) developed the model abstraction structure for representing models as a feasible basis for developing model management systems. Dolk (1993) attempts to connect both artificial intelligence and database management to evolve a theory of model management via model integration relying heavily upon the relational database theory. In the model processing area, Blanning (1982) investigated important issues in the design of relational model bases and presented a framework for the development of a relational algebra for the specification of join implementation in model bases.

New Approaches to Model Management

During the 1990s, there has been a variety of new approaches to manage models as organizational resources. Some of notable approaches include the development of graph-based modeling (Jones, 1995), object-oriented approach (Muhanna & Pick, 1994), mdeling by analogy (analogical modeling) and cased-based learning (Liang & Konsynski, 1993), modeling by example (Angehrn, 1991), active modeling systems (Dolk & Kridel, 1991), knowledge-based model construction (Murphy & Stohr, 1986).

Model integration is another active line of DSS research. There has been several new approaches to model integration such as integrating simulation modeling and inductive learning in an adaptive decision support system (Piramuthu, Raman, Shaw, & Park, 1993), Object-oriented model integration (Dempster & Ireland, 1991), model integration using metagraph (Basu & Blanning, 1994 p. 195).

User-Interface/Implementation

Factor 4, User Interface/Implementation, is a new factor that emerged in this study. Design, implementation, and evaluation of information systems are an integrated and inseparable process, as shown by several studies, which reported the identification of DSS implementation success factors and the linkage between those factors and DSS effectiveness (Ramamurthy, King, & Premkumar, 1992). Because of the inseparable nature of system design and implementation, factor 4 includes authors in these two areas. DSS implementation research over the past three decades (1970-1999) has been primarily concerned with investigating the relationship between various factors (user related and other factors) and implementation success. User-related factors include cognitive style (the characteristic ways individuals process and utilize information to solve problems), personality (the cognitive structures maintained by individuals to facilitate adjustment to events and situations), demographics (age, sex, and education), and user-situation variables (training, experiences, and user involvement) (Alavi & Joachimsthaler, 1992), (Zmud, 1979).

DSS implementation research aims at systematically identifying factors which will influence the implementation success of DSS so that those critical factors can be managed effectively to increase the successful implementation of DSS in organizations. Several studies reported the identification of DSS implementation success factors and the linkage between those factors and DSS effectiveness (Alavi & Joachimsthaler, 1992; Igbaria & Guimaraes, 1994; Ramamurthy, King, & Premkumar, 1992). Like so many empirical studies in other fields, no direct comparison of these studies is possible due to different methodologies, samples, etc. For example, Ramurthy, King, and Prekumar (1992 p.469) concluded that "user's domain-related expertise, systems experience, gender, intelligence, and cognitive style have important influence on one or more dimensions of DSS effectiveness. However, their relative importance vary with the outcome measure of choice." An empirical study of Igbaria and Guimaraes (1994) strongly confirmed that user involvement and DSS friendliness are critical DSS success factors.

Despite conflicting and confusing findings in the area of implementation research, a systematic integration and assessment of a large set of DSS implementation research seems to suggest that DSS implementation research has been accumulating its research findings. The review of DSS implementation research over the period of 1991 through 1997 led us to conclude the following:

1. Over the past three decades, MIS/DSS researchers have identified numerous factors affecting MIS/DSS. They are classified into external environments (environmental characteristics), organizational environment (characteristics), task environment (characteristics), other factors including human factors (user's personal characteristics and MIS staff characteristics) and non-human factors (MIS policies). Of these, the majority of implementation researchers focus on the user related factors (cognitive styles, personality, demographics, user-situational variables).
2. Based on the meta analysis of prior implementation research, we are further readjusting the focus of DSS implementation research on the user-situational variables. The user-situational variables (involvement, training and experience) are more important than cognitive styles and personality and the DSS implementation success rate can be improved by as much as 30 percent by manipulating user-situational variables(Alavi & Joachimsthaler, 1992). Fur-

ther their study concluded that "future research should be directed toward developing causal models of DSS implementation that weave these key factors together in a form that makes their interrelationship explicit."

DSS EVALUATION

Decision support systems management consists of an ongoing, inseparable process of designing, implementing, and evaluating DSS. Decision support systems are to assist managers in their decision making processes to improve the effectiveness of decision making rather than its efficiency. Evaluation of DSS is concerned with analyzing costs and benefits of decision support systems before and after DSS development and implementation. The unique nature of DSS evaluation is that although some DSS provides substantial cost saving and profit increase, measurements of benefits of DSS have been a difficult problem, due to the fact that quantification of the positive impacts of improved decision process is difficult (Peter G. W. Keen & Scott Morton, 1978).

Evaluating DSS is concerned with determining the value of DSS. The value of DSS can be measured by a smorgasbord of eight methodologies: (1) decision outputs; (2) changes in the decision process; (3) changes in managers' concepts of the decision situation; (4) procedural changes; (5) classical cost/benefit analysis; (6) service measures; (7)managers' assessment of the system's value; and (8) anecdotal evidence (Peter G. W. Keen & Scott Morton, 1978).

Aldag and Power(1986) reviewed a number of prior research. Most of these papers are concerned with the use of multiple attribute decision making model based decision aids to improve the effectiveness of decision making. An interactive heuristic program was used to solve a case problem by two groups of students. The result of the Aldag and Power study detected no significant differences in performances although user attitudes toward the computer-based decision aid were favourable and user gained more decision confidence. Their study suggested that user affect should not be used as a proxy for decision quality and DSS must be adopted with caution in that if such decision aids result in positive user affect and heightened decision confidence without corresponding improvements in decision quality, DSS may be dysfunctional.

Sharda and others (1988) conducted an experimental investigation using an executive decision making game played by two sections of a business strategy course to find out that the groups with DSS made significantly more effective decisions and examined more alternatives and exhibited a higher confidence level in their decisions than their non-DSS counterparts.

REFERENCE DISCIPLINES OF DSS RESEARCH

Cognitive Science

Factor 5 seems to represent ***Cognitive science***, which is an interdisciplinary field which studies diverse human cognitive activities such as language understanding, thinking, visual cognition and action, etc. The focus of cognitive science research is on how cognition typically works in normal adults, how it varies across individuals/different populations/cultures, how it develops, etc.

To discuss specific findings in the literature of a psychological decision theory, Einhorn and Hogarth (1981) decomposed the processes of judgment and choice into the subprocesses of information acquisition (information search and storage), evaluation, action, and feedback/learning and discussed several issues related to each subprocess. In regard to evaluating a set of alternatives, Einhorn and Hogarth discussed the need for finding principles underlying choice processes including a possibility of an over-riding cost/benefit analysis, which may induce suboptimal behavior in some circumstances.

Behavioral decision theorists proposed a variety of mechanisms that influence strategy selection. Of these, the cost-benefit framework of Payne (1982) and his colleagues provided a basis for DSS researchers for understanding the behavior of decision makers using DSS as to the selection of their strategy and the relationship between the use of DSS and decision quality. The cost-benefit framework of cognition aims at maximizing accuracy and/or decision quality as well as minimizing cognitive effort. A study of Todd and Benbasat (Todd & Benbasat, 1992), for example, concluded that the amount of information processed does not necessarily increase with the use of a decision aid. Instead, it appears in the study that decision makers use decision aids to reduce the amount of efforts required to complete a task.

Decision makers use different processes in different types of tasks. Decision processes are sensitive to seemingly minor changes in the task-related factors. Tversky and Kahneman (1982) described three heuristics in making judgment under uncertainty (representativeness, availability, and adjustment and anchoring), which lead to systematic and predictable errors. The findings of Tversky and Kahneman (1982) have contributed to controlling bias in user assertions in DSS and provide a guiding principle for overcoming the user's poor capabilities to calculate probabilities when designing DSS. Their findings provided a theoretical basis for reaching an important conclusion that the cognitive styles of users should not be the basis of information systems design in that "predispositions are often dysfunctional" (Huber, 1983).

CSCW

Factor 6 seems to define CSCW. The term computer-supported cooperative network (CSCN) has often been used interchangeably with computer-supported cooperative work (CSCW). According to Johansen (1988), CSCW has been used synonymously with terms such as groupware, teamware, technological support for workgroup collaboration, group decision support systems, workgroup computing, collaborative computing, interpersonal computing, computer assisted communication, augmented knowledge workshop, interfunctional computing, computer-supported groups, and flexible interactive technologies for multi-person tasks. However, there seems to be a consensus that groupware is a generic term for information systems that make it possible for a network of collaborating individuals to carry out cooperative activities such as designing products, writing project reports, developing computer software, making decisions, etc. This network of collaborating individuals across time and space, not only performing work but also creating knowledge, is often termed a computer-supported cooperative network (CSCN).

CSCN may be also defined as the multidisciplinary academic field that investigates how the computer and related technologies affect group behaviour and group productivity and how to design, develop and implement the information systems to support people working together. CSCN is rooted in computer science, cognitive science, psychology, sociology, anthropology, ethnography, organizational behaviour and management information systems. CSCN systems consist of hardware, software, data, procedures and people, like any other type of information system. Working together with other people inevitably necessitates communication, sharing data and information, and solving problems together.

Many earlier pioneering researchers in the computing area had a vision of using computers for supporting a group of people working together. Doug Engelbart was one of them. He envisioned that future computers may enhance/augment human intellect 'through technologically mediated and facilitated formal collaboration'. His augmented knowledge workshop laboratory worked on the several fundamental topics that are essential elements for understanding today's CSCW. They include collaborative dialogue, teleconferencing, recorded dialogue, group writing, group programming, group databases and multimedia communications (Johansen, 1988). The two most widely used items of groupware (e-mail, computer conferencing) emerged in the 1970s. It was in the 1980s that we witnessed the emergence of key terms and ideas in the CSCW area, such as groupware, group decision support systems, computer-supported cooperative work and electronic meeting systems. Today, group support systems and electronic meeting systems have emerged as the most important fields of decision support systems. Readers are referred to Part 1 of Greif (1988) and Chapter 1 of Johansen (1988) for a concise history of the field.

A group of artificial intelligence researchers at Xerox Palo Alto Research Center explored the idea of putting computers as an collaborative problem solving tool in face-to-face meeting to replace, extend, and augment the chalkboard (Stefik et al., 1987). Using an experimental meeting room equipped with personal computers which are linked together over a local area network that supports a distributed database, they created several tools for collaboration such as Boardnoter (imitating the functionality of a chalkboard), Cognoter (a cognition noter for brainstorming, organizing, and evaluating ideas), and Argnoter (argument noter for organizing and evaluating arguments).

MCDM/MCDSS

Factor 8 seems to represents two groups of authors—MCDM and MCDSS. The first group, as explained in an earlier section, represents the management scientist group. The second group represents information systems researcher as defined by (Tung X. Bui, 1984; Tung X. Bui, 1987; Jelassi & Foroughi, 1989; Jelassi, Jarke, & Stohr, 1985). A multiple criteria decision support system can be defined as an MCDM model-embedded DSS to solve various semistructured and unstructured decisions involving multiple attributes or multiple objectives or both. According to Zeleny (1982, p. 17), the term "Multiple-Criteria decision making" (MCDM) indicates a concern with the general class of problems that involve multiple attributes, objectives, and goals. Attributes are characteristics of objects in the real world. These attributes can be specified "in relative independence from the decision maker's needs or desires." (Zeleny, 1982, p. 15). Objectives, which are "closely identifiable with a decision maker's needs and desires," specify (1) a set of attributes (or an attribute) and (2) "directions of improvement or preferences along individual attributes or complexes of attributes." (Zeleny, 1982, p. 15). Various techniques have been introduced for MCDM, such as multiattribute utility theory, multicriteria linear programming, heuristic search methods, goal programming, compromise programming, and learning models (see, (MacCrimmon, 1973),(Zeleny, 1982), for a description of MCDM models).

Multiple criteria DSS/Negotiation Support Systems represents MCDM model-embedded decision support systems (Tung X. Bui & Jarke, 1984; H. B. Eom, Lee, Snyder, & Ford, 1987-1988); (Jarke, Jelassi, & Shakun, 1987); (Jelassi, Jarke, & Stohr, 1985). They can be broadly categorized into a generalized data-oriented MCDSS which is based on multiattribute decision making models (Jelassi, Jarke, & Stohr, 1985), a model-oriented MCDSS which is based on multiple objective decision making models (H. B. Eom, Lee, Snyder, & Ford, 1987-1988), and data-oriented MCDM Group DSS (Tung X. Bui & Jarke, 1984), and negotiation support systems (Jarke, Jelassi, & Shakun, 1987).

MCDSS intend to provide the necessary computerized assistance to decision makers in such a way that the decision maker is encouraged to explore the support tools available in an iterative fashion with the aim of further defining and refining the nature of the problem (Minch & Sanders, 1986). The ultimate success of DSS lies in its ability to help decision makers solve ill-structured problems through direct interaction with analytical models. Such an ability can be enhanced by intermingling the various features of MCDM with DSS. These features include (1) the multiple-objective goal structure designed to handle quantitative and qualitative information crucial for ill-structured problems, (2) the interactive solution search procedure designed to analyze continuous trade-offs among various alternatives until the best available solution is attained, and (3) the emphasis on the decision maker's judgment or bounded rationality which better reflects his/her actual cognitive behaviors.

Among the numerous individuals contributing to give rise to the field of MCDM, Keeney and Raiffa (1976) have provided us with an excellent and complete overview of multiple attribute utility theory, along with numerous examples of practical applications. By the nature of multiple criteria decision making, usually there are numerous non-dominated solutions in MCDM problems. Integration of MCDM into DSS has long been advocated by the researchers in the DSS and MCDM areas. Keen and Scott Morton (1978 p.48) believe that the multiple criteria decision problem is at the core of decision support and "a marriage between MCDM and DSS promises to be practically and intellectually fruitful." An important reason for the emergence of MCDM model-embedded DSS (MCDSS) is that MCDM complements DSS and vice versa due to the differences in underlying philosophies, objectives, support mechanisms, and relative support roles.

Some features of MCDM include (1) the multiple-objective goal structure designed to handle quantitative and qualitative information crucial for ill-structured problems, (2) the interactive solution search procedure designed to analyze continuous trade-offs among various alternatives until the best available solution is attained, and (3) the emphasis on the decision maker's judgment or bounded rationality which better reflects his/her actual cognitive behaviors.

Integration of MCDM into DSS has long been advocated by the researchers in both areas. Founding fathers of DSS such as Keen and Scott Morton (1978, p.48) believe that the multiple criteria decision problem is at the core of decision support and "A marriage between MCDM and DSS promises to be practically and intellectually fruitful." The emergence of MCDM model-based DSS was predicted in the early 1980s (Zeleny, 1982). A series of studies (Sean B. Eom & Min, 1992); (Sean B. Eom, 1995); (Sean B. Eom, 1996) reached the compelling conclusion that the MCDM model-embedded DSS have positioned themselves at the core of DSS. An important reason for the emergence of the MCDM model-embedded DSS is that MCDM complements DSS and vice versa due to the differences in underlying philosophies, objectives, support mechanisms, and relative support roles (Nazareth, 1993).

Since MCDM inherently necessitates a simultaneous comparison of the large number of decision criteria and alternatives which demand a complex array of information, an integration of MCDM with DSS is inevitable. The rationale may be that MCDM models can be useful tools for DSS problem solving, while DSS helps maintain and retrieve MCDM models for the repeated usage. Indeed, Weistroffer and Narula (1997) argued that efficient and user-friendly DSSs are crucial for successfully solving real-world MCDM problems. Recognizing such complementary nature of MCDM and DSS, we briefly synthesize the current status of MCDM utilization within the DSS framework. Since two prior studies (Sean B. Eom & Min, 1992); (Nazareth, 1993) have already summarized various attempts of integrating MCDM with DSS up until 1991, this section primarily focuses on the review of more recent efforts (1992-1998) of embedding MCDM within model bases of DSS along with some omissions made by the two prior studies. Hereafter, MCDM-method-embedded DSS will be abbreviated as MCDSS. In general, MCDSS is a DSS that helps implement MCDM models with an objective of aiding decision makers in analyzing, exploring, and comparing a set of incompatible alternatives. Some characteristics of MCDSS that is absent from other types of DSS includes analysis of multiple criteria, involvement of MCDM methods, and the integration of user-input in the modeling process (Dyer, Fishburn, Steuer, Wallenius, & Zionts, 1992). A vast majority of the past MCDSS literature, however, has been focused on the design of independent DSSs to solve a specific set of related MCDM problems. Readers are referred to (Sean B. Eom & Min, 1999) for an in-depth discussion on the contribution of MCDM on many DSS subsystems such as user-interface, model management, data management, foundations, etc.

Organization Science

Factor 9 appears to represent organizational science. Earlier, Davis and Olson (1985) claimed that management information systems were influenced significantly by four major academic areas: managerial accounting, operations research, management and organizational theory, and computer science. Our bibliometrics-based empirical research shows the visibility of only multiple criteria decision making and organizational science to the development of DSS research subspecialties. This study, however, should not necessarily be interpreted as disproof of Davis and Olson's claim that the theory of managerial accounting was an important *MIS* reference discipline. This study shows that the *DSS* research has not been significantly influenced by the authors in management accounting, although our citing and cited database files did not exclude accounting papers. In addition to the result of this study, the result of the previous two studies (M. J. Culnan, 1986; Mary J. Culnan, 1987) also failed to identify management accounting as a reference discipline of MIS. Readers are referred to the previous chapter and Eom and Farris (Sean B. Eom & Farris, 1996) for a detailed discussion on the contributions of organization science to the development of DSS research subspecialties.

Social Psychology

Factor 10 seems to define social psychology represented by Dubrovsky, Siegel, McGuire, and Kiesler. (Dubrovsky, Kiesler, & Sethna, 1991; Siegel, Dubrovsky, Kiesler, & McGuire, 1986). Social psychologists such as Siegel, Dubrovsky, Kiesler, and McGuire explored the effects of computer-mediated communications on communication efficiency (group members' capabilities to function/communicate among themselves in the least wasteful manner), participation (distribution of communication in the group), interpersonal behavior (overt expressive behavior), and group decision making. An earlier experiment showed that when compared to face-to-face groups, computer-mediated group members made fewer remarks but revealed more uninhibited behavior, took longer to make their group decisions, exhibited higher social equalization in that group members participated more equally in discussions.

A laboratory experiment by Dubrovsky, Kiesler, and Sethna investigated status effects in computer-mediated and face-to-face decision making groups by varying the external status of group members, and by varying the decision tasks to manipulate expertise. In face-to-face meetings, the high-status member dominated with the low-status members and more often was a "first advocate" in the discussion. These status inequalities in the face-to-face meeting were clearly apparent when the high status member's expertise was relevant to the decision task. But status and expertise inequalities in participation were reduced when the same group members made similar decisions via e-mail conferencing.

CONCLUSION

This study investigates an ongoing change in the intellectual structure of DSS research fields through a comparison of the bibliometrical study between the first two decades of DSS research (1969-1990) and the recent ten years (1990-1999). For further discussions on the intellectual structure of the DSS area up to 2004, readers are referred to (Sean B. Eom, 2007). We are now in a better position to understand the dynamic dimension of the intellectual structure of DSS research. This study identified a dynamic dimension of DSS research areas to account for the ongoing changes in its "disciplinary matrix" —the four emerging areas (Implementation, Design, and Cognitive science); continuously growing areas (GDSS, Model management, MCDM, and Organization science); and dying areas (Individual difference and Foundations).

In a nutshell, the factors in this study appear to reflect the maturing stages of DSS research, aiming at providing a macro view of DSS research literature with a goal of providing a basis for developing coherent DSS theories to sort out a confusing body of a variety of DSS literature. The changes in the intellectual structures in decision support systems have been profound over the 1990s. Focus of DSS research appears to be shifting from the study of DSS components (data, model, individual differences of decision makers) during the periods of 1970 through 1990 to the design, implementation, and user-interface management (which have not been shown to be substantive DSS research subspecialties in the previous research), to provide useful guiding principles for practitioners in the integrated processes of design, implementation, and evaluation of decision support systems. The model management and group decision support systems areas have been continuously researched over the two periods of investigation. In the area of model management, much progress has been made in the sub-areas of model representation, model base processing, model integration, and the application of artificial intelligence to model management.

REFERENCES

Abdolmohammadi, M. J. (1985). Bayesian Inference Research in Auditing: Some Methodological Suggestions. *Contemporary Accounting Research, 2*, 76-94.

Aiken, M., & Carlisle, J. (1992). An Automated Idea Consolidation Tool for Computer Supported Cooperative Work. *Information & Management, 23*(6), 373-382.

Aiken, M. W., Liu Sheng, O. R., & Vogel, D. R. (1991). Integrating Expert Systems with Group Decision Support Systems. *ACM Transactions On Information Systems, 9*(1), 75-95.

Alavi, M., & Joachimsthaler, E. A. (1992). Revisiting DSS Implementation Research: A Meta-Analysis of the Literature and Suggestions for Researchers. *MIS Quarterly, 16*(1), 95-116.

Aldag, R. J., & Power, D. J. (1986). An Empirical Assessment of Computer-Assisted Decision Analysis. *Decision Sciences, 17*(4), 572-588.

Angehrn, A. A. (1991). Modeling by Example: A Link between Users, Models and Methods in DSS. *European Journal of Operational Research, 55*(3), 296-308.

Ariav, G., & Ginzberg, M. J. (1985). DSS Design: A Systemic View of Decision Support. *Communications of The ACM, 28*(10), 1045-1052.

Arinze, B. (1991). A Contingency Model of DSS Development Methodology. *Journal of Management Information Systems, 8*(1), 149-166.

Basu, A., & Blanning, R. W. (1994). Model Integration Using Metagraphs. *Information Systems Research, 5*(3), 195-218.

Blanning, R. W. (1982). A Relational Framework for Model Management in Decision Support Systems. In G. W. Dickson (Ed.), *Decision Support Systems-82 Transactions* (pp. 16-28). San Francisco, CA.

Bui, T. X. (1984). Building Effective Multiple Criteria Decision Support Models: A Decision Support System Approach. *Systems, Objectives, Solutions, 4*(1), 3-16.

Bui, T. X. (1987). *Co-Op: A Group Decision Support System for Cooperative Multiple Criteria Decision Making* (Vol. 290). Berlin: Springer-Verlag.

Bui, T. X., & Jarke, M. A. (1984). A DSS for Cooperative Multiple Criteria Group Decision Making. In *Proceedings of the 5th International Conference on Information Systems* (pp. 101-113). Tucson, AZ.

Culnan, M. J. (1986). The Intellectual Development of Management Information Systems, 1972-1982: A Co-Citation Analysis. *Management Science, 32*(2), 156-172.

Culnan, M. J. (1987). Mapping the Intellectual Structure of MIS, 1980-1985: A Co-Citation Analysis. *MIS Quarterly, 11*(3), 341-353.

Dempster, M. A. H., & Ireland, A. M. (1991). Object-Oriented Model Integration in a Financial Decision Support System. *Decision Support Systems, 7*(4), 329-340.

Dennis, A. R., Daniels, R. M., Jr., Hayes, G., & Nunamaker, J. F., Jr. (1993-1994). Methodology-Driven Use of Automated Support in Business Process Re-Engineering. *Journal of Management Information Systems, 10*(3), 117-138.

Dolk, D. R. (1993). An Introduction to Model Integration and Integrated Modeling Environments. *Decision Support Systems, 10*(3), 249-254.

Dolk, D. R., & Konsynski, B. R. (1984). Knowledge Representation for Model Management Systems. *IEEE Transactions On Software Engineering, SE-10*(6), 619-628.

Dolk, D. R., & Kridel, D. J. (1991). An Active Modeling System for Econometric Analysis. *Decision Support Systems, 7*(4), 315-328.

Dubrovsky, V. J., Kiesler, S., & Sethna, B. N. (1991). The Equalization Phenomenon: Status Effects in Computer-Mediated and Face-to-Face Decision Making Groups. *Journal of Human Computer Interaction, 6*(2), 119-146.

Dyer, J. S., Fishburn, P. C., Steuer, R. E., Wallenius, J., & Zionts, S. (1992). Multiple Criteria Decision Making, Multiattribute Utility Theory: The Next Ten Years. *Management Science, 38*(5), 645-654.

Einhorn, H. J., & Hogarth, R. M. (1981). Behavioral Decision Theory: Processes of Judgment and Choice. *Annual Review of Psychology, 32*(1), 53-88.

Eom, H. B., Lee, S. M., Snyder, C. A., & Ford, N. F. (1987-1988). A Multiple Criteria Decision Support System for Global Financial Planning. *Journal of Management Information Systems, 4*(3), 94-113.

Eom, S. B. (1995). Decision Support Systems Research: Reference Disciplines and a Cumulative Tradition. *Omega: The International Journal of Management Science, 23*(5), 511-523.

Eom, S. B. (1996). Mapping the Intellectual Structure of Research in Decision Support Systems through Author Cocitation Analysis (1971-1993). *Decision Support Systems, 16*(4), 315-338.

Eom, S. B. (2003). *Author Cocitation Analysis Using Custom Bibliographic Databases--an Introduction to the SAS Approach.* Lewiston, New York: The Edwin Mellen Press.

Eom, S. B. (2007). *The Development of Decision Support Systems Research: A Bibliometrical Approach.* Lewiston, NY: The Edwin Mellen Press.

Eom, S. B., & Farris, R. (1996). The Contributions of Organizational Science to the Development of Decision Support Systems Research Subspecialties. *Journal of the American Society for Information Science, 47*(12), 941-952.

Eom, S. B., Lee, S. M., Kim, E. B., & Somarajan, C. (1998). A Survey of Decision Support System Applications (1988-1994). *Journal of The Operational Research Society, 49*(2), 109-120.

Eom, S. B., & Min, H. (1992). The Changing Role of Multiple Criteria in Decision Support Systems. *Human Systems Management, 11*(3), 137-144.

Eom, S. B., & Min, H. (1999). The Contributions of Multi-Criteria Decision Making to the Development of Decision Support Systems Subspecialties: An Empirical Investigation. *Journal of Multi-Criteria Decision Analysis, 8*(5), 239-255.

Fjermestad, J., & Hiltz, S. R. (1998-1999). An Assessment of Experimental Research of Group Support Systems: Methodology and Results. *Journal of Management Information Systems, 15*(3), 7-149.

Geoffrion, A. M. (1987). An Introduction to Structured Modeling. *Management Science, 33*(5), 547-588.

Greif, I. (Ed.). (1988). *Computer-Supported Cooperative Work: A Book of Readings.* San Mateo, CA: Morgan Kaufmann Publishers, Inc.

Huber, G. P. (1983). Cognitive Style as a Basis for MIS and DSS Design: Much Ado About Nothing? *Management Science, 29*(5), 567-579.

Igbaria, M., & Guimaraes, T. (1994). Empirically Testing the Outcomes of User Involvement in DSS Development. *Omega, 22*(2), 157-172.

Jarke, M. A., Jelassi, M. T., & Shakun, M. F. (1987). Mediator: Towards a Negotiation Support System. *European Journal of Operational Research, 31*(3), 314-334.

Jelassi, M. T., & Foroughi, A. (1989). Negotiation Support Systems: An Overview of Design Issues and Existing Software. *Decision Support Systems, 5*(2), 167-181.

Jelassi, M. T., Jarke, M. A., & Stohr, E. A. (1985). Designing a Generalized Multiple Criteria Decision Support System. *Journal of Management Information Systems, 1*(4), 24-43.

Johansen, R. (1988). *Groupware: Computer Support for Business Teams.* New York: The Free Press, A Division of Macmillan, Inc.

Jones, C. V. (1995). Development in Graph-Based Modeling for Decision Support. *Decision Support Systems, 13*(1), 61-74.

Keen, P. G. W. (1980). MIS Research: Reference Disciplines and a Cumulative Tradition. In E. R. McLean (Ed.), *Proceedings of the First International Conference on Information Systems* (pp. 9-18). Philadelphia, PA.

Keen, P. G. W., & Scott Morton, M. S. (1978). *Decision Support Systems: An Organizational Perspective.* Reading, MA: Addison-Wesley.

Keeney, R. L., & Raiffa, H. (1976). *Decisions with Multiple Objectives: Preferences and Value Tradeoffs.* New York: John Wiley and Sons.

Konsynski, B. R., & Dolk, D. R. (1982). Knowledge Abstractions in Model Management. In *Decision Support Systems-82 Transactions* (pp. 187-202). San Francisco, CA.

Kuhn, T. S. (1970). *The Structure of Scientific Revolutions* (2nd ed.). Chicago: The University of Chicago Press.

Liang, T.-P., & Konsynski, B. R. (1993). Modeling by Analogy: Use of Analogical Reasoning in Model Management Systems. *Decision Support Systems, 9*(1), 113-125.

MacCrimmon, K. R. (1973). An Overview of Multiple Objective Decision Making. In J. L. Cochrane & M. Zeleny (Eds.), *Multiple Criteria Decision Making* (pp. 18-44). Columbia, SC: University of South Carolina Press.

Minch, R. P., & Sanders, G. L. (1986). Computerized Information Systems Supporting Multicriteria Decision Making. *Decision Sciences, 17*(3), 395-413.

Muhanna, W. A., & Pick, R. A. (1994). Meta-Modeling Concepts and Tools for Model Management: A Systems Approach. *Management Science, 40*(9), 1093-1120.

Murphy, F. H., & Stohr, E. A. (1986). An Intelligent System for Formulating Linear Programs. *Decision Support Systems, 2*(1), 39-47.

Nazareth, D. L. (1993). Integrating Mcdm and DSS: Barriers and Counter Strategies. *INFOR, 31*(1), 1-15.

Payne, J. W. (1982). Contingent Decision Behaviour. *Psychological Bulletin, 92*(2), 382-402.

Piramuthu, S., Raman, N., Shaw, M. J., & Park, S. C. (1993). Integration of Simulation Modeling and Inductive Learning in an Adaptive Decision Support System. *Decision Support Systems, 9*(1), 127-142.

Ramamurthy, K., King, W. R., & Premkumar, G. (1992). User Characteristics-DSS Effectiveness Linkage: An Empirical Assessment. *International Journal of Man-Machine Studies, 36*(3), 469-505.

Reisman, S., Johnson, T. W., & Mayes, B. T. (1992). Group Decision Program: A Videodisc-Based Group Decision Support System. *Decision Support Systems, 8*(2), 169-180.

Remus, W. E. (1984). An Empirical Investigation of the Impact of Graphical and Tabular Data Presentation on Decision Making. *Management Science, 30*(5), 533-542.

Sharda, R., Bar, S. H., & McDonnell, J. C. (1988). Decision Support System Effectiveness: A Review and Empirical Test. *Management Science, 34*(2), 139-159.

Shaw, M. J., & Fox, M. S. (1993). Distributed Artificial Intelligence for Group Decision Support: Integration of Problem Solving, Coordination, and Learning. *Decision Support Systems, 9*(4), 349-367.

Siegel, J., Dubrovsky, V. J., Kiesler, S., & McGuire, T. W. (1986). Group Processes in Computer-Mediated Communication. *Organizational Behaviour and Human Decision Processes, 37*(2), 157-187.

Singh, M. G., Bennavail, J. C., & Chen, Z. J. (1992). A Group Decision Tool for Combining Subjective Estimates Based on an Optimization Approach. *Decision Support Systems, 8*(6), 541-549.

Sprague, R. H., Jr. , & Carlson, E. D. (1982). *Building Effective Decision Support Systems*. Englewood Cliffs, NJ: Prentice Hall.

Stabell, C. B. (1983). A Decision Oriented Approach to Building Decision Support Systems. In J. L. Bennett (Ed.), *Building Decision Support Systems* (pp. 221-260). Reading, MA: Addison-Wesley.

Stefik, M., Foster, G., Bobrow, D., Kahn, K., Lanning, S., & Suchman, L. (1987). Beyond the Chalkboard: Computer Support for Collaboration and Problem Solving in Meetings. *Communications of The ACM, 30*(1), 32-47.

Todd, P. A., & Benbasat, I. (1992). The Use of Information in Decision Making: An Experimental Investigation of the Impact of Computer-Based Decision Aids. *MIS Quarterly, 16*(3), 373-393.

Turoff, M., Hiltz, S. R., Bahgat, A. N. F., & Rana, A. R. (1993). Distributed Group Support Systems. *MIS Quarterly, 17*(4), 399-417.

Tversky, A., & Kahneman, D. (1982). Judgment under Uncertainty: Heuristics and Biases. In D. Kahneman, P. Slovic & A. Tversky (Eds.), *Judgment under Uncertainty: Heuristics and Biases* (pp. 1124-1131). London: Cambridge University Press.

Weistroffer, H. R., & Narula, S. C. (1997). The State of Multiple Criteria Decision Support Software. *Annals of Operations Research, 72*(1), 299-313.

Zeleny, M. (1982). *Multiple Criteria Decision Making*. New York: McGraw-Hill.

Zmud, R. W. (1979). Individual Differences and MIS Success: A Review of the Empirical Literature. *Management Science, 25*(10), 966-979.

ENDNOTE

1 An earlier version of this paper was presented at The 2004 IFIP International Conference on Decision Support Systems (DSS2004): Decision Support in an Uncertain World, Prato, Italy, July 1-3, 2004.

APPENDIX

Table 1. Factor structure correlations (1990-1999) (factor loading at .40 or higher; rotation method: oblique; number of factors =10)

Factor 1			Factor 2			Factor 3		
GSS			**Design**			**Model Management**		
DENNIS	X32	0.963	SPRAGUE	X142	0.952	GEOFFRION	X47	0.978
JESSUP	X70	0.959	TURBAN	X154	0.918	KRISHNAN	X88	0.962
VALACICH	X157	0.958	KEEN	X77	0.903	DOLK	X37	0.947
NUNAMAKER	X114	0.956	CARLSONED	X23	0.901	GREENBERG	X52	0.940
GALLUPE	X46	0.955	SCOTT MORTON	X132	0.886	MUHANNA	X110	0.936
GEORGE	X48	0.953	WATSONHJ	X163	0.865	MURPHYFH	X111	0.934
POOLE	X120	0.952	ELAM	X41	0.860	JONESCV	X74	0.923
VOGEL	X160	0.952	HOLSAPPLE	X62	0.860	PICK	X118	0.914
MCGRATHJE	X103	0.948	WHINSTON	X166	0.859	STOHR	X147	0.909
DESANCTIS	X33	0.943	HENDERSONJ	X57	0.856	LIANG	X93	0.896
CONNOLLY	X25	0.941	ALTER	X4	0.849	BINBASIOGLU	X17	0.894
WATSONRT	X164	0.936	BENNETT	X15	0.839	FOURER	X44	0.880
NORTHCRAFTGB	X113	0.914	BONCZEK	X20	0.832	MEERAUS	X107	0.868
HILTZ	X58	0.913	ARIAV	X7	0.824	BASU	X12	0.867
TUROFF	X155	0.913	GORRY	X50	0.817	MA	X96	0.866
BOSTROM	X21	0.907	GINZBERG	X49	0.815	KIMBROUGH	X82	0.861
KRAEMER	X87	0.907	EOM	X42	0.793	BHARGAVA	X16	0.859
ZIGURS	X169	0.906	LEESM	X90	0.782	KERNIGHAN	X80	0.859
STEINERID	X146	0.905	WATKINS	X162	0.775	BLANNING	X18	0.842
TANSIK	X151	0.905	SIMONHA	X140	0.763	KENDRICK	X79	0.830
SHAWME	X136	0.905	HURT	X64	0.747	DUTTA	X39	0.790
HACKMAN	X55	0.903	SHAWMJ	X137	0.690	SHAWMJ	X137	0.697
SEIBOLD	X133	0.901	KOTTEMANN	X86	0.678	HOLSAPPLE	X62	0.633
SAMBAMURTHY	X130	0.898	BLANNING	X18	0.673	WHINSTON	X166	0.631
GALEGHER	X45	0.897	DUTTA	X39	0.667	JARKE	X67	0.620
MCLEODPL	X106	0.894	COURTNEY	X27	0.648	BONCZEK	X20	0.617
PINSONNEAULT	X119	0.894	BARIFF	X8	0.644	ELAM	X41	0.577
KIESLER	X81	0.892	MARCHJG	X98	0.623	KOTTEMANN	X86	0.539
RAOVS	X125	0.885	LIANG	X93	0.614	HENDERSONJ	X57	0.525
LIMLH	X94	0.873	ZIONTS	X170	0.604			
WYNNEBE	X168	0.872	WALLENIUS	X161	0.602			
HOLLINGSHEAD	X60	0.871	NEWELL	X112	0.600			
COOPERWH	X26	0.871	ALAVI	X2	0.589			
RAMAN	X124	0.870	STOHR	X147	0.583			
CHIDAMBARAM	X24	0.866	SILVERMS	X139	0.578			
JANIS	X66	0.866	KINGWR	X84	0.565			
WILLIAMSE	X167	0.865	JARKE	X67	0.557			
BASTIANUTTI	X11	0.865	BASU	X12	0.537			
ANSON	X5	0.863	DOLK	X37	0.533			
DUBROVSKY	X38	0.861	ROCKART	X128	0.524			
JOHANSEN	X71	0.859	REMUS	X126	0.522			
DICKSON	X35	0.859	SANDERSGL	X131	0.514			

continued on following page

Table 1. continued

HOLMES	X61	0.856	JONESCV	X74	0.510			
MCGUIRETW	X104	0.853	ZMUD	X171	0.504			
SIEGELJ	X138	0.853	JELASSI	X69	0.479			
DAFT	X28	0.851	SWANSON	X150	0.471			
MARTZ	X99	0.847	MINTZBERG	X108	0.450			
JOHNSONK	X72	0.846	MALONE	X97	0.427			
OSBORNAF	X116	0.846	GEOFFRION	X47	0.426			
MCGOFF	X102	0.840	HOGARTH	X59	0.419			
DELBECQ	X30	0.829	ROBEY	X127	0.411			
LEWISLF	X92	0.829	LUCAS	X95	0.410			
SPROULL	X143	0.828	SHARDA	X135	0.409			
HEMINGER	X56	0.828	ALDAG	X3	0.408			
LENGEL	X91	0.827	KONSYNSKI	X85	0.408			
VAN DE VEN	X158	0.824	RAIFFA	X122	0.401			
GRAYP	X51	0.824						
DIEHL	X36	0.822						
GROHOWSKI	X53	0.821						
ACKOFF	X1	0.820						
STROEBE	X148	0.817						
BEAUCLAIR	X13	0.816						
JARVENPAA	X68	0.812						
KINGJL	X83	0.810						
ROHRBAUGH	X129	0.802						
APPLEGATE	X6	0.799						
JOHNSTONSC	X73	0.794						
HUBER	X63	0.793						
STEEB	X144	0.792						
GUSTAFSON	X54	0.779						
WEICK	X165	0.738						
KONSYNSKI	X85	0.710						
BUI	X22	0.666						
MALONE	X97	0.658						
FOSTER	X43	0.653						
MCLEANER	X105	0.642						
BENBASAT	X14	0.625						
SUCHMAN	X149	0.609						
KAHNK	X76	0.599						
LANNING	X89	0.597						
MITROFF	X109	0.597						
MASONRO	X100	0.591						
BOBROW	X19	0.590						
DAVISFD	X29	0.588						
STEFIK	X145	0.543						
JELASSI	X69	0.501						
MINTZBERG	X108	0.495						
ALAVI	X2	0.446						
SHAKUN	X134	0.442						
Variance		62.592			29.400			24.327
% of Variance		36.6%			17.2%			14.2%

continued on following page

Table 1. continued

Factor 4			Factor 5			Factor 6		
User Interface/Implementation			**Cognitive Science**			CSCW		
IVES	X65	0.941	PAYNEJW	X117	0.930	STEFIK	X145	0.915
ZMUD	X171	0.910	HOGARTH	X59	0.901	SUCHMAN	X149	0.900
BAROUDI	X9	0.900	EINHORN	X40	0.897	KAHNK	X76	0.899
LUCAS	X95	0.893	TVERSKY	X156	0.881	BOBROW	X19	0.896
OLSONMH	X115	0.887	SLOVIC	X141	0.876	FOSTER	X43	0.893
ROBEY	X127	0.873	KAHNEMAN	X75	0.857	LANNING	X89	0.892
SWANSON	X150	0.871	TODD	X153	0.842	MALONE	X97	0.708
SANDERSGL	X131	0.850	VESSEY	X159	0.805	KINGJL	X83	0.704
COURTNEY	X27	0.821	NEWELL	X112	0.792	APPLEGATE	X6	0.683
KINGWR	X84	0.813	REMUS	X126	0.748	KONSYNSKI	X85	0.682
ROCKART	X128	0.810	SILVERMS	X139	0.745	BEAUCLAIR	X13	0.673
ALAVI	X2	0.779	SIMONHA	X140	0.738	GRAYP	X51	0.668
GINZBERG	X49	0.698	DEXTER	X34	0.714	STEEB	X144	0.657
DAVISFD	X29	0.680	BENBASAT	X14	0.691	ROHRBAUGH	X129	0.655
DEXTER	X34	0.676	MARCHJG	X98	0.672	BUI	X22	0.653
MCLEANER	X105	0.673	BARIFF	X8	0.611	JOHNSTONSC	X73	0.635
BENBASAT	X14	0.661	LUCAS	X95	0.541	LEWISLF	X92	0.620
KEEN	X77	0.660	COURTNEY	X27	0.494	HUBER	X63	0.613
WATSONHJ	X163	0.656	DAVISFD	X29	0.490	JELASSI	X69	0.610
ALDAG	X3	0.618	KOTTEMANN	X86	0.486	JOHANSEN	X71	0.601
POWERDJ	X121	0.613	ZMUD	X171	0.485	KRAEMER	X87	0.577
DELONG	X31	0.611	ALDAG	X3	0.481	JARKE	X67	0.543
SIMONHA	X140	0.594	POWERDJ	X121	0.481	RAOVS	X125	0.534
REMUS	X126	0.580	KEEN	X77	0.476	NUNAMAKER	X114	0.532
MINTZBERG	X108	0.577	KINGWR	X84	0.473	DESANCTIS	X33	0.530
HUBER	X63	0.563	ROBEY	X127	0.459	TUROFF	X155	0.526
BARIFF	X8	0.561	MINTZBERG	X108	0.443	ACKOFF	X1	0.524
SCOTT MORTON	X132	0.546	GINZBERG	X49	0.442	HILTZ	X58	0.517
MARCHJG	X98	0.543	SCOTT MORTON	X132	0.438	GALLUPE	X46	0.514
SILVERMS	X139	0.541	ARIAV	X7	0.433	DICKSON	X35	0.510
VESSEY	X159	0.522	HURT	X64	0.426	MARTZ	X99	0.508
ALTER	X4	0.519	RAIFFA	X122	0.412	VOGEL	X160	0.507
TODD	X153	0.517	SHARDA	X135	0.403	ZIGURS	X169	0.505
MCDONNELLJC	X101	0.493				GUSTAFSON	X54	0.505
BARSH	X10	0.492				JARVENPAA	X68	0.504
MASONRO	X100	0.481				MCGOFF	X102	0.501
JARVENPAA	X68	0.481				HEMINGER	X56	0.488
GORRY	X50	0.473				GROHOWSKI	X53	0.487
DICKSON	X35	0.461				WATSONRT	X164	0.483
SPRAGUE	X142	0.460				POOLE	X120	0.472
SHARDA	X135	0.456				GEORGE	X48	0.462
ARIAV	X7	0.456				JESSUP	X70	0.456
HENDERSONJ	X57	0.454				DENNIS	X32	0.450
NEWELL	X112	0.448				DELBECQ	X30	0.439
TURBAN	X154	0.426				VAN DE VEN	X158	0.437
HURT	X64	0.416				SHAKUN	X134	0.434

continued on following page

Table 1. continued

WEICK	X165	0.404				MASONRO	X100	0.429
						MCGUIRETW	X104	0.417
						PINSONNEAULT	X119	0.417
						SIEGELJ	X138	0.417
						KIESLER	X81	0.416
						MITROFF	X109	0.416
						BENNETT	X15	0.416
Variance		25.252			18.252			24.01
% of Variance		14.8%			10.7%			14.0%
Factor 7			Factor 8			Factor 9		
Evaluation			MCDM/MCDSS			Organization Science		
MCDONNELLJC	X101	0.907	RAIFFA	X122	0.839	THEORET	X152	0.887
BARSH	X10	0.904	WALLENIUS	X161	0.838	RAISINGHANI	X123	0.880
SHARDA	X135	0.895	KEENEY	X78	0.832	MINTZBERG	X108	0.837
POWERDJ	X121	0.844	ZIONTS	X170	0.822	MITROFF	X109	0.736
ALDAG	X3	0.837	JELASSI	X69	0.821	MASONRO	X100	0.673
BARIFF	X8	0.749	SHAKUN	X134	0.802	SIMONHA	X140	0.569
KINGWR	X84	0.737	JARKE	X67	0.748	MARCHJG	X98	0.547
DEXTER	X34	0.732	BUI	X22	0.727	WEICK	X165	0.518
REMUS	X126	0.703	KONSYNSKI	X85	0.529	NEWELL	X112	0.492
BENBASAT	X14	0.599	HUBER	X63	0.526	COURTNEY	X27	0.441
COURTNEY	X27	0.597	LEESM	X90	0.502	HUBER	X63	0.425
LUCAS	X95	0.596	EOM	X42	0.488	JANIS	X66	0.419
VESSEY	X159	0.550	BEAUCLAIR	X13	0.478	OSBORNAF	X116	0.407
ZMUD	X171	0.545	KINGJL	X83	0.465	KONSYNSKI	X85	0.407
TODD	X153	0.543	HENDERSONJ	X57	0.462	SILVERMS	X139	0.398
KEEN	X77	0.516	MASONRO	X100	0.447			
GINZBERG	X49	0.513	MALONE	X97	0.445			
KOTTEMANN	X86	0.504	HURT	X64	0.445			
ROBEY	X127	0.503	APPLEGATE	X6	0.443			
HURT	X64	0.501	ROHRBAUGH	X129	0.443			
ALAVI	X2	0.489	STEEB	X144	0.441			
SWANSON	X150	0.485	MITROFF	X109	0.440			
HUBER	X63	0.483	GRAYP	X51	0.434			
SANDERSGL	X131	0.475	BENNETT	X15	0.433			
SIMONHA	X140	0.471	JOHNSTONSC	X73	0.426			
ALTER	X4	0.465	ELAM	X41	0.425			
SILVERMS	X139	0.463	KEEN	X77	0.414			
JARVENPAA	X68	0.451	KRAEMER	X87	0.403			
IVES	X65	0.436	GUSTAFSON	X54	0.396			
DICKSON	X35	0.432	DESANCTIS	X33	0.394			
HENDERSONJ	X57	0.429						
SCOTT MORTON	X132	0.420						
SPRAGUE	X142	0.414						

continued on following page

Table 1. continued

BAROUDI	X9	0.407						
ELAM	X41	0.406						
PAYNEJW	X117	0.400						
Variance		17.116			17.916			13.586
% of Variance		10.0%			10.5%			7.9%

Factor 10		
Social Psychology		
DUBROVSKY	X38	0.411
SIEGELJ	X138	0.407
MCGUIRETW	X104	0.407
KIESLER	X81	0.398
Variance		2.598
% of Variance		1.5%

About the Author

Sean Eom is a professor of management information systems (MIS) and had been appointed as a Copper Dome Faculty Fellow in Research at the Harrison College of Business, Southeast Missouri State University during the academic years 1994-1996 and 1998-2000. In recognition of his continuing research contributions to the information systems area, he is listed as one of outstanding intellectuals in the first edition of *2000 Outstanding Intellectuals of the 21st Century*, *Top 100 Educators 2005*, *Who's Who in American Education*, and *59th Edition of Marquis Who's Who in America.*

He received his PhD in management science from the University of Nebraska–Lincoln in 1985. His other degrees are from the University of South Carolina at Columbia (MS, international business), Seoul National University (MBA, international management), and Korea University (BA). His research areas include decision support systems (DSS), expert systems, and global information systems management.

He is the author/editor of six books including *Decision Support Systems Research (1970-1999): A Cumulative Tradition and Reference Disciplines*, *Author Cocitation Analysis using Custom Bibliographical Databases: An Introduction to the SAS Systems*, *Inter-Organizational Information Systems in the Internet Age*, and *Encyclopedia of Information Systems*. He has published more than 50 refereed journal articles and 60 articles in encyclopedia, book chapters, and conference proceedings.

Index

J

K

L

M

N

O

P

R

S

T

U

V

W

www.ingramcontent.com/pod-product-compliance
Ingram Content Group UK Ltd.
Pitfield, Milton Keynes, MK11 3LW, UK
UKHW051028041026
15470UKWH00015B/1697